Advanced Principles of Wireless Sensor Networks

Advanced Principles of Wireless Sensor Networks

Edited by Arthur Nelson

CLANRYE
INTERNATIONAL
www.clanryeinternational.com

Clanrye International,
750 Third Avenue, 9th Floor,
New York, NY 10017, USA

ISBN: 978-1-63240-927-0

Cataloging-in-Publication Data

Advanced principles of wireless sensor networks / edited by Arthur Nelson.
    p. cm.
Includes bibliographical references and index.
ISBN 978-1-63240-927-0
1. Wireless sensor networks. 2. Sensor networks. 3. Wireless communication systems. I. Nelson, Arthur.
TK7872.D48 A38 2020
681.2--dc23

For information on all Clanrye International publications
visit our website at www.clanryeinternational.com

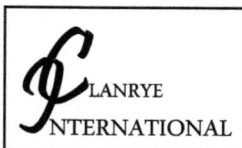

$\mathcal{CL}$ LANRYE
INTERNATIONAL

Contents

Preface

This book has been a concerted effort by a group of academicians, researchers and scientists, who have contributed their research works for the realization of the book. This book has materialized in the wake of emerging advancements and innovations in this field. Therefore, the need of the hour was to compile all the required researches and disseminate the knowledge to a broad spectrum of people comprising of students, researchers and specialists of the field.

Wireless sensor network is a group of dedicated and spatially distributed sensors used to monitor and record the physical conditions of the environment. It also organizes the collected data at a central location. It helps in measuring the environmental conditions such as temperature, pollution levels, sound, humidity, and wind. They rely on wireless connectivity and spontaneously form a network to ensure the wireless transportation of sensor data. Modern wireless sensor networks are bi-directional that enable the control of sensor activity. It plays an important role in military applications such as battlefield surveillance. Such networks are also used in many industrial and consumer applications such as industrial process monitoring and control and machine health monitoring. This book elucidates the concepts and innovative models around prospective developments with respect to wireless sensors network. Some of the diverse topics covered herein book address the varied branches that fall under this category. The book is appropriate for those seeking detailed information in this area.

At the end of the preface, I would like to thank the authors for their brilliant chapters and the publisher for guiding us all-through the making of the book till its final stage. Also, I would like to thank my family for providing the support and encouragement throughout my academic career and research projects.

Editor

RSHSC-Routing Algorithm based on Simplified Harmony Search and Coding for UWSNs

Meiju Li,[1,2,3] **Xiujuan Du**◉,[1,2] **and Chunyan Peng**◉[1,2]

[1]*Computer Department, Qinghai Normal University, Xining 810008, China*
[2]*Key Laboratory of the Internet of Things of Qinghai Province, Xining 810008, China*
[3]*College of Physics and Electronic Information Engineering, Qinghai Nationalities University, Xining 810007, China*

Correspondence should be addressed to Xiujuan Du; 124111397@qq.com

Guest Editor: Mohammed T. Ghoneim

With the development of wireless networks and increasingly interest of people in underwater resources and environment, UWSNs are being paid more and more attention. Because of the characteristics of underwater channel and acoustic signal, the protocols used in the terrestrial networks cannot be directly used in UWSNs. In this paper, a reliable and energy-efficient routing protocol based on SHS and coding, called RSHSC, is proposed. Firstly, regular nodes are assigned to cluster heads according to simplified harmony search algorithm. Secondly, partial network encoding is introduced and the next two-hop information is considered when data packets are transmitted to sink nodes from the source node. Only the best next-hop forwards data packets. All data packets from neighbor nodes are used for decoding. Thirdly, two schemes of updating routing are designed and compared. Lastly, extensive simulations prove RSHSC is effective in improving reliability and decreasing energy consumption.

1. Introduction

In recent years, UWSNs (underwater wireless sensor networks) attracted more and more attentions because of the wide application such as marine source exploration and pollution monitor [1–3]. Like other wireless networks, it is necessary to improve the communication performance for UWSNs. However, UWSN is different from other wireless networks. UWSN has some characteristics that other wireless networks have not. Firstly, UWSN is provided energy by batteries. Once the battery is exhausted, the node is considered as dead [4]. Meanwhile, the acoustic communication consumes more energy than radio signal and optical signal. Secondly, underwater channel is complex because of shipping, marine organisms, and so on [5]. Besides, the communication quality is affected by multipath resulting from the reflections, water salinity, temperature, and so on [6]. Thirdly, mobile topology [7]. The position of the node is forced to change with the water currents and marine biological activities [8]. Fourthly, the propagation delay is longer than terrestrial networks. The propagation speed of acoustic signal under water is just 1500 m/s, which is five magnitude lower than radio signal on land (3×10^8 m/s) [9]. Because of those factors of UWSNs, the protocols used on the land cannot be directly applied in UWSNs, especially for routing protocols. Therefore, designing a routing protocol special for UWSNs is necessary [10–12].

Like most wireless networks, the performances of delay, energy consumption, reliability, and throughput and bandwidth utilization are important in UWSNs [13]. In recent years, more and more routing protocols are proposed. Some of them are to decrease energy consumption, some of them are to shorten delay. They are effective in certain scenarios but cannot take into account multiple performance.

The contributions of this paper are showed as follows:

(1) The regular nodes are assigned to the cluster heads based on simplified harmony search algorithm (SHS). Two schemes are designed

(2) Partial network coding is introduced when routing data from source cluster head to the sink node. Meanwhile, the data from all neighbor nodes are used

for decoding, which makes full use of the broadcast characteristics of UWSNs

(3) Two schemes are compared for maintaining routing timely: when the speed of water flow is constant, scheme 1 is adopted; otherwise, the scheme 2 is used

(4) Through simulations, the best SHS and routing maintaining algorithm are determined and the performance of RSHSC is evaluated

2. Relative Works

Researchers have designed some routing protocols for UWSNs in recent years. In 2008, Yan et al. proposed DBR routing algorithm [14]. In DBR, neighbors with less depth need to wait a certain time to forward data packets. The lengths of waiting time of neighbors are set according to the depth. The more depth, the shorter waiting time. The neighbor will cease to forward if hearing the same data packets during waiting time. DBR has some defects. Therefore, some improved protocols are provided. Residual energy is introduced to select next-hop in EEDBR [15]. DBR-NC uses the coding technique [16]. To avoid selecting next-hop in the void area, next-hop of the next-hop is considered in WDFAD-DBR [17]. Meanwhile, the forwarding area is adjusted according to node dense. LMPC is proposed in [18]. In LMPC, a data packet is delivered along multiple paths according to binary tree. FLMPC improved LMPC in [19]. Nodes are classified into cross nodes and regular nodes. Only the cross nodes generate binary tree. Hao et al. proposed GPNC algorithm [20]. Data packets are forwarded to sink greedily using the sensor nodes' location information. GPNC incorporates partial network coding. In L2-ABF, the sender selects next-hop by calculating the forwarding angle. CoUWSN uses cost function of distance and SNR of the link to decide the next-hop [21]. Sink mobility pattern is given in DEADs [22]. AEDG is proposed in [23]. In AEDG, AUV is used to collect data. To prevent gateway overloading, the number of associated node with the gateway node is limited. The residual energy and number of neighbors are considered. Hubcode is proposed in [24]. Hubcode is an algorithm based on cluster. Hubcode exchanges coefficient matrix and gets initial data via calculating inverse matrix. In BLOAD [25], the weight of nodes is calculated according to the distance among nodes, the more overlap area in the coverage range of the node, the less importance for the node. In TSBNC, the coding is introduced into the time-slot algorithm [26]. Network coding and cross-layer are used in NCRP [27]. The node with the least weight is selected as the cluster head. In QDAR [28], end-to-end delay and residual energy are introduced into the Q-Learning algorithm. In LB-AGR, the best next-hop is determined based on density, available energy, location, and level difference between neighbor nodes [29]. The characteristics of these protocols are showed as Table 1.

3. Model Analysis

In this sector, network model, energy and propagation model are analyzed.

3.1. Network Model. The network model is showed in Figure 1. In our network, nodes are deployed in the 3D underwater environment and divided into three types: regular nodes, advanced nodes, and sink nodes. The first two nodes are defined as follows:

(i) Regular nodes: the nodes have ordinary amount of initial energy, which can communicate with advanced nodes and sink nodes

(ii) Advanced nodes: the nodes have more initial energy than regular nodes. Advanced nodes can communicate with each other, regular nodes and sink nodes. The advanced node is also called as the cluster head (CH). The advanced nodes with almost depth are divided into the same levels

In the network, each sink node is equipped with both acoustic modem and RF modem, which are responsible for communicating with underwater nodes (regular nodes and advanced nodes) and data centers, respectively. Each underwater node is equipped with an acoustic modem, which is responsible for communicating with each other and sink nodes. All nodes mobile with water flow. Each node can get the location information of itself.

3.2. Channel Model. To design a routing algorithm with high bandwidth utilization, energy efficient and reliable for UWSNs, we must learn about the channel model. In UWSNs, channel is affected by absorption loss and spreading loss. The attenuation $A(d,f)$ is showed as (1) [21].

$$A(d,f) = A_0 d^k \alpha(f)^d. \tag{1}$$

Here, k is the spreading factor, and the value is 1.5 for practical applications. d is the spreading distance. A_0 is a normalization constant. $\alpha(f)$ is presented as (2).

$$10 \log \alpha(f) = \frac{0.11f^2}{1+f^2} + \frac{44f^2}{4100+f^2} + \frac{2.75f^2}{10^4} + 0.003 [\text{dB/km}]. \tag{2}$$

There are great many types of noise under water. The main noise include turbulence (N_t), shipping (N_s), waves (N_w), and thermal noise (N_{th}). These noises can be modeled by the power spectral and Gaussian statistics as (3), (4), (5), (6), and (7) [22].

$$N(f) = N_t(f) + N_s(f) + N_w(f) + N_{th}(f), \tag{3}$$

$$10 \log N_t(f) = 17 - 30 \log f, \tag{4}$$

$$10 \log N_s(f) = 40 + 20(s-0.5) + 26 \log f - 60 \log (f+0.03), \tag{5}$$

$$10 \log N_w(f) = 50 + 7.5\sqrt{w} + 20 \log f - 40 \log (f+0.4), \tag{6}$$

$$10 \log N_{th}(f) = -15 + 20 \log f. \tag{7}$$

TABLE 1: Protocols characteristics.

Protocol name	Characteristics	Advantage	Disadvantage
DBR	Select next-hop only based on depth	Independent location information, decrease part of redundant forwarding	Long delay, high energy consumption
EEDBR	Residual energy is considered	Prolong network life	Long delay, redundant forwarding
DBR-NC	Coding is introduced	Reliable	Long delay, energy is not considered
WDFAD-DBR	Next two-hop information is considered. Adaptively adjust the forwarding area according to node dense	Avoid selecting the node in the void area as the next-hop	Long delay in sparse network delivery ratio is low
iAMCTD	Courier node is used	Courier node decreases delay, optimized threshold decreases redundant forwarding	Energy unbalance
LMPC	Establish binary tree	Reliable, high delivery ratio	High energy consumption, void node is not considered
FLMPC	Establish binary tree from the sensor nodes which reside near the layer	Copies in the cross node increase the delivery ratio, decrease the retransmission	Routing update is not considered, multiple routing wastes energy
GPNC	Network coding is used based on geographic location information	Reliable, decrease energy consumption, shorten delay	Void node is not considered, energy unbalance
CoUWSN	Uses multiple input multiple output	Save transmit power, increase data rate, extend the communication range	Consume more energy, long delay
DEADs	Cooperative routing is joined with sink mobility	High throughput, prolong network life	Waste energy, energy unbalance
AEDG	Use shortest path tree to assign nodes to gate way	Network lifetime prolongation, throughput maximization	Long delay
Hubcode	Use hubs as relay and encode multiple messages address to the same destination	Reducing the forwarding overheads, increasing the delivery ratio	Energy balance long delay
BLOAD	Addressing energy hole, mixed routing scheme (including directly and multiple hop communication) is used	Balance energy, avoid energy hole	The node with longer distance to sink dies quickly
TSBNC	Theory of network coding is introduced into time-slot based routing algorithm	Decrease energy consumption and collisions	Energy unbalance, long delay
NCRP	Network coding and cross-layer design are used	High delivery ratio, save energy	Void node is not considered, energy unbalance
QDAR	Q-learning algorithm is introduced	Extending network lifetime and short delay	In the mobile network, energy consumption is high. Link quality is not considered
LB-AGR	Nodes are divided into different levels, upstream and downstream are considered when selecting the best-next-hop	Comprehensive factors are considered	The void node is not considered, greedy routing

Here, s is the shipping factor (varies from 0 to 1), W is the wind velocity (varies from 0 to 10 m/s), and f denotes the carrier frequency.

Therefore, the signal to noise (SNR) is showed as (8), which is related to the propagation distance (d) and the carrier frequency (f). B is the bandwidth.

$$SNR(d,f) = \frac{P}{A(d,f)N(f)B}. \quad (8)$$

According to Shannon theory, channel capacity is calculated as (9) [21].

$$C(d,f) = B\log_2(1 + SNR(d,f)). \quad (9)$$

4. Protocol Design

The implementation of RSHSC consists of four steps: initialization, cluster construction, intercluster routing, and routing

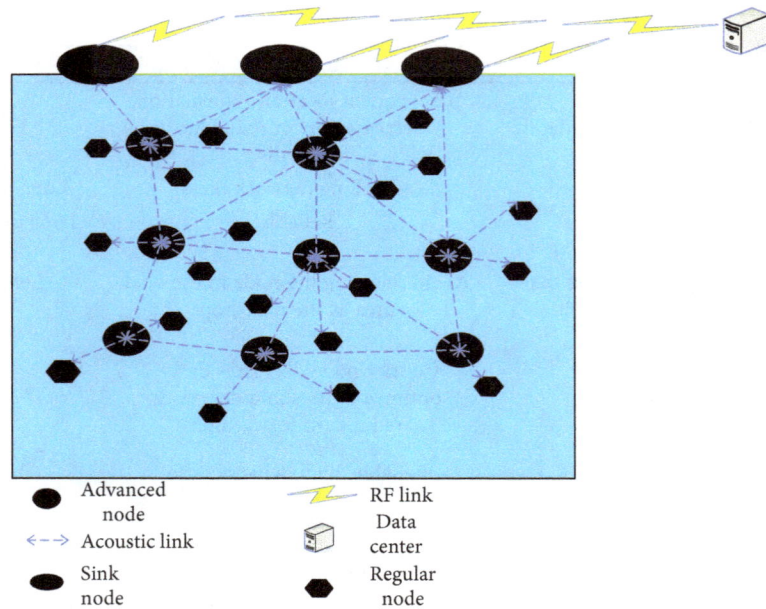

FIGURE 1: Network model.

TABLE 2: Format of hello packet.

ID	Seq number	ID1	ID2	ID3	Reserve	RE_ENERGY	Address

maintaining. In initialization phase, sink nodes broadcast beacon signal with the location information of themselves to underwater nodes. Cluster heads broadcast hello packets. Regular nodes broadcast information packets. The details of initialization phase are showed in Section 4.1. In cluster construction phase, regular nodes are assigned to the cluster head based on SHS algorithm, which is detailed in Section 4.2. In the interclusters routing phase, the cluster head encodes original packets and selects the best next-hop according to the information of neighbors and next two hop, which is detailed in Section 4.3. Because nodes mobile with the water flow, the routing needs to be updated timely. In routing maintaining phase, two algorithms are provided to decrease the energy consumption and collision results from sending too many nondata packets, which is detailed in Section 5.

4.1. Initialization.
Sink nodes broadcast beacon signal including location and ID of themselves. After receiving beacon signal, each cluster head calculates and saves the distance between itself and sink nodes. Then, the cluster head broadcasts hello packet, whose format is showed in Table 2. "ID" is the source node ID of the hello packet. Seq number is the sequence number of the hello packet. "ID1," "ID2," and "ID3" are the ID of member nodes. Initially, the field of "ID1", "ID2," and "ID3" are filled with "0." Address is the location information of the cluster head. "RE_ENERGY" is the residual energy of the cluster head. The regular node sends information packets after receiving a hello packet. The format of information packet is showed in Table 3. "ID1" is the ID number of the cluster head with the highest priority that the regular node joins in. "Distance 1" is the

TABLE 3: Format of information packet.

ID	ID1	Distance1	ID2	Distance 2

distance to the cluster head with ID1. "ID2" and "Distance 2" are as the ID1 and Distance1. Before receiving less than three information packets from different regular nodes, the corresponding ID is filled with "0." Analogously, the ID2 and Distance2 are filled with "0" when receiving hello packets with only ID address.

In this phase, on one hand, regular nodes and cluster heads get information from each other; on the other hand, cluster heads get the information of neighbor cluster heads.

4.2. Cluster Constructing.
In the process of cluster constructing, the harmony search algorithm is introduced.

4.2.1. Harmony Search (HS) Algorithm.
HS is a heuristic global search algorithm. The I instruments ($I = 1, 2, \ldots, m$) are analogous as the I variables to solve optimization problems, and the harmonic R_j of each musical instrument tone ($j = 1, 2, \ldots, m$) is equivalent to the j solution vectors of the optimization problem, and the evolution is analogous to the objective function.

The procedure of HS is as follows:

(1) The algorithm generates m initial solutions into the harmony memory (HM) and searches for new solutions in HM with a probability HR

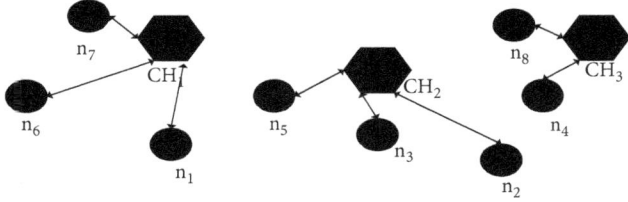

FIGURE 2: Regular node assignment.

(2) The algorithm generates local perturbations to the new solution with probability P_r. If the new solution objective function value is better than the worst solution in HM, replace it

(3) Iterate iteratively until the expected number of iterations is T_{max}

(4) As above, by introducing HR and P_r, HS algorithm is expected to the balance of exploration ability of the solution space, but there is no the recital basis for how to choose the value

4.2.2. Simplified HS. According to the characteristics of UWSNs, the average number of regular nodes managed by each cluster head is small, and it is difficult to get the global information of the whole networks. Therefore, the simplified HS algorithm is proposed for UWSNs.

(A) Taking regular nodes as research objects, the assignment is determined by regular nodes. We simplify the HS (HSA) as follows:

(1) The objective function is

$$F_1 = \alpha \sum_{i=1}^{\tilde{CH}_i} \frac{\left|d_{n_i-CH_i}\right|}{R} + \beta \sum_{i=1}^{i=N} \frac{\left|\tilde{CH}_i - \bar{N}\right|}{\bar{N}}. \quad (10)$$

Here, $\alpha + \beta = 1$. $\left|d_{n_i-CH_i}\right|$ presents the distance between the regular node n_i and the cluster head CH_i. $\tilde{CH}_i$ presents the number of members for cluster head CH_i. R is the communication radius. When $\tilde{CH}_i$ is constant, the less F_1, the better assignment. The first part reflects the balance among regular nodes, and the second part reflects the balance among cluster heads.

(2) Because any regular node is assigned to only the cluster head in its communication range, we use the local selection of regular nodes being assigned to the cluster head to replace the global assignment. So the $X_i = (x_{i,1}, x_{i,2} \ldots x_{i,k} \ldots x_{i,n})$. n is the number of member nodes of cluster heads CH_i and adjacent to CH_i. X_i is the HM, $x_{i,k}$ is the ID of the cluster head the regular node n_i joining in. For example, in Figure 2, the regular nodes are $R = (n_1, n_2, n_3, n_5, n_6, n_7)$, $X_1 = (1, 2, 2, 2, 1, 1)$.

While $R = (n_1, n_2, n_3, n_4, n_5, n_6, n_7, n_8)$, $X_2 = (1, 2, 2, 2, 3, 2, 1, 1, 3)$

(3) According to the information of cluster heads recorded in the regular nodes, the HM is adjusted at probability P. P is calculated by (11). Here, $p'_i = 1/d_i$, $p' = \sum_{i=1}^{i=n_i^{\cdot}} (1/d_i(1/d_1))$, $n_i^{\cdot}$ is the number of cluster head in the communication range of regular node n_i, d_i is the distance between the node n_i and the cluster head. If the adjusted objective function is bigger than the value before, replace it. Otherwise, the HM is kept

$$P = \frac{p'_i}{p'}. \quad (11)$$

(B) Taking cluster heads as research objects, the assignment is determined by cluster heads. We simplify the HS (HSB) as follows:

(1) The objective function is the same as HSA

(2) Similar to HSA, $X_i = (x_{i,1}, x_{i,2} \ldots x_{i,k} \ldots x_{i,n})$. $x_{i,k}$ is the aggregate of member nodes in the cluster head CH_i. As is showed in Figure 2, $R = (CH_1, CH_2, CH_3)$, then $X_i = (\{n_1, n_6, n_7\}, \{n_2, n_3, n_5,\}, \{n_3, n_4, n_8\})$

(3) According to the information of regular nodes recorded in the cluster heads, the HM is adjusted at probability P. P is calculated by (12), p_i^- is the probability each member n_i having only one cluster head, and p'_i is the probability n_i being adjusted into another cluster head. p'_i can be gotten from (13). R_i and R_j denote the residual energy of the cluster heads in a regular node communication range, $\tilde{CH}_i$ and $\tilde{CH}_j$ are the number of members in a cluster, respectively, $\gamma_1 + \gamma_2 = 1$

$$P = 1 - \prod_{i=1}^{i=\tilde{CH}_i} \left(1 - (1 - p_i^-)p'_i\right), \quad (12)$$

$$p'_i = \gamma_1 \frac{R_i}{R_i + R_j} + \gamma_2 \frac{1/\tilde{CH}_i}{\left(1/\tilde{CH}_i\right) + \left(1/\tilde{CH}_j\right)}. \quad (13)$$

4.3. Routing Construction. How are data packets transmitted from a source node to the sink node? In this section, the procedure is analyzed.

4.3.1. Review of Routing. According to the initialization procedure, each CH saves the information of neighbor CHs as showed in Table 4. After receiving a hello packet, the CH

TABLE 4: Neighbors' information table of cluster head.

ID	Address	Residual energy	Address of next-hop	Next-hop residual energy	Priority

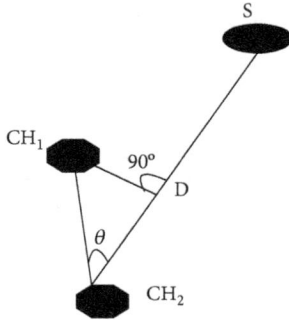

FIGURE 3: Advanced distance.

firstly checks the "address" field and calculates the distance between the source node of hello packet and the sink node. Only the cluster head with shorter distance to the sink node has opportunity to be the next-hop. The priority of neighbors is calculated by formula (14). $d(t)$ is the advanced distance to the sink node. As showed in Figure 3, CH_2 is the source cluster head, CH_1 is the neighbor of CH_2, S is the sink node. The $\overline{CH_2D}$ is the advanced distance to node S. We set the coordinate of CH_2, CH_1, and S is (x_0, y_0, z_0), (x_1, y_1, z_1), and $(0,0,0)$, respectively. From the coordinate, we can get $\overline{CH_2S} = \sqrt{x(t)_0^2 + y(t)_0^2 + z(t)_0^2}$, $\overline{CH_1S} = \sqrt{x(t)_1^2 + y(t)_1^2 + z(t)_1^2}$, $\overline{CH_2CH_1} = \sqrt{(x(t)_0 - x(t)_1)^2 + (y(t)_0 - y(t)_1)^2 + (z(t)_0 - z(t)_1)^2}$, $\cos \theta = \overline{CH_2CH_1}^2 + \overline{CH_2S}^2 - \overline{CH_1S}^2 / 2\overline{CH_2CH_1} \times \overline{CH_1S}$. So $d(t) = \overline{CH_2CH_1} \cos \theta$. The more $p(t)$, the higher priority. Only the neighbor cluster head with the highest priority forwards the data packet.

$$p(t) = k_1 \frac{E_R(t)}{E_I(t)} + k_2 \frac{d(t)}{R}. \qquad (14)$$

As above, the node with bigger advanced distance to sink and more residual energy is selected as the best next-hop. Therefore, the selected next-hop may have far distance to the previous hop, which leads to the big probability of failing to deliver. To enhance the reliability of the network, in our algorithm, coding is introduced.

4.3.2. Network Coding. In most of conventional routing algorithms, the forwarder is responsible for relaying data without any processing. However, in UWSNs, like other wireless communications, broadcast is adopted. Meanwhile, the bandwidth is limited and delay is long. In order to make full use of the characteristics of broadcast and limited bandwidth, network coding is an efficient measure. The basic procedure of network coding is showed in Figure 4.

The data packet a is from node A. The data packet b is from node B. The node C is a relay.

In the network without coding as Figure 4(a), the necessary number of time slots to exchange data packets a and b is 4. While in Figure 4(b), the number is 3.

Furthermore, we compare the partial network coding with the full network coding. In Figure 5, the data packets d_1, d_2, and d_3 are sent from the node A to node B. Here, we make

$$[d] = \begin{matrix} d_1 \\ d_2 \\ d_3. \end{matrix} \qquad (15)$$

Node A encode data packets $[d]$ into $[d']$.

$$[d'] = \begin{bmatrix} \gamma_{11} & \gamma_{12} & \gamma_{13} \\ \gamma_{21} & \gamma_{22} & \gamma_{23} \\ \gamma_{31} & \gamma_{32} & \gamma_{33} \end{bmatrix} [d]. \qquad (16)$$

To decode the encoded packets into original data packets, node B needs all data packets. So, the delay to transmit these three data packets is 3T (T is one slot time). In Figure 5(b), the partial network coding is showed. Here, $d_1' = \gamma_1 d_1$, $d_2' = \gamma_2 d_1 + \gamma_3 d_2$, and $d_3' = \gamma_4 d_1 + \gamma_5 d_2 + \gamma_6 d_3$. So, in the partial network coding, recovering d_1', d_2', and d_3' into original packets needs T, 2T, and 3T, respectively. The average delay to transmit these three packets is $(T + 2T + 3T)/3 = 2T$. Comparing the partial network coding with full network coding, we can get the partial network coding is better than full network coding.

In our algorithm, partial network coding is adopted. Each relay receives the encoded packets and decodes them. Then the next-hop of relay continues to encode the original data packets until the sink node receives the packets. If the sink node is in the communication range of the relay, the decoded packets are directly forwarded to the sink node.

As showed in Figure 6, during the routing procedure, the CH receives original data packets from member nodes and encodes the original data packets. The CH hears the data packets from regular nodes which are not its member node, drops them. When it hears a data packet from the neighbor with longer distance to the sink node, the CH checks weather it is helpful to decode if the next-hop of data packets is not the current ID. If received encoded packet have been decoded, CH drops it. If there is no association with the encoded packets saved, CH saves them and broadcasts data packets for a certain time. Otherwise, the received data packet is used to decode. For example, CH_1 successively hears encoded packets P_1, P_2, P_3 and P_4, $P_1 = d_1 \oplus d_2$, $P_2 = d_2 \oplus d_3$, $P_3 = d_1 \oplus d_3$, and $P_4 = d_4 \oplus d_5$. We assume d_1 have been decoded in CH_1, then d_2 is decoded using P_1, d_3 is decoded by P_2, however, P_3 is helpless for decoding any new original

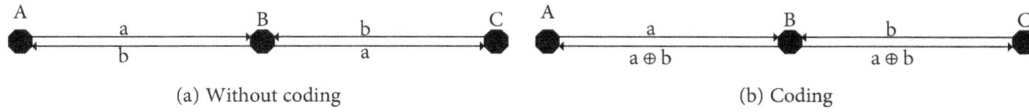

(a) Without coding   (b) Coding

FIGURE 4: Coding theory.

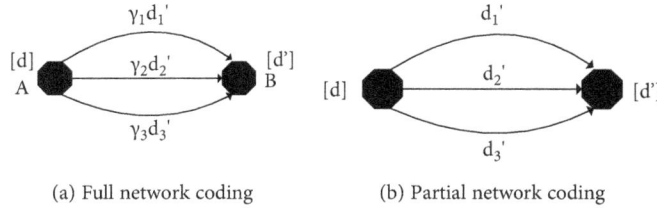

(a) Full network coding   (b) Partial network coding

FIGURE 5: Full network coding and partial network coding.

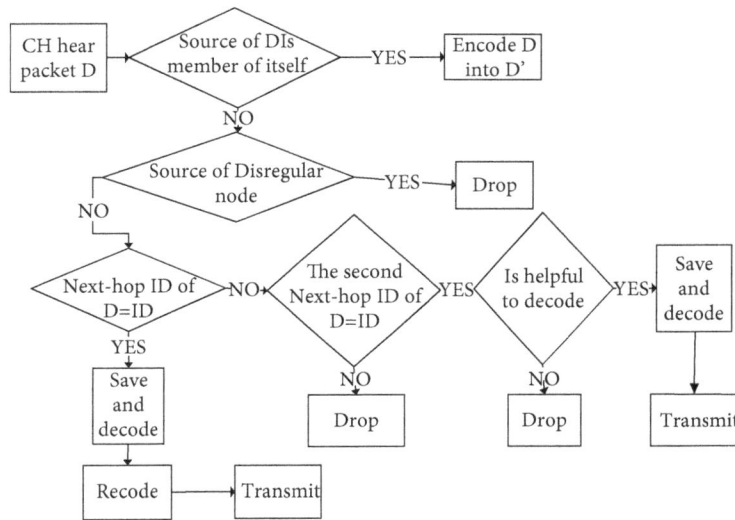

FIGURE 6: Encoding algorithm flow chart.

packet, CH_1 will drop P_3. While there is no association between data packets are encoded into P_4 and the encoded packets saved in CH_1, so CH_1 saves P_4 and broadcasts P_4 after a certain time.

The procedure of encoding algorithm is showed in Figure 6.

4.3.3 Routing Procedure. The detailed routing procedure is showed in Figure 7. After receiving a hello packet, cluster head checks three conditions including: (1) the hello packet is from a neighbor nearer to the sink node. (2) The residual energy is more than a preset value Re_{-th}. (3) The source node of D exits next-hop. When these three conditions meet requirements, the node calculates the priority $p(t)$ of the neighbor as (14). If the $p(t)$ is bigger than the priority value P_0 of the next-hop saved before, $p(t)$ replaces P_0. Otherwise, drop the hello packet.

5. Routing Maintained and Update

Because of the topology is mobile as the time goes by, routing needs to be updated timely. Here, there are two schemes to update route.

5.1. Scheme 1. In our protocol, data is transmitted block by block [27]. In scheme 1, control packets are introduced. After receiving the last packets in a block, the next-hop node reply ACK packets including the number and ID of recovered packets and unrecovered packets. When the ratio of recovered packets to transmitted packets is lower than 70%, the previous hop resends the unrecovered original packets. Meanwhile, the cluster head sends request packets to update routing. The neighbor nodes send reply packets after receiving hello packets according to the requirements in Figure 7. The node calculates the priority and updates the best next-hop.

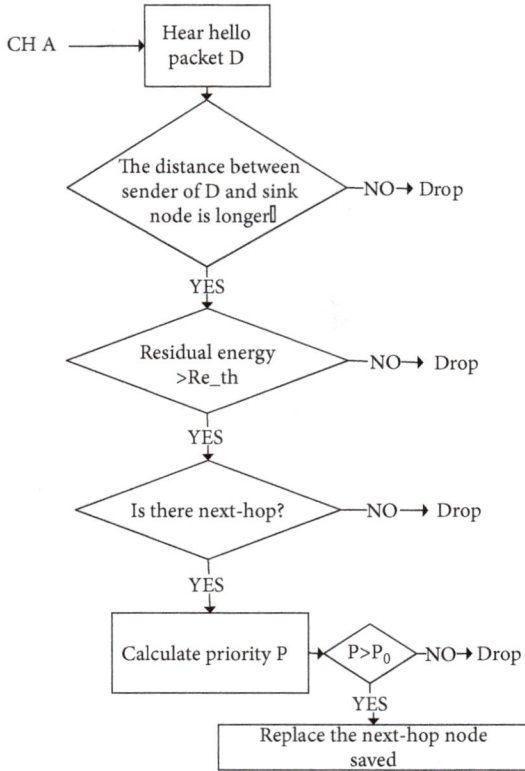

FIGURE 7: Routing algorithm flow chart.

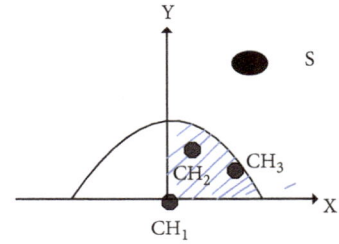

FIGURE 8: Expected time.

time is $\overline{E(t)}$ as showed in (18). Here, v is the water flow speed. The period of updating routing is set as $\overline{E(t)}$.

$$\overline{E(d)} = \int_0^R \int_0^R \left(\sqrt{R^2 - y^2} - x \right) \mathrm{d}x \mathrm{d}y, \qquad (17)$$

$$\overline{E(t)} = \frac{\overline{E(d)}}{v}. \qquad (18)$$

6. Performance Evaluation

6.1. Theoretical Analysis

6.1.1. The Size of a Block. To get the information of recovered data packets in next-hop, an ACK packet is necessary from the next-hop. To decrease the number of ACK packets, an ACK packet is sent for a block. The bigger the block, the less ACK to transmit constant data packets. However, because of the harsh environment under water, the bigger the block, the less probability of a successful delivery, which leads to too many re-transmission of original data packets. We know, the more original packets being transmission, the more energy consumption.

For the size of block is n_b packets and the packets is l bits, we assume the probability failing to deliver one bit is p. The probability of a block successfully delivery to next-hop is P as (19).

$$P = \left((1-p)^l \right)^{n_b} = (1-p)^{n_b l}. \qquad (19)$$

Therefore, we should select moderate block size to balance the delivery ratio and energy consumption.

6.1.2. The Number of Encoded Packets. From the basic theory of coding, we know the number of sending encoded packets should be more than original packets to decode them into initial packets. In fact, the more linearly independent encoded packets are sent, the easier they are to decode. However, too many packets being transmitted wastes energy and results in collision. The effective transmission is defined as (20). μ can be adjusted according to ACK from the next-hop. When the ratio recovered packets to total transmitted original packets is lower than α_0, the number of sending encoded packets must be adjusted according to (21). $\alpha(t)$ is

The scheme 1 is simple to realize, and period broadcasting control packets is avoided. Extra control packets are decreased compared with the previous routing update algorithms, which decreases the collisions and energy consumption. Meanwhile, only the next-hop with lower delivery ratio is updated, and other nodes are not affected.

However, the back and forth time of the control packets is too long because of the long propagation delay in UWSNs. The unstable stage results in the failing to deliver data packets. Therefore, the scheme 2 of updating routing is proposed.

5.2. Scheme 2. We know, in our algorithm, the location information of nodes can be gotten. The updating algorithm can be designed according to the location information. As Figure 8, CH_1 is the sender and CH_2 and CH_3 are the neighbors of CH_1 with the shorter distance to the sink node. With the water flow by, CH_2 and CH_3 may run out of the communication range of CH_1 in some time. Because of the distance between CH_1 and its neighbors is different, the time varies. To get the average time of running out of the current cluster head for each neighbor, the expected value of time is calculated.

Comparing with the horizontal movement, the motion of vertical direction has a small range and can be ignored. The neighbors with shorter distance to the sink node are set as uniform distribution. The expected distance, $\overline{E(d)}$, running out of communication range is showed as (17). The expected the number ratio of recovered packets to total transmitted original packets at time t.

$$\mu = \frac{\text{number of sending encoded packets}}{\text{number of original data packets}}, \quad (20)$$

$$\mu(t+1) = \mu(t) + \mu(t)(\alpha_0 - \alpha(t)). \quad (21)$$

6.1.3. The Energy Consumption. We assume the total number of original packets is n_0 and the average ratio sending encoded packets to original packets is μ_0. Because the number ratio of recovered packets to total transmitted original packets fluctuates near α_0, the average value is set as α_0. The average hop is n_h. Therefore, the total number of sending data packets N' is showed as (22).

$$N' = n_0(1 + \mu_0) + n_0(1 - \alpha_0). \quad (22)$$

In addition, the ACK packets are sent, and the number of ACK packets is relative with the size of block. So, the number of ACK packets n_a is showed in (23)

$$n_a = \frac{n_0}{n_b}. \quad (23)$$

During initial phase and routing updating phase, the control packets are sent. The number of control packets is increased as the number of nodes. We assume the value is constant n_c. During routing update phase, in scheme 1, only the recovered ratio is lowered than 70%, the node sends request packet, and the neighbor sends reply packets. We set the number of control packets as n_{cu}. In scheme 2, the node update routing according to $\overline{E(t)}$. The number of sending control packets is $n_u = (T_l/\overline{E(t)})n_c$.

In our algorithm, for simplicity, the energy consumption of receiving packets is ignored, the energy consumption of sending a data packets and a control packet is set as e_d and e_c. Therefore, the energy consumption E in scheme 1 and scheme 2 are showed, respectively, as (24) and (25).

$$E = N' e_d + (n_c + n_{cu})e_c, \quad (24)$$

$$E = N' e_d + \left(n_c + \frac{T_l}{\overline{E(t)}}n_c\right)e_c. \quad (25)$$

6.2. Simulation Analysis. Because the cost of arranging the UWSNs is too high, simulations are adopted to evaluate the performance of the designed protocol. NS3 is a popular simulator to simulate UWSNs. NS3 offers some characteristics that other simulators have not, such as underpinning to discrete-event-driven networks, simulation of high-fidelity UWSNs channels, complete protocol stack, and mobile 3D networks. Firstly, the schemes of constructing cluster and updating routing are evaluated. Secondly, the influence of the node density and the number ratio of regular nodes to cluster heads on the performance of the system is tested. Lastly, we compare RSHSC with NCRP and VBF. NCRP uses network coding to greedily forward data packets to the sink.

VBF constructs a vector pipe from the source node to the sink node. Only the nodes in the pipe forward data.

In our simulations, the nodes are deployed in a 3D area with $3000\,\text{m}*3000\,\text{m}*2000\,\text{m}$. The number of data packets is set as 60 in a block. When a node energy is exhausted, the simulation of this round is over. We show the average value of 50 runs. The detailed parameters are set as follows:

The data rate is 10 kbps. The center frequency is 12 KHZ. The bandwidth is 10 KHZ. Packet error rate model is ns3: UanPhyPerNoCode. Mode type is FSK. Signal noise model is ns3 :: Uan Phy Calc Sinr Default. Acoustic propagation speed is 1500 m/s. UAN Propagation model is ns3 :: Uan Prop Model Thorp. Energy model is acoustic modem energy model. MAC model is CWMAC. The mobility model is random walk 2D mobility model (speed: 2 ~ 4 m/s, directions are chosen randomly). The payload of DATA is 64 bytes, and the number of data packets in each block is 60. Deployment region is 3D region of $3 \times 3 \times 2\,\text{km}^3$. Node number is 20–75. The initial energy of advanced nodes is set as 100 J, and regular nodes is set as 25 J.

6.2.1. Performance at Different Cases. In our design, the ways of constructing clusters include SHSA and SHSB. Maintaining routing includes scheme 1 and scheme 2. Here, the number ratio of regular nodes to advanced node is 2 : 1. To evaluate the performance of each scheme, four cases are studied as Table 4. Delivery ratio and network life are two important parameters to evaluate the performance of the protocol. Delivery ratio is the ratio of number of sink nodes received data packets to the number of regular nodes sent data packets. Network life is total performing time until the first node drains its energy. For simplicity, we replace network life with the ratio of network life to evaluate the performance of the protocol.

From Figure 9(a) (Supplementary material (available here)), the delivery ratio of SHSA is almost the same with SHSB. The SHSA is based on regular based during constructing cluster, which can make each regular node assigned to a cluster head. Some regular nodes are kicked out by all cluster heads in its coverage in SHSB, which leads to the data packets collected by these nodes cannot be delivered to the sink node. The delivery ratio of scheme 2 is lower than scheme 1. On the one hand, a large number of control packets cause collisions. On the other hand, the instability of the whole network leads to the failure of delivery. The instable stage in scheme 1 also leads to failure of delivery. However, the successful reception of an original packet may increase the delivery rate. Because the reception of an original packet can help to decode multiple encoding packets. The different cases are showed in Table 5. In Figure 9(b), the network life ratio is not almost affected by the SHSA and SHSB (Supplementary material). The network life of case 1 and case 2 is longer than case 3 and case 4. Because scheme 2 produces many control packets, which wastes energy. Combining the above simulation results, we continue to study the performance of the protocol based on case 1.

6.2.2. The Affection of Number Ratio of Regular Nodes to Advanced Nodes (RRTA). In Figure 10 (Supplementary

(a) The delivery ratio

(b) The network life ratio

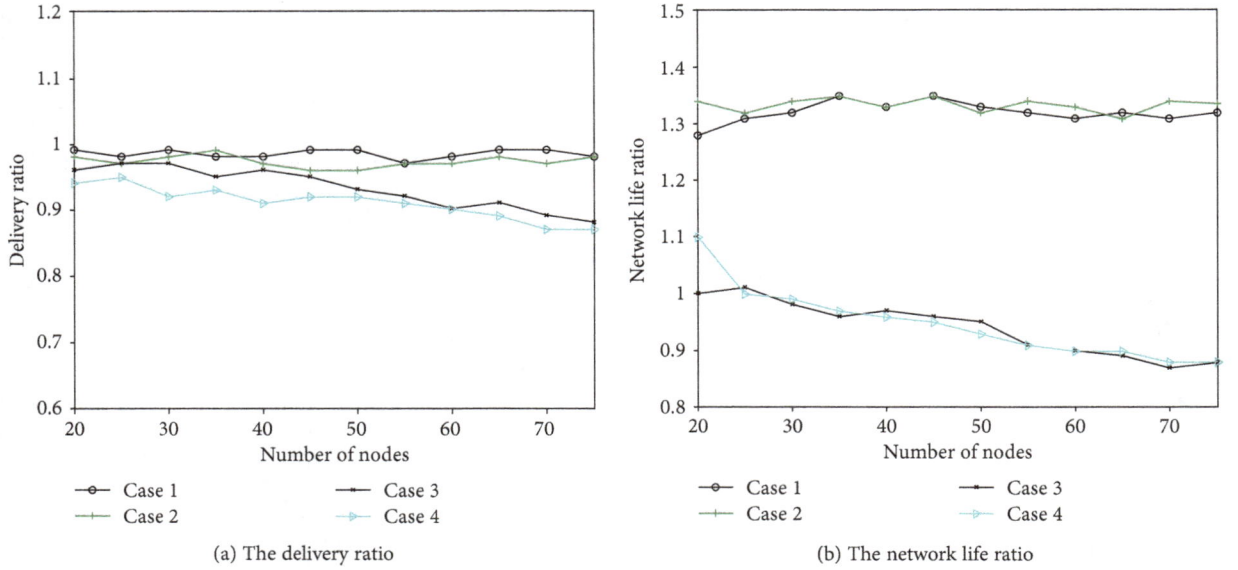

FIGURE 9: Affection of different schemes.

TABLE 5: Four cases.

Case number	Constructing cluster	Maintaining routing
Case 1	SHSA	Scheme 1
Case 2	SHSB	Scheme 1
Case 3	SHSA	Scheme 2
Case 4	SHSB	Scheme 2

material), N presents the layer numbers of advanced node being layout along vertical direction. We can see that the network life ratio decreases with the RRTA increasing, which is because the node first drains energy is regular node when the ratio is small. The regular nodes are in charge of only collecting data. The lifetime of regular nodes is related with the frequency of producing data packets and the initial energy. Meanwhile, the network life ratio decreases with the layer numbers increasing, because of the more layers, upper cluster heads consuming more energy than below nodes. In our simulation, the deliver ratio is about 60% when the communication range of nodes is 1000 m [27], and the vertical depth of the research area is 2000 m. Therefore, we select $N = 3$ and R RTA = 3.

6.2.3. Comparation with NCRP and VBF.
NCRP is a cross-layer routing protocol based on network coding (NCRP) for UWSNs, which utilizes network coding and cross-layer design to greedily forward data packets to sink nodes efficiently [27]. We set 60 data packets in each transmission block. The degree distribution is degree value = [1 4 6], degree probability distribution = [0.500 0.075 0.425], and average degree = 3.35. In VBF, a vector pipe is created from the source to the sink node [30]. Only the node in the pipe is qualified to be the next-hop. To limit the number of forwarding nodes, the delay time of forwarding data for each node is set. In simulations, the routing pipe radius is set as 300 m, the time interval of forwarding a packet is set as $T_{adaption} = \sqrt{\alpha} \times$

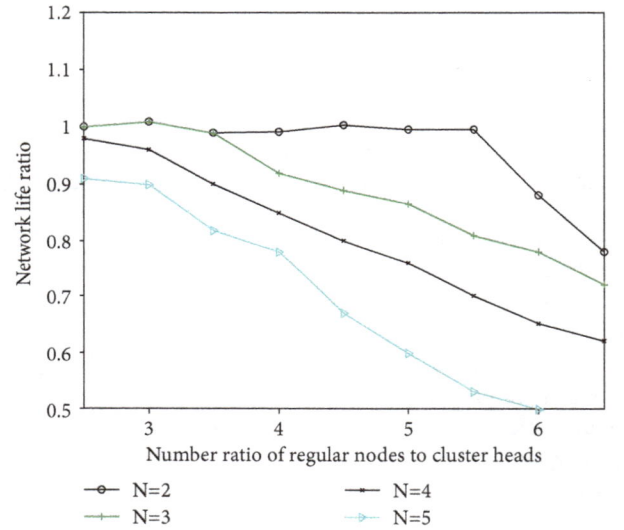

FIGURE 10: Affection of RRTA.

$T_{delay} + (R - d/v_0)$, α is calculated according to [30], T_{delay} is set as $0.02\sqrt{\bar{d}}$, $\bar{d}$ is the average distance among all nodes. R is the transmission radius. d is the distance between the current node and the forwarder. The other simulation parameters of VBF and NCRP are the same with the RSHSC.

From Figure 11(a) (Supplementary material), the delivery ratio of RSHSC is higher than VBF. Because in VBF, only the nodes in the pipe can forward data packets. However, in the sparse network, there may be no nodes in the pipeline, which leads to the failure of delivery. The delivery ratio of RSHSC is higher than NCRP. Although both the two protocols use coding techniques, the information of the next two hop is considered in RSHSC and avoids selecting void node as the next-hop. Figure 11(b) shows the comparison of energy consumption (Supplementary material). The energy

(a) Delivery ratio

(b) Energy consumption per packet

(c) End-to-end delay

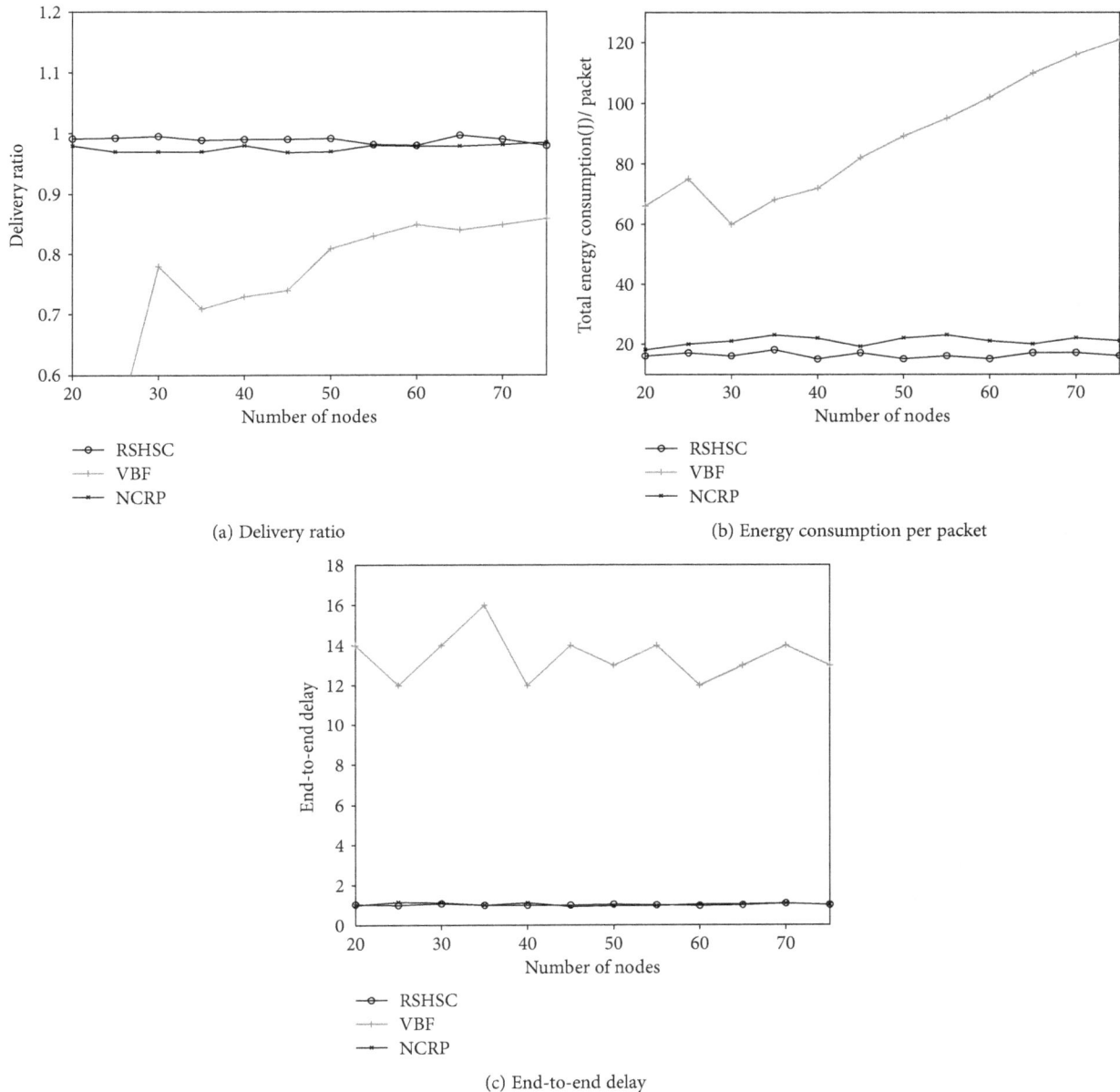

FIGURE 11: Comparison of performance.

consumption of VBF is higher than others, because VBF needs to send a large number of control packets. In addition, redundant forwarding exists in VBF. The energy consumption of RSHSC is lower than NCRP. In NCRP, the secondary nodes join to forward data packets, which wastes energy. While in RSHSC, only the best next-hop forwards data packets, and the data packets from other neighbors just join to decode. In addition, simplified harmony search algorithm is helpful to save energy of regular nodes and balance the energy among cluster heads. Figure 11(c) indicates the end-to-end delay of RSHSC is almost equal to NCRP and lower than VBF. Because VBF needs to wait to forward data packets, while NCRP and RSHSC transmit data block by block, the node immediately forward once data

received a data packet. After transmitting a block, the acknowledgement is sent, which shortens waiting time.

7. Conclusion

In this paper, RSHSC, an energy effective and reliable routing protocol based on harmony search algorithm and coding, is proposed for UWSNs. The process of RSHSC can be mainly divided into four parts: initialization, constructing clusters, intercluster routing and routing maintenance. In the initialization phase, the control packets are sent and nodes got the information of each other. In the process of constructing routing, each regular node is assigned to a cluster head according to simplified harmony search algorithm. The energy balance between cluster heads and the energy between ordinary nodes are taken into account when assigning. During the process of constructing routing, data packets are encoded. The best next-hop is responsible for forwarding data packets. The data packets from other neighbors is used to decode. The utilization of data packets from other neighbors increases the channel utilization. Meanwhile, the next two-hop is considered when selecting the best next-hop, which avoids the void node being selected as the next-hop. Two schemes are compared in the phase of maintaining routing. We get that scheme 1 is effective to save energy via theoretical analysis. Lastly, extensive simulations are conducted based NS-3. The results show that RSHSC is more effective than VBF and NCRP in decreasing energy consumption and increasing delivery ratio. In the future work, we will explore the implement of RSHSC and the better topology for improving the performance of UWSNs.

Conflicts of Interest

The authors declare that they have no conflicts of interest.

Authors' Contributions

Meiju Li designed the experiments; Meiju Li and Chunyan Peng analyzed the data; Xiujuan Du contributed analysis tools; and Meiju Li and Xiujuan Du wrote the paper.

Acknowledgments

This work is supported by the National Natural Science Foundation of China (61751111), Qinghai Office of Science and Technology (2018-SF-143), Key Lab of IoT of Qinghai (2017-ZJ-Y21), Hebei Engineering Technology Research Center for IOT Data Acquisition and Processing, CERNET Innovation Project (NGII20160307).

References

[1] E. Felemban, F. K. Shaikh, U. M. Qureshi, A. A. Sheikh, and S. B. Qaisar, "Underwater sensor network applications: a comprehensive survey," *International Journal of Distributed Sensor Networks*, vol. 11, no. 11, Article ID 896832, 2015.

[2] M. Murad, A. A. Sheikh, M. A. Manzoor, E. Felemban, and S. Qaisar, "A survey on current underwater acoustic sensor network applications," *International Journal of Computer Theory and Engineering*, vol. 7, no. 1, pp. 51–56, 2015.

[3] J. Lloret, "Underwater sensor nodes and networks," *Sensors*, vol. 13, no. 9, pp. 11782–11796, 2013.

[4] X. Du, C. Peng, and K. Li, "A secure routing scheme for underwater acoustic networks," *International Journal of Distributed Sensor Networks*, vol. 13, no. 6, 2017.

[5] C. Peng, X. Du, K. Li, and M. Li, "An ultra-lightweight encryption scheme in underwater acoustic networks," *Journal of*

[6] S. Climent, A. Sanchez, J. Capella, N. Meratnia, and J. Serrano, "Underwater acoustic wireless sensor networks: advances and future trends in physical, MAC and routing layers," *Sensors*, vol. 14, no. 1, pp. 795–833, 2014.

[7] M. Li, X. Du, K. Huang, S. Hou, and X. Liu, "A routing protocol based on received signal strength for underwater wireless sensor networks (UWSNs)," *Information*, vol. 8, no. 4, p. 153, 2017.

[8] S. Cai, N. Yao, and Z. Gao, "A reliable data transfer protocol based on twin paths and network coding for underwater acoustic sensor network," *EURASIP Journal on Wireless Communications and Networking*, vol. 2015, no. 1, 2015.

[9] X. Du, K. Li, X. Liu, and Y. Su, "RLT code based handshake-free reliable MAC protocol for underwater sensor networks," *Journal of Sensors*, vol. 2016, Article ID 3184642, 11 pages, 2016.

[10] M. R. Jafri, N. Javaid, N. Amjad, M. Akbar, Z. A. Khan, and U. Qasim, "Impact of acoustic propagation models on depth-based routing techniques in underwater wireless sensor networks," in *2014 28th International Conference on Advanced Information Networking and Applications Workshops*, pp. 479–485, Victoria, BC, Canada, 2014, IEEE.

[11] I. F. Akyildiz, D. Pompili, and T. Melodia, "Underwater acoustic sensor networks: research challenges," *Ad Hoc Networks*, vol. 3, no. 3, pp. 257–279, 2005.

[12] R. Headrick and L. Freitag, "Growth of underwater communication technology in the U.S. Navy," *IEEE Communications Magazine*, vol. 47, no. 1, pp. 80–82, 2009.

[13] S. Misra, S. Dash, M. Khatua, A. V. Vasilakos, and M. S. Obaidat, "Jamming in underwater sensor networks: detection and mitigation," *IET Communications*, vol. 6, no. 14, pp. 2178–2188, 2012.

[14] H. Yan, Z. J. Shi, and J. H. Cui, "DBR: depth-based routing for underwater sensor networks," in *NETWORKING 2008 Ad Hoc and Sensor Networks, Wireless Networks, Next Generation Internet*, vol. 4982 of Lecture Notes in Computer Science, pp. 72–86, Springer, Berlin, Heidelberg, 2008.

[15] A. Wahid, S. Lee, H. J. Jeong, and D. Kim, "EEDBR: energy-efficient depth-based routing protocol for underwater wireless sensor networks," in *Advanced Computer Science and Information Technology*, pp. 223–234, Springer, Berlin, Heidelberg, 2011.

[16] B. Diao, Y. Xu, Q. Wang et al., "A reliable depth-based routing protocol with network coding for underwater sensor networks," in *2016 IEEE 22nd International Conference on Parallel and Distributed Systems (ICPADS)*, pp. 270–277, Wuhan, Hubei, China, 2017, IEEE.

[17] H. Yu, N. Yao, T. Wang, G. Li, Z. Gao, and G. Tan, "WDFAD-DBR: weighting depth and forwarding area division DBR routing protocol for UASNs," *Ad Hoc Networks*, vol. 37, Part 2, pp. 256–282, 2016.

[18] J. Xu, K. Li, G. Min, K. Lin, and W. Qu, "Energy-efficient tree-based multipath power control for underwater sensor networks," *IEEE Transactions on Parallel and Distributed Systems*, vol. 23, no. 11, pp. 2107–2116, 2012.

[19] B. Ali, A. Sher, N. Javaid, S. Islam, K. Aurangzeb, and S. Haider, "Retransmission avoidance for reliable data delivery

in underwater WSNs," *Sensors*, vol. 18, no. 2, p. 149, 2018.

[20] K. Hao, Z. Jin, H. Shen, and Y. Wang, "An efficient and reliable geographic routing protocol based on partial network coding for underwater sensor networks," *Sensors*, vol. 15, no. 6, pp. 12720–12735, 2015.

[21] S. Ahmed, N. Javaid, F. A. Khan et al., "Co-UWSN: cooperative energy-efficient protocol for underwater WSNs," *International Journal of Distributed Sensor Networks*, vol. 11, no. 4, Article ID 891410, 2015.

[22] A. Umar, N. Javaid, A. Ahmad et al., "DEADS: depth and energy aware dominating set based algorithm for cooperative routing along with sink mobility in underwater WSNs," *Sensors*, vol. 15, no. 6, pp. 14458–14486, 2015.

[23] N. Ilyas, T. A. Alghamdi, M. N. Farooq et al., "AEDG: AUV-aided efficient data gathering routing protocol for underwater wireless sensor networks," *Procedia Computer Science*, vol. 52, pp. 568–575, 2015.

[24] S. Ahmed and S. S. Kanhere, "HUBCODE: message forwarding using hub-based network coding in delay tolerant networks," in *Proceedings of the 12th ACM international conference on Modeling, analysis and simulation of wireless and mobile systems - MSWiM '09*, pp. 288–296, Tenerife, Canary Islands, Spain, 2009, ACM.

[25] I. Azam, N. Javaid, A. Ahmad, W. Abdul, A. Almogren, and A. Alamri, "Balanced load distribution with energy hole avoidance in underwater WSNs," *IEEE Access*, vol. 5, no. 99, pp. 15206–15221, 2017.

[26] H. Wu, M. Chen, and X. Guan, "A network coding based routing protocol for underwater sensor networks," *Sensors*, vol. 12, no. 4, pp. 4559–4577, 2012.

[27] H. Wang, S. Wang, R. Bu, and E. Zhang, "A novel cross-layer routing protocol based on network coding for underwater sensor networks," *Sensors*, vol. 17, no. 8, p. 1821, 2017.

[28] Z. Jin, Y. Ma, Y. Su, S. Li, and X. Fu, "A Q-learning-based delay-aware routing algorithm to extend the lifetime of underwater sensor networks," *Sensors*, vol. 17, no. 7, p. 1660, 2017.

[29] X. J. Du, K. J. Huang, S. L. Lan, Z. X. Feng, and F. Liu, "LB-AGR: level-based adaptive geo-routing for underwater sensor network," *The Journal of China Universities of Posts and Telecommunications*, vol. 21, no. 1, pp. 54–59, 2014.

[30] P. Xie, J. H. Cui, and L. Lao, "VBF: vector-based forwarding protocol for underwater sensor networks," in *NETWORKING 2006. Networking Technologies, Services, and Protocols; Performance of Computer and Communication Networks; Mobile and Wireless Communications Systems*, pp. 1216–1221, Springer, Berlin, Heidelberg, 2006.

2

Heuristic Localization Algorithm with a Novel Error Control Mechanism for Wireless Sensor Networks with Few Anchor Nodes

Yujia Sun[iD]**, Xiaoming Wang**[iD]**, Jiyan Yu, and Yu Wang**

Ministerial Key Laboratory of ZNDY, Nanjing University of Science and Technology, Nanjing 210094, China

Correspondence should be addressed to Xiaoming Wang; 202xm@163.com

Academic Editor: Bruno Andò

A novel iterative localization algorithm with high accuracy and low anchor node dependency for large-scale wireless sensor networks is proposed in this paper. At each iteration, blind nodes are located using a weighted linear least squares-based algorithm. To prevent errors in the blind nodes from propagating and accumulating throughout the network, an anchor geometric feature-based error control mechanism is used to select the nodes that participate in the localization and to estimate the localization confidence. The simulation results show that the algorithm can be used when only a few anchor nodes are involved. This algorithm is more advanced than traditional methods, which often require a large number of well-placed anchor nodes to operate appropriately. By optimizing the decision parameter v of the algorithm, the average localization error of the algorithm is approximately 0.43 meters. When the ratio of anchor nodes (the ratio of the number of anchor nodes to the number of sensor nodes in the network) is 1.25% (i.e., 5 anchor nodes for 400 sensor nodes), the received signal strength indicator (RSSI) variance is 8 dBm, and the radio range is 50 meters. A comparison of the proposed algorithm with global localization methods, including multidimensional scaling (MDS), semidefinite programming (SDP), and shortest-path access (SPA), shows that the proposed algorithm achieves higher location accuracy and stability when the number of anchor nodes is varied. The efficiency of the proposed localization algorithm is evaluated in a real sensor network, and the accuracy is high and robust to radio channel variance.

1. Introduction

Wireless sensor networks (WSNs) are a basic component of the Internet of Things. Based on the development of mobile computing and embedded technologies, WSNs have been widely implemented in daily applications, such as health care monitoring, natural disaster prevention, and surveillance [1–3]. Sensor network services, such as mobile sinks, geographic routing, and location-based multicasting, rely on the physical location information of sensor nodes or the phenomena of interest. One method of accessing the locations of nodes is to use the Global Positioning System (GPS). However, installing a GPS chip on every sensor node is expensive. Moreover, the GPS is not always available because of environmental or hardware limitations, such as shadowing, energy, volume, or cost issues. Therefore, efficient and inexpensive localization technologies must be developed [4, 5].

Various WSN localization approaches have been presented in the literature. Depending on the usage of the range information between nodes, node localization approaches can be classified into range-based localization and range-free localization. In a range-based localization method, the blind nodes estimate the distance to each anchor node and then estimate the location based on the distance to and position of the anchor nodes. Anchor nodes are a special type of sensor node with known locations, with the help of GPS, or are predeployed. Blind nodes are sensor nodes with unknown positions. To estimate distance, blind nodes use radio channel information, such as the received signal strength indicator (RSSI) [6, 7], time of arrival (ToA) [8, 9], time difference of arrival (TDoA) [10, 11], angle of arrival (AoA) [12], or some combination of these methods. The RSSI-based distance estimation is attractive because of its low cost, long range, and simplicity. After determining the distance between blind nodes and anchor nodes, multilateration technology is used for position estimation [13, 14]. Trilateration is a particular form of multilateration that utilizes three anchor nodes to calculate the location of a blind node in two dimensions. For a given WSN, to estimate the position of all blind nodes, at least three anchor nodes are required in the communication range of the blind node. For a large-scale outdoor randomly deployed WSN, a large number of anchor nodes should be deployed to guarantee that every blind node is located. However, additional anchor nodes correspond to additional costs associated with hardware, deployment, maintenance, and so on.

Iterative node localization is a novel localization method that can decrease the dependence on anchor nodes by importing a blind node leveling-up scheme [15, 16]. For iterative node localization, the located blind node is upgraded to new anchor nodes (pseudoanchor) in which the location information of the initial anchor nodes is progressively propagated to other nodes. Theoretically, all blind nodes in a network can be located with three anchor nodes under an iterative localization method. In practical localization systems, measurement noise during node ranging is inevitable, and the location of a pseudoanchor is uncertain. As other blind nodes locate themselves and refer to the pseudoanchors with uncertain locations, the localization error increases as more pseudoanchors are introduced into the localization. This process is called the error propagation problem in iterative node localization [17]. Various methods have been proposed to mitigate the propagation error. Liu et al. [18] proposed an error control mechanism that tracks the estimation error and quantifies each location estimation with a given level of uncertainty. Based on the location certainty information, the algorithm introduces a neighbor selection procedure and an update criterion. The simulation results show that the localization accuracy and robustness are significantly improved with this approach. Yang and Liu [19] proposed the concept of the quality of trilateration (QoT), which considers both the geometric relationship and ranging errors. Based on the QoT, a confidence-based iterative localization (CIL) scheme can be designed. At each stage, the CIL selectively utilizes trilaterations and reduces the likelihood of using low-confidence references, which effectively halts error propagation. Wu et al. [20] considered the error control problem for non-Gaussian noise measurements. The proposed error control algorithm estimates the location error based on nonlinear least squares residuals. Additionally, a robust formulation of error control is proposed that can reduce the accumulated error. The simulations show that more anchors, less noise, and fewer iterations reduce the localization error. Hu et al. [21] derived an upper bound of error propagation for iterative localization that works well when the precise environmental noise distribution is unavailable. The minimum upper bound is adopted to evaluate the localization result. With this method, the algorithm constructs the proper linear least squares localization with high probability. The main concept underlying these methods is the estimation of the ranging error. The algorithms exhibit good localization accuracy when the noise is precisely estimated.

In this paper, we analyze the localization problem to reduce anchor node dependency in a large-scale WSN. A novel iterative localization algorithm with an anchor geometric feature-based error control mechanism is proposed. Compared with previous research that precisely modeled the localization error during each iteration, we formulate the error control problem as a classification problem. Based on node localization data under varying ratios of anchor nodes, RSSI variances, and radio communication ranges in a randomly deployed node environment, we characterize the localization condition by considering the number of anchor nodes, the anchor positions, the distance between the anchors and blind nodes, and the spatial distribution of anchors. Additionally, a one-class support vector decision-maker is trained based on the selected features. The simulation results show that the algorithm can be used when only a few anchor nodes are involved, which is more advanced than traditional methods that often require a large number of well-placed anchor nodes to work well. We evaluate the performance of our algorithm with different numbers of anchor nodes and RSSI variances. The results show that the algorithm efficiently solves the error propagation problem. The average localization error of the algorithm is approximately 0.43 meters when the percentage of anchor nodes is 1.25%, the RSSI variance is 8 dBm, and the radio range is 50 meters. The location accuracy is better than those of certain global localization methods, including MDS, SDP, and SPA. The efficiency of the proposed localization algorithm is evaluated in a real sensor network, and the accuracy is high and robust to radio channel variance.

The remainder of this paper is organized as follows. In Section 2, we present the RSSI channel model and the weighted linear least squares location estimator. In Section 3, we present the anchor geometric features and the proposed error control mechanism. In Section 4, we evaluate the proposed localization algorithm with simulation experiments. In Section 5, we conclude the paper.

2. Localization Method

In this paper, we classify wireless sensor nodes as blind nodes, anchor nodes, pseudoanchor nodes, and reference nodes based on their localization duty. These names will be discussed in the following sections, and their respective duties are explained at the beginning of the paper. Anchor nodes are nodes that can access their absolute location at the beginning of the deployment. Blind nodes are nodes that do not know their location and attempt to estimate it based on the locations of the anchor nodes. A pseudoanchor node is a blind node that has an estimated location and is upgraded to an anchor node. The location of the pseudoanchor node always has an error, and the algorithm should be designed to stop the error from propagating throughout the network. The reference node is a node used by the blind node to locate itself. Both the anchor nodes and the pseudoanchor nodes can be used as reference nodes.

2.1. RSSI Signal Model. Many channel models have been proposed in the literature for indoor or outdoor localization applications [22]. In this paper, the log-normal shadowing model is used for RSSI-based localization because of its simplicity. Let P_{RX} denote the received power and d denote the distance between the transmitter and the receiver. Then, the log-normal shadowing model is expressed as follows:

$$P_{RX}(\text{dBm}) = A - 10\eta \log \frac{d}{d_0} + N, \qquad (1)$$

where A is the path loss at distance d_0, η is the path loss exponent, and $N \sim N(0, \sigma^2)$ is a zero mean random variable with variance σ. In general, d_0 is equal to 1 meter, and η is in the range of 2 to 4. The parameters A and η should be measured through real experiments. In this paper, we conduct the experiment based on the TI CC2650 sensor node in an outdoor environment. The results are shown in Figure 1. This channel model is used in the following section.

2.2. Weighted Linear Least Squares Estimator. The distance $\tilde{d}_i$ between a blind node and each reference i $(i = 1, 2, \ldots, m)$ is estimated based on the radio channel model. Let (x, y) represent the position of the blind node and (x_i, y_i) represent the position of the reference. The distance estimation error is expressed as follows:

$$\varepsilon = \sum_{i=1}^{m} \left(\sqrt{(x_i - x)^2 + (y_i - y)^2} - d_i \right)^2. \qquad (2)$$

The localization problem finds the optimal (x, y) that minimizes ε. Formula (2) is a nonlinear function, and the optimal position is iteratively calculated using a searching method, such as a gradient search, as follows:

$$\begin{bmatrix} x \\ y \end{bmatrix}_{k+1} = \begin{bmatrix} x \\ y \end{bmatrix}_k - \alpha \begin{bmatrix} \dfrac{\partial \varepsilon}{\partial x} \\ \dfrac{\partial \varepsilon}{\partial y} \end{bmatrix}_{x=x_k, y=y_k}. \qquad (3)$$

Searching methods require an initial estimate of the position, which is difficult in real environmental applications. Therefore, we prefer to translate the problem into a linear problem that can be solved with a least squares estimator. Based on the positions of the blind node and the references, the distance between them can be expressed as follows:

$$d_i^2 = (x_i - x)^2 + (y_i - y)^2, \qquad (4)$$

where d_i is the real distance, which is different from the noisy estimation distance $\tilde{d}_i$. By setting reference node 1 as an origin, we obtain $d_1^2 = (x_1 - x)^2 + (y_1 - y)^2$. Then, for the other references, we obtain the following equation:

$$d_i^2 - d_1^2 = (x_i - x)^2 + (y_i - y)^2 - (x_1 - x)^2 - (y_1 - y)^2, \qquad (5)$$

where $i > 1$. Formula (5) can be expressed in matrix form as follows:

$$G - 2WP = 0, \qquad (6)$$

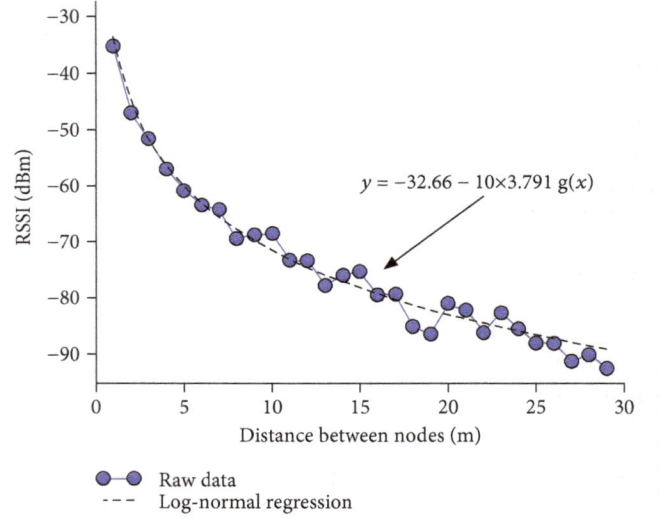

FIGURE 1: RSSI model in the outdoor environment.

where

$$G = \begin{bmatrix} d_2^2 - d_1^2 - x_2^2 - y_2^2 + x_1^2 + y_1^2 \\ d_3^2 - d_1^2 - x_3^2 - y_3^2 + x_1^2 + y_1^2 \\ \vdots \\ d_m^2 - d_1^2 - x_m^2 - y_m^2 + x_1^2 + y_1^2 \end{bmatrix},$$

$$W = \begin{bmatrix} x_1 - x_2 & y_1 - y_2 \\ x_1 - x_3 & y_1 - y_3 \\ \vdots & \vdots \\ x_1 - x_m & y_1 - y_m \end{bmatrix}. \qquad (7)$$

In a real RSSI-based localization system, the real distance between nodes is never obtained. Instead, a noisy distance estimation $\tilde{d}_i$ is obtained. Then, matrix G can be formulated as follows:

$$\tilde{G} = \begin{bmatrix} \tilde{d}_2^2 - \tilde{d}_1^2 - x_2^2 - y_2^2 + x_1^2 + y_1^2 \\ \tilde{d}_3^2 - \tilde{d}_1^2 - x_3^2 - y_3^2 + x_1^2 + y_1^2 \\ \vdots \\ \tilde{d}_m^2 - \tilde{d}_1^2 - x_m^2 - y_m^2 + x_1^2 + y_1^2 \end{bmatrix}. \qquad (8)$$

Now, the position of the blind node is calculated as the least squares solution of (6), which is expressed as follows:

$$P = \frac{1}{2} \left(W^T W \right)^{-1} W^T \tilde{G}. \qquad (9)$$

The RSSI signal is sensitive to environmental interference. The linear least squares localization solution exhibits poor accuracy when the RSSI variance is increasing. To improve the localization accuracy, a weighted linear least squares estimator is introduced, in which the solution is weighted based on the variance of the distance estimation.

The weighted linear least squares solution is expressed as follows:

$$P = \frac{1}{2}\left(\mathbf{W}^T\mathbf{S}^{-1}\mathbf{W}\right)^{-1}\mathbf{W}^T\mathbf{S}^{-1}\tilde{\mathbf{G}}, \tag{10}$$

where matrix $\mathbf{S}$ is the covariance matrix of $\tilde{\mathbf{G}}$. Assuming that the positions of the references are constant and that the distance measurements are independent, matrix $\mathbf{S}$ is expressed as follows:

$$\mathbf{S} = \begin{bmatrix} V\left(\tilde{d}_1^2\right) + V\left(\tilde{d}_2^2\right) & V\left(\tilde{d}_1^2\right) & \cdots & V\left(\tilde{d}_1^2\right) \\ V\left(\tilde{d}_1^2\right) & V\left(\tilde{d}_1^2\right) + V\left(\tilde{d}_3^2\right) & \cdots & V\left(\tilde{d}_1^2\right) \\ \vdots & \vdots & \ddots & \cdots \\ V\left(\tilde{d}_1^2\right) & V\left(\tilde{d}_1^2\right) & \cdots & V\left(\tilde{d}_1^2\right) + V\left(\tilde{d}_m^2\right) \end{bmatrix}, \tag{11}$$

where $V()$ is the variance of each distance measurement.

The weighted linear least squares localization displays increased accuracy and robustness and less complexity than traditional methods.

3. Error Control Method

3.1. Geometric Features of Localization. Multilateration is infeasible when three references are collinear, and it is feasible when three references are not collinear [21]. In a large-scale randomly deployed WSN, the references are rarely collinear. However, this does not mean that the blind node has a good localization result for any type of anchor node position. Therefore, we conduct a simulation to investigate the relationship between the reference node geometry and localization accuracy. In the simulation, the network includes 400 wireless sensor nodes that are randomly deployed in a square area with an edge equal to 200 m. The blind node is located based on (10), and the RSSI measurement suffers from Gaussian noise with a mean of zero and a variance of 2 dBm. Additional simulation details are presented in Table 1.

Based on the localization results of the simulation, six typical localization situations for a single blind node with different numbers of anchor nodes are plotted in Figure 2. The red triangle represents the position of the blind node, and the blue dots and yellow stars represent the positions of the anchor nodes and the estimated positions of the blind node, respectively. The red circle is the communication range of the blind node. Figures 2(a)–2(c) are the situations where the blind node is associated with a bad localization result, with numbers of anchor nodes equal to 6, 10, and 16, respectively. The distribution of the estimated positions becomes concentrated when there are more anchor nodes, which means that the localization accuracy improves when the number of anchor nodes increases. Figures 2(d)–2(f) are situations in which the blind node is associated with a good localization result, with the same number of anchor nodes. Compared with Figure 2(a), the localization result is much more accurate in Figure 2(d). As the number of anchor nodes

TABLE 1: Description of simulation parameters.

Parameter	Description	Default
M	Number of blind nodes	380
N	Number of anchor nodes	20
L	Number of initial pseudoanchors	5
R	Radio communication range	30 m
σ	RSSI variance	1 dBm
A	Path loss at one meter	32.6 dBm
η	Path loss exponent	3.79
ϕ	OC-SVM kernel type	Linear
v	OC-SVM parameter nu	0.1
e	OC-SVM parameter epsilon	0.001

and the network configuration are the same for the situations in Figures 2(a) and 2(d), the geometry of the anchor nodes in Figure 2(d) can help improve the localization accuracy. Furthermore, a comparison of Figures 2(c) and 2(d) reveals that this improvement is significant. Figure 2(d) is more accurate than Figure 2(c) with a smaller number of anchor nodes (6 in Figure 2(d) and 16 in Figure 2(c)). In this paper, offset and collinear are used to describe the geometric pattern of the anchor nodes, and the quantified features are discussed in Section 3.2. The pattern offset means that the centroid of the anchor nodes is far from the blind node, as Figures 2(b) and 2(c) show. In this situation, the estimated position tends to be in the direction of the vector centroid based on the real blind node position. The collinear pattern is not a real collinear condition, but the anchor nodes are almost located in a single line, as Figure 2(a) shows. In this pattern, the centroid of the anchor node may be close to the real position of the blind node, but the estimated position has a large variance in the direction from the centroid to the real blind node position.

As Figure 2 shows, the number of anchor nodes affects the localization accuracy of the blind node, and the result is more accurate when the number of anchor nodes is high. We collect the number of references of each blind node and the associated localization error during the simulation. The data are plotted in Figure 3. The number of references has a large range from 3 to 25. Additionally, for most blind nodes, the number of references ranges from 10 to 15. The localization error decreases as the number of references increases. Moreover, the localization results exhibit considerable uncertainty when the number of references is less than 5, as the anchor nodes tend to display a bad geometric pattern. In this way, the number of references can be used as a feature for predicting localization errors that are larger than a specified threshold.

Our observations show that the number of references and the reference geometric pattern significantly affect the localization accuracy. When the references are not evenly distributed around the blind node or when the number of references is less than the threshold, the localization results will exhibit considerable uncertainty. Therefore, we believe

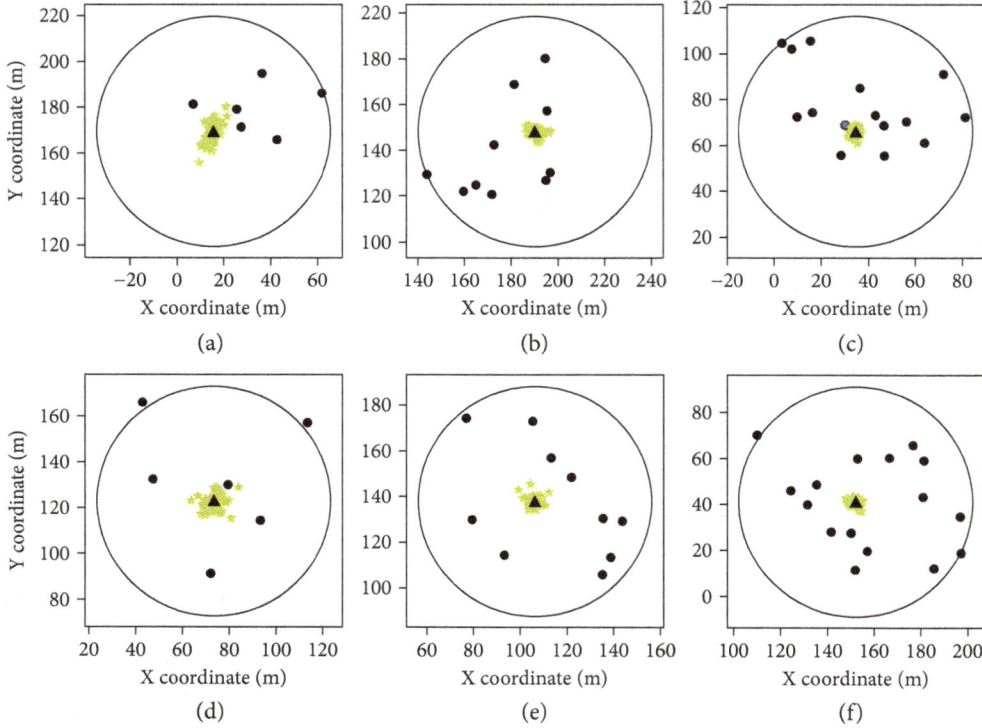

FIGURE 2: Estimation layouts for different anchor (neighbor) geometries. (a) Uncertain with 6 references; (b) uncertain with 10 references; (c) uncertain with 16 references; (d) stable with 6 references; (e) stable with 10 references; and (f) stable with 16 references.

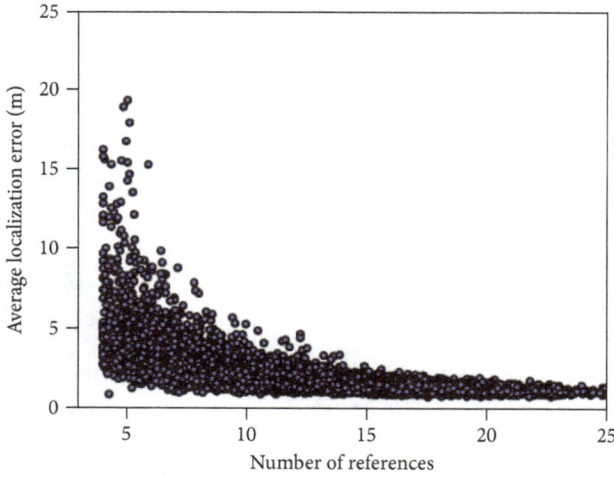

FIGURE 3: Localization error for different numbers of references.

that the localization error can be predicted by characterizing the reference geometry.

3.2. Feature Selection.

3.2. Feature Selection. The uncertain locations of the references, the distance estimation errors, and the reference geometry are the factors that affect the localization accuracy. These features can be used to predict the localization accuracy and inform the decision-making in the blind upgrading process.

The covariance matrix is used to describe the data distribution in a higher dimension, which is an extension of the variance. In our implementation, we construct the covariance matrix based on the coordinates of the references in a 2-dimensional plane. Given a blind node, let $X = \{x_1, x_2, \dots, x_m\}$ denote the x coordinates of the references and $Y = \{y_1, y_2, \dots, y_m\}$ denote the y coordinates of the references, where m represents the number of references. The covariance matrix of X and Y is given as follows:

$$C = \begin{bmatrix} \text{cov}(X, X) & \text{cov}(X, Y) \\ \text{cov}(Y, X) & \text{cov}(Y, Y) \end{bmatrix}, \tag{12}$$

where $\text{cov}(X, Y) = E[(X - (EX))(Y - (EY))]$, and $E(X)$ is the expectation of X. Our geometric features are based on the eigenvalue of C. We denote the larger eigenvalue as λ_1 and the smaller one as λ_2. Then, we define a feature ALC to describe the linearity of a set of references because this linearity causes unstable localization. ALC can be expressed as follows:

$$\text{ALC} = \frac{\lambda_1 - \lambda_2}{\lambda_1}. \tag{13}$$

When the spatial location of a set of references is located on a line, ALC is equal to 1. However, when the set of references is located on a circle, ALC is equal to 0.

To describe the uniformity of the references, we define two features ALR and ALS. ALR is expressed as follows:

$$ALR = \left| \frac{R^2 - \lambda_1 \lambda_2}{R^2 (d\,max - d\,min)} \right|. \qquad (14)$$

ALS is expressed as follows:

$$ALS = \frac{\sum_{i=1}^{m} \left| \sqrt{(x_c - x_i)^2 + (y_c - y_i)^2} - d_i \right|}{m}, \qquad (15)$$

$$(x_c, y_c) = \left(\frac{\sum_{i=1}^{m} x_i}{m}, \frac{\sum_{i=1}^{m} y_i}{m} \right), \qquad (16)$$

where (x_c, y_c) represents the centroid of the references, R is the communication range of the blind node, and d_i is the distance from the blind node to each reference. When the references are uniformly located around the blind node, the values of ALR and ALS tend to be 0; however, the values here are 1 and R, respectively.

In addition to the previously proposed features, several basic data features used in classification applications are shown in Table 2. These basic features work on three signal channels, which are the X coordinates of the references, the Y coordinates of the references, and the distance measurement from each reference. Based on these features, a feature space is constructed, and the importance of the features is analyzed in the following section.

In total, 25 features are extracted, but some are irrelevant or redundant features. Therefore, feature selection and feature ranking are applied to identify the important features in the feature space. Feature selection methods can be classified as the wrapper method, the embedded method, and the filter method. In this paper, we apply a two-stage feature selection procedure. First, the maximal information coefficient (MIC) is used to score each feature. Then, the 10 best features are selected to form a feature subset, and a recursive feature elimination (RFE) method is used to select the 4 best features from the subset. The MIC is a measure of the strength of the linear relationship between two variables X and Y. The basic concept of the MIC is to use the relevant mutual information (MI). To obtain the MIC score, the values of two variables are separated into different numbers of bins to construct rectangular grids with different resolutions. Thus, the distribution of the cells in each grid is the fraction of points that fall into that cell. Then, the MI statistics for each grid are calculated, and the maximum is chosen as the MIC score. The MIC of two variables X and Y is defined as follows:

$$MIC(X, Y) = \max_{x_n y_n < N^{0.6}} \left[\frac{MI_{x_n, y_n}(x, y)}{\log_2 \min \{x_n, y_n\}} \right], \qquad (17)$$

where N is the sample size; x_n and y_n denote the number of bins for the x- and y-axes, respectively; and MI_{x_n, y_n} are the MI statistics between the variables for an $x_n - by - y_n$ rectangular grid. We calculate the MIC between the feature and the localization error based on (17). The result is used to score the corresponding feature, as shown in Figure 4.

The 10 best features with their corresponding ranks are shown in Table 3. The features proposed in this paper yield

TABLE 2: Signal features from the literature.

Feature	Description
Len()	Number of references
Mean()	Mean value
Std()	Standard deviation
Max()	Largest value in a series
Min()	Smallest value in a series
Sma()	Signal magnitude area
Energy()	Sum of squares divided by the number of values
Entropy()	Signal entropy

high MIC scores and are ranked 1, 2, and 4. In addition, the feature pLen is ranked 3, which means that the number of references is highly related to the localization error. For all possible combinations of up to four of these features, we test the classification accuracy based on a support vector machine (SVM). Table 4 details the accuracy of the three best and the three worst feature combinations. The feature combination with the most accurate results (ALC, ALR, ALS, and pLen) is used for further analysis.

3.3. Error Control Algorithm. As previously noted, in a large-scale WSN localization problem, the percentage of anchor nodes is expected to be small, which results in a situation in which a small percentage of blind nodes can directly communicate with the anchor nodes. However, none of the blind nodes in the network may be able to find at least three anchor nodes to derive a location. To overcome this problem, a two-phase localization algorithm that includes the initialization phase and the error control phase is proposed.

Considering the small number of anchor nodes, the initialization algorithm will find a series of pseudoanchor nodes and estimate their positions based on a distance vector-hop (DV-hop) [23] distance estimation. Initially, the anchor nodes in the network broadcast a packet that contains the unique ID and position of the node and hop information (equal to 0). All the neighbors surrounding the anchor node receive the packets. The RSSI, position of the anchor node, and hop information of 1 are recorded by the one-hop neighbors, and they then broadcast another packet containing the anchor position and hop information of 1. The nodes that receive the second packet and did not receive the first packet will record the anchor position and hop information of 2, and they then begin a new broadcast. The anchor nodes that receive a packet with hop information of 1 record the unique ID and the RSSI value of the packet to form a one-hop neighbor list. This process continues until each anchor position and the associated hop information have been spread to every node in the network. The nodes that receive hop information multiple times search for the smallest number of hops for each anchor node. This process improves the accuracy of a DV-hop distance estimation and reduces network traffic.

By determining the hop information associated with each anchor node, a blind node estimates the distances by multiplying the hop information by a shared average hop distance metric. The DV-hop distance estimation suffers from large

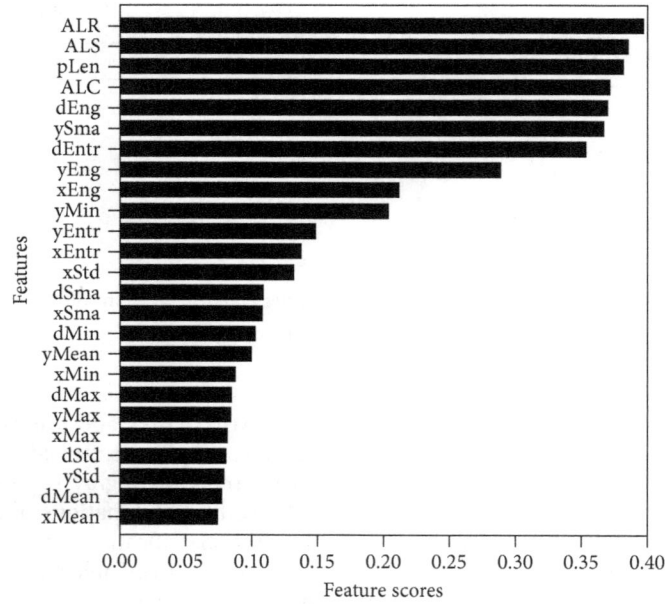

FIGURE 4: MIC scores for 25 different features.

TABLE 3: Rank of the 10 best features.

Rank	1	2	3	4	5
Name	ALR	ALS	pLen	ALC	dEng
Rank	6	7	8	9	10
Name	ySma	dEntr	yEng	xEng	yMin

TABLE 4: Classification accuracy for different feature combinations.

Accuracy	1	2	3	4	5	6	7	8	9	10
0.843	×	×	×	×						
0.839		×		×		×	×			
0.835	×	×	×			×				
0.636							×	×	×	×
0.645				×	×				×	×
0.649					×	×	×			×

error compared with those of channel model-based methods because the localization error of the DV-hop estimation process is large. Therefore, multiple blind nodes are selected as pseudoanchor nodes in the initialization phase. The blind nodes in the one-hop range of each anchor node are the best choices of pseudoanchor nodes because one reference distance can be estimated based on the channel model. Given the one-hop neighbors of an anchor, the blind nodes are sorted by their distance to the anchor node, and the nearest n_init blind nodes are the pseudoanchor nodes. Then, the positions of these pseudoanchor nodes are estimated with (10). The initialization phase is summarized in Algorithms 1 and 2.

In an iterative localization scenario, the position of a pseudoanchor node is uncertain. This positional error may further affect the localization of the other nodes and increase

```
Input: Number of initial Pseudo-anchor n_init
Define: HopInfo Packet HIP = {ID, position, hopInfo}
        NeighborInfo Packet NIP = {ID_1,···, ID_{n_init}}
        Neighbor RSSI List NRL = [ID_i, RSSI_i]
        Neighbor Distance List NDL = [ID_i, d_i]
Procedures:
1: send an HIP
2: while not time up:
3:    if received an HIP:
4:       parse the HIP to obtain ID, position, hopInfo
5:       obtain the RSSI
6:       if hopInfo is equal to 1:
7:          append [ID, RSSI] to NRL
8:          reset the timer
9:       else:
10:         pass
11: end while
12: for i from 1 to sizeof(NRL):
13:    estimate d_i based on RSSI_i by Eq. (1)
14:    append [ID_i, d_i] to NDL
15: endfor
16: sort NDL by d_i in ascending
17: if sizeof (NDL) > n_init:
18:    set NIP equal to the first n_init IDs in NDL
19: else:
20:    set NIP equal to the IDs in NDL
21: send an NIP
```

ALGORITHM 1: Initialization phase for anchor nodes.

the localization error, which represents the problem of error propagation in an iterative localization scenario. Therefore, an error control procedure must be implemented to improve the localization performance. In this paper, we use an error control method based on the neighboring geometric features

---

Input: The unique ID of current node ID_c
Define: HopInfo Packet $HIP = \{ID, position, hopInfo\}$
      NeighborInfo Packet $NIP = \{ID_1, \cdots, ID_{n\_init}\}$
      Anchor HopInfo List
              $AHL = [ID_i, position_i, hopInfo_i]$
      Neighbor Distance List $NDL = [ID_i, d_i]$
Procedure:
1: **if** received an HIP:
2:    parse the HIP to obtain $ID, position, hopInfo$
3:    **if** ID in $AHL.ID$ and ID is an anchor:
4:       **if** $hopInfo < AHL\ [ID].hopInfo$
5:          $AHL\ [ID].hopInfo = hopInfo$
6:    **else if** ID is an anchor:
7:       append $[ID, position, hopInfo]$ to AHL
8:       send an HIP
9:       **if** $hopInfo == 1$:
10:          record ID and $RSSI$
11: **if** received an NIP:
12:    parse the NIP
13:    **if** ID_c in NIP and sizeof $(AHL) > 2$:
14:       **for** i from 1 to sizeof (AHL):
15:          estimate d_i to each anchor based on hopInfo and the RSSI
16:          estimate the position of ID_c based on position in AHL and a series of d_i
17: change node type to pseudo-anchor

ALGORITHM 2: Initialization phase for blind nodes.

of the target. The features used to estimate the localization confidence are discussed in Section 3.2.

Blind node j begins localization by broadcasting a packet containing a localization request and its unique ID. The neighbor of the target that received the request packet will return a location packet that contains the position of the node and the node type. The target collects all the location packets and the associated RSSI during the communication. Then, the target counts the number of anchor nodes in its neighborhood, which is denoted as pLen. If the value of pLen is greater than or equal to 3 (which is the minimum requirement for multilateration), the target calculates its geometric features (ALC_j, ALR_j, and ALS_j). Before calculating the features, the target should first estimate the distance to the anchors based on the RSSI values via (1). After determining the distances and positions of the anchor nodes, the geometric features of the current target node are estimated using (13), (14), and (15).

After discovering the current geometric features, a one-class support vector classifier is used for the localization decision. Conventional binary classification methods require both positive and negative training samples. However, guaranteeing that most of the negative states are included in the training is difficult. The OC-SVM differs from traditional classifiers in that only one class of training samples is required. The OC-SVM has demonstrated good performance for anomaly and fault detection.

By detecting anomalies during location estimation, we treat the error control problem as a one-class classification problem in which the OC-SVM is used with a linear kernel to detect good reference geometries and reject the others. The OC-SVM maps the input samples to a high-dimensional feature space with a kernel function and then searches for a hyperplane to separate the mapping points from the origin (in this feature space) [24, 25]. We take the training set $D = \{p_i\}$, $p_i \in R^K$, AND $1 \le i \le n$ where p_i represents the ith training sample, K is the number of visual features, and n denotes the total number of samples. A kernel function ϕ is assumed, and it maps from the original space R^K to the infinite dimensional space χ, which satisfies $\phi(p_i) \in \chi$. Thus, the hyperplane is used to perform the following classification:

$$\min \quad \frac{1}{2}\|\mathbf{w}\|^2 + \frac{1}{vn}\sum_{i=1}^{n}\xi_i - \rho, \tag{18}$$

$$\text{s.t.} \quad \mathbf{w} \cdot \phi(p_i) \ge \rho - \xi_i \xi_i \ge 0,$$

where $\mathbf{w}$ is the normal vector of the hyperplane, ρ represents the interval between the hyperplane and the original point, ξ_i is the slack variable corresponding to the ith sample that punishes the points that deviate from the hyperplane, and $v \in [0, 1]$ indicates the compromise between the maximum interval and the penalty term. A Lagrange function is used to derive the hyperplane, and the decision equation is obtained as follows:

$$f(p) = \text{sgn}\left((\mathbf{w} \cdot \phi(p)) - \rho\right). \tag{19}$$

If $f(p) \ge 0$, then p is assigned to the targets; otherwise, p is assigned to false alarms.

4. Localization Simulation

In this section, the localization algorithm described above is validated with a simulation experiment. First, we conduct a simulation in a grid sensor deployment that directly assesses the performance of the error control. Then, we further examine the algorithm performance in different environmental conditions for a randomly deployed sensor arrangement, including the number of anchor nodes, radio communication range, and RSSI variance. Finally, we compare the proposed algorithm to several global localization methods.

In the simulation, 400 sensor nodes are deployed in a 2D square environment with an edge length equal to 200 meters. The details of the simulation parameters are shown in Table 1. Not all the simulations in this section follow the value settings in the table. The differences will be described in the specific simulations.

4.1. Performance Metrics. The localization error in a 2D plane is defined as follows:

$$e = \sqrt{(x_i - \widehat{x}_i)^2 + (y_i - \widehat{y}_i)^2}, \qquad (20)$$

where (x_i, y_i) is the true ground position of the blind node, and $(\widehat{x}_i, \widehat{y}_i)$ is the estimated position.

The relative localization error is defined as follows:

$$\bar{e} = \frac{1}{MR} \sum_{i=1}^{M} \sqrt{(x_i - \widehat{x}_i)^2 + (y_i - \widehat{y}_i)^2}, \qquad (21)$$

where M is the number of blind nodes and R is the radio communication range. This relative localization error is used to analyze the algorithm performance for different radio communication ranges R.

The Cramer-Rao lower bound derived in [26] is used to evaluate the algorithm. Based on the log-normal channel model, the Fisher information matrix is as follows:

$$F = \begin{bmatrix} F_{xx} & F_{xy} \\ F_{xy} & F_{yy} \end{bmatrix}, \qquad (22)$$

and

$$\begin{aligned} F_{xx} &= b \sum_{i=1}^{m} \frac{(x - x_i)^2}{d_i^4}, \\ F_{yy} &= b \sum_{i=1}^{m} \frac{(y - y_i)^2}{d_i^4}, \\ F_{xy} &= b \sum_{i=1}^{m} \frac{(x - x_i)(y - y_i)}{d_i^4}, \\ b &= \left(\frac{10\eta}{\sigma \ln 10} \right)^2, \end{aligned} \qquad (23)$$

where m is the number of anchor nodes. Then, the Cramer-Rao lower bound is given as follows:

$$\text{CRLB} = E\left[(x - x)^2 + (y - y)^2\right] \geq \frac{F_{xx} + F_{yy}}{F_{xx}F_{yy} - F_{xy}^2}. \qquad (24)$$

4.2. Localization Results in Grid Deployment. Figure 5 shows a heat map of the direct localization results. In the figure, each square represents a sensor node, and the center of the square is the true ground position of the node. The square with a white background and a triangle in its center is an anchor node, and the other colored squares are blind nodes. Different colors are used to represent the localization error of each blind node, as indicated by the color bar on the right. Large localization errors are reflected by colors close to magenta, which represents an error of two meters, and the black-colored squares have a localization error larger than two meters. The two subfigures display the localization error when the number of anchor nodes is equal to 3 and 10. Additionally, in each anchor node scheme, we present four stages of localization results, in which the percentage of located blind nodes is 25%, 50%, 75%, and 100%.

As the figure shows, in the 100% stage, all the blind nodes in the network are located for each scheme, and the localization error is less than two meters, with the exception of a node in the corner of scheme (a). The localization starts from the neighborhood of the anchor nodes and then spreads all over the network. Certain blind nodes that have large localization error in the early stage exhibit a good localization result by the end of the localization. During the localization, a blind node with a bad localization result can be found and relocated with a better reference combination, thereby representing the effect of the proposed error control method. This method recognizes a bad localization condition based on the geometric features of the anchor node and improves the localization result by selecting a good reference combination.

4.3. Evaluation of Decision Parameter v. The localization results shown in Section 4.2 can be further improved by tuning the localization parameter v, as discussed in Section 3.3. Simulation experiments are conducted to determine the effect of parameter v on the localization accuracy of the proposed algorithm. Figure 6 shows the localization error for parameter v when the RSSI variance is $\sigma = 8$. Additionally, four different anchor node schemes denoted as NA (number of anchor nodes) are considered to examine parameter v. Figure 6 shows that the error has a minimum value when v varies from 0.1 to 1. When $v = 0$, the algorithm makes looser localization decisions, which means that more blind nodes will be located, even if the associated location confidence is bad. When $v = 1$, the algorithm makes strict decisions. Figure 6 shows that the algorithm obtains the minimum localization error when parameter v is equal to 0.5 and that this value is stable for four different numbers of anchor nodes when the RSSI variance is equal to 8 dBm.

4.4. Comparison with the Nonerror Control Method. Figure 7 plots the cumulative distribution function of localization errors for different numbers of anchor nodes and shows that the proposed error control algorithm (the curve with upward triangular markers) can largely increase the localization accuracy compared with the iterative localization algorithm

(a) 3 anchor nodes

(b) 10 anchor nodes

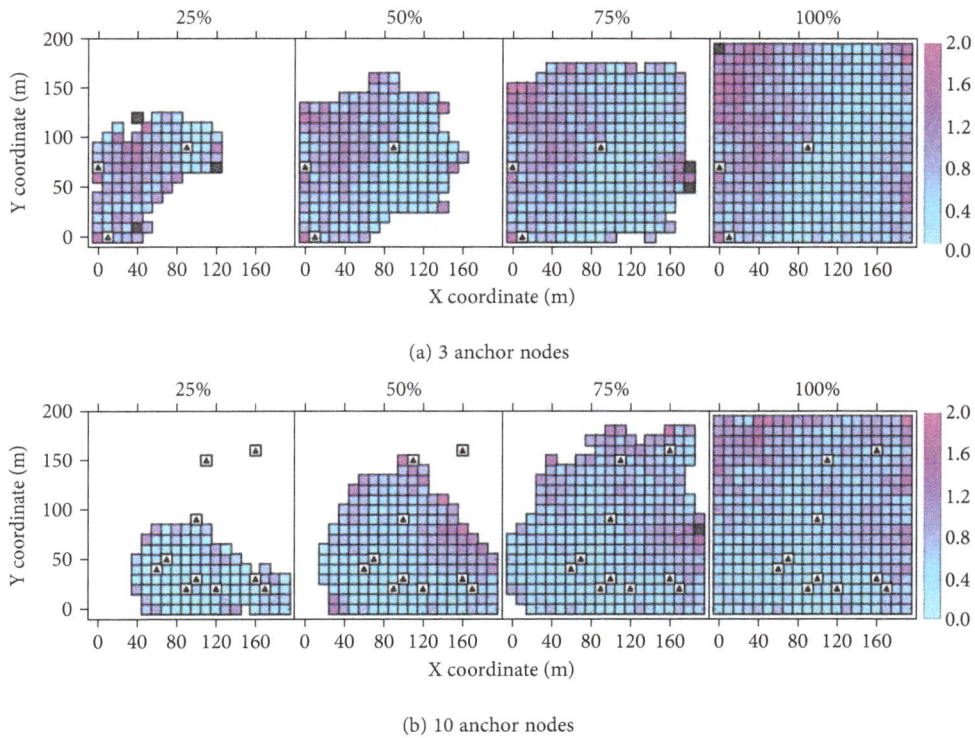

FIGURE 5: Localization error heat map with four different numbers of anchor nodes: (a) 3 anchor nodes are set and (b) 10 anchor nodes are set.

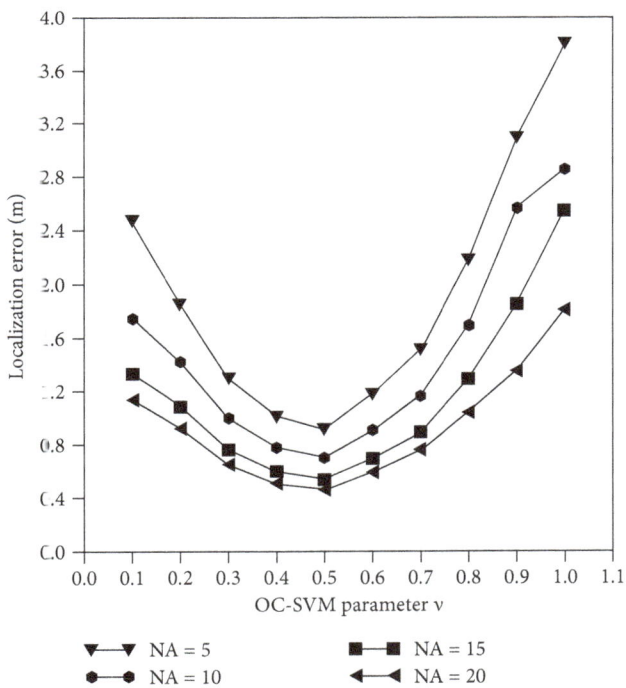

FIGURE 6: Performance of parameter ν for different numbers of anchor nodes when $\sigma = 8$.

FIGURE 7: CDF for different numbers of anchor nodes when error control is used or not used.

without error control (the curve with dot markers). The value in the legend is the ratio of anchor nodes (1.25% represents 5 anchor nodes for 400 sensor nodes). The localization error decreases with an increase in the number of anchor nodes.

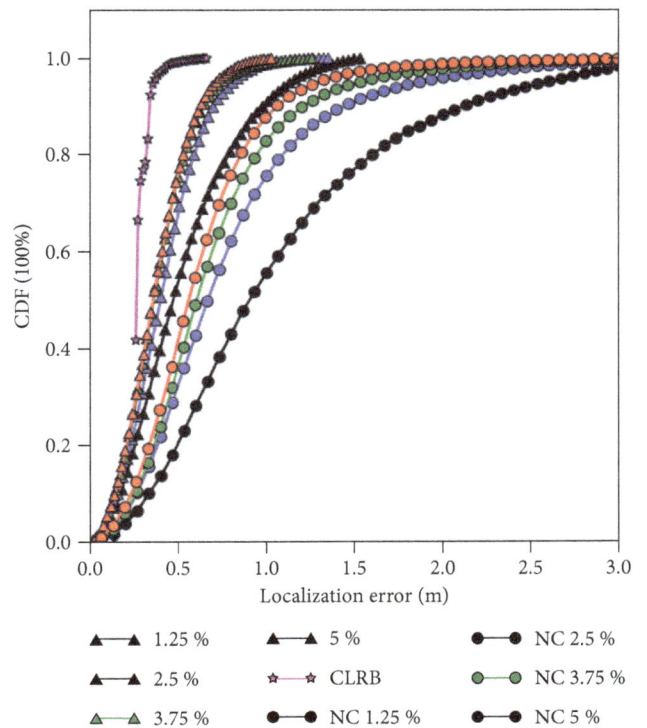

However, when the ratio of anchor nodes is larger than 2.5% (the number of anchor nodes is larger than 10), the performance gradually improves as the number of anchor nodes further increases. The green curve and the red curve with upward triangular markers are almost identical. Additionally, the localization results are similar to those of the CRLB (calculated by (24)) when the ratio of anchor nodes is equal to 5% (20 anchor nodes).

4.5. Evaluation of the RSSI Variance σ. Equation (8) displays that the variances in localization errors are caused by the variances in the ranging method. The ranging method used in the proposed algorithm is RSSI based. Therefore, the RSSI variances affect the uncertainty of the localization results. Simulation experiments are conducted, and the results are shown in Figure 8. We set the RSSI variances to range from 1 dBm to 10 dBm and set the means of the RSSI errors to zero. We also vary the number of anchor nodes (indicated by NA) and the radio communication range (indicated by R). The results show that the accuracy of the algorithm gradually decreases as the value of σ increases. Increases in the environmental variance will cause variance in the RSSI signal, produce noise in the ranging results, and affect the localization results. When the number of anchor nodes (NA = 20) and the radio communication range (R = 50 m) are large, the algorithm is more stable relative to cases with fewer anchor nodes and a shorter radio range. More anchor nodes provide more localization references, which means that the proposed algorithm has a greater chance of selecting a better localization condition to minimize the localization error. A longer radio communication range can similarly increase the number of anchor nodes and improve the localization performance.

4.6. Comparison with Global Localization Methods. Finally, we compare the proposed algorithm with several global localization algorithms for varying numbers of anchor nodes. These localization algorithms are MDS [27], SDP [28], and SPA [29], which are global in nature, although heuristic algorithms have been used to calculate the associated distributions. Among these algorithms, SPA is simple and easy to implement in real sensor networks; thus, it is used as a baseline comparison. To compare the performances, 400 sensor nodes are randomly deployed in a 200 m × 200 m square area. The number of anchor nodes is varied from 3 to 10. The RSSI variance is set to 2 dBm, and the mean of the RSSI error is set to zero. The decision parameter v of our algorithm is set to 0.5. The values of the other simulation parameters are the same as those in Table 1. The results of the simulation are shown in Figure 9. The localization error obviously decreases with an increase in the number of anchor nodes. When the number of anchor nodes is small, the DSP and SPA algorithms perform poorly, and the MDS algorithm performs relatively well. Figure 10 plots the accuracy of each localization method when the number of anchor nodes is set to 10. The proposed algorithm has the best localization accuracy compared with the other three algorithms.

5. Performance Results for Experimental Data

In this section, we conduct some experimental analyses to evaluate the efficiency of the proposed localization algorithm based on a real WSN deployed in an outdoor football field.

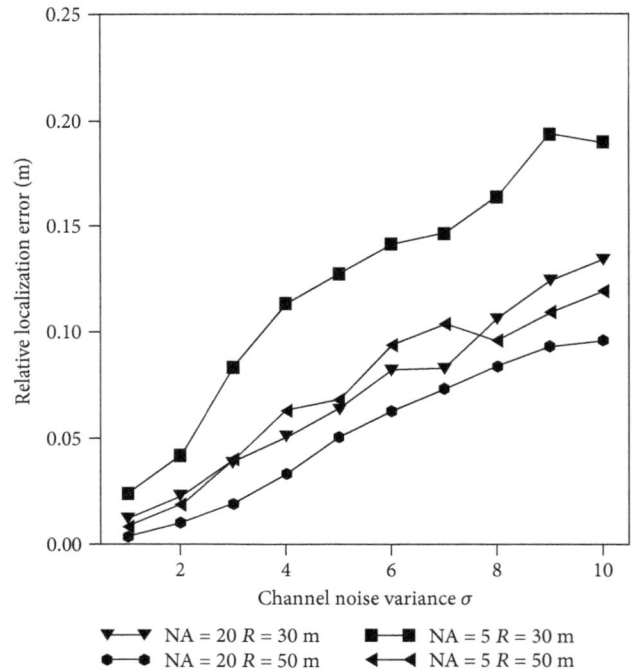

FIGURE 8: Performance of RSSI variance σ for different numbers of anchor nodes and radio ranges.

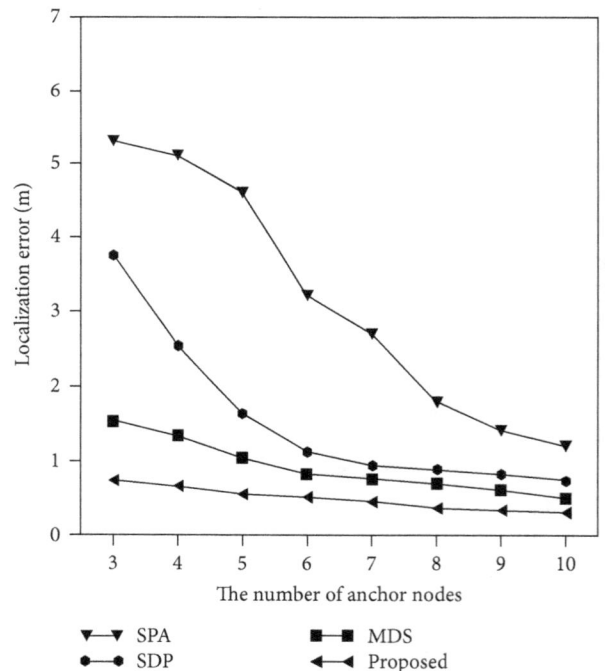

FIGURE 9: Comparison of the performance of different algorithms with different numbers of anchor nodes.

There are two stages in the experiment. In the first stage, we deploy a sensor network with only one blind node to test the efficiency of the weighted least squares position estimator. In the second stage, a multihop sensor network is

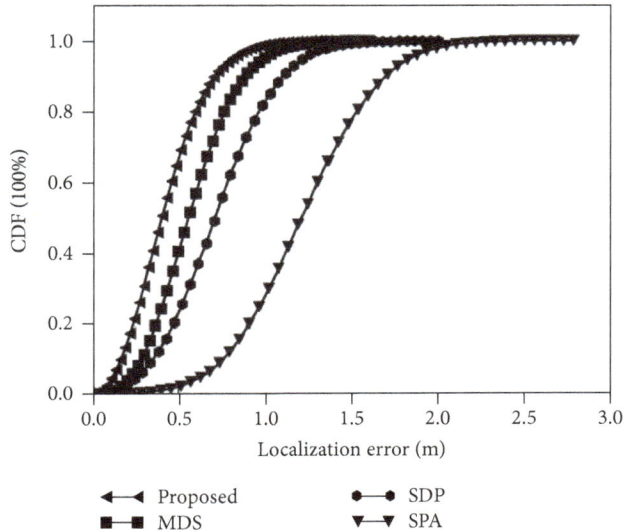

FIGURE 10: Accuracy of the different localization algorithms.

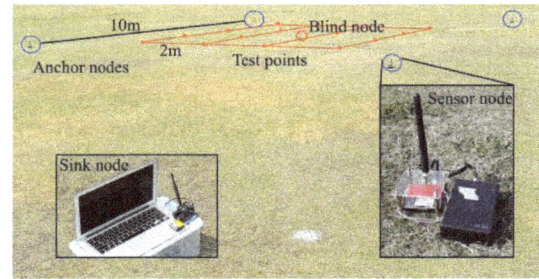

FIGURE 11: Deployment area in the position estimation experiment.

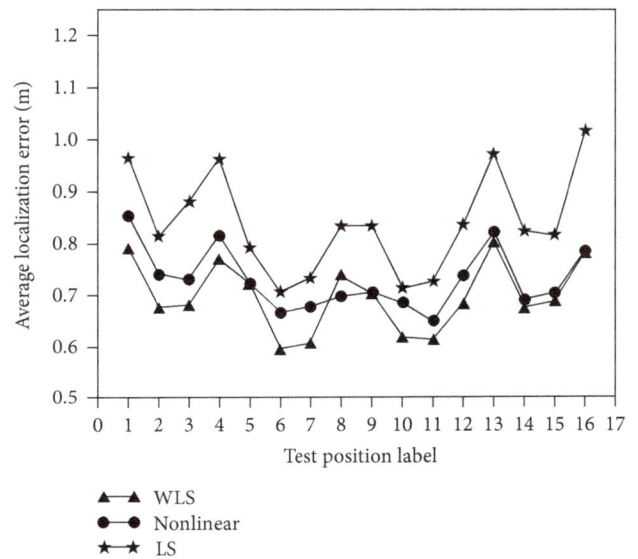

FIGURE 12: Average localization error for each test point with different estimation methods.

deployed to test the efficiency of the proposed localization algorithm with error control.

The sensor nodes are based on a TI CC2650 wireless radio transceiver, which runs a TI-RTOS real-time embedded system and is powered by a 4800 mAh battery bank. A 2.4 GHz whip antenna is attached to the sensor nodes, and the nodes are synchronized by an FTSP time synchronization algorithm.

5.1. Position Estimation Experiment. In this experiment, four sensor nodes working as anchor nodes are deployed at the vertexes of a square, which are indicated by the blue circles in Figure 11. The edge of the square is 10 meters, and 16 positions are selected as the test points in the square, which are indicated by the red dots. The test points are set in a grid with a 2-meter edge length, as the figure shows. Additionally, a blind node is established among the test points. There is also a sink node, which is set 10 meters from the test area; a laptop is connected to the sink node by a USB cable to store the data collected. During the experiments, the anchor nodes will send a message to each blind node. After receiving the message from the reference nodes, the blind node extracts the RSS value of the communication and stores the data in a local buffer for further use. When 10 RSS values from each anchor node are collected, the blind node estimates its position using linear least squares (LS) and weighted least squares (WLS) methods. Then, the RSS data and the position estimation results are uploaded to the sink node and stored in the laptop. On the laptop, a nonlinear position estimation method is implemented to estimate the position of the blind node, and the results are printed with the results from the blind node. For each test point, the procedure described above is repeated 50 times; then, the radio parameters are changed, and the experiment is repeated. After all the radio parameters have been tested, the blind node will be moved to the next position.

Figure 12 shows the localization error for each test point with the different position estimation methods: WLS, LS, and

nonlinear methods. The WLS position estimator exhibits the lowest error for all 16 test points (the average localization errors are 0.68 m for WLS, 0.72 m for nonlinear, and 0.85 m for LS); furthermore, the variance of the error is only 0.1 m. Moreover, the errors for different groups of test points display different trends. For test points 6, 7, 10, and 11, the localization errors are similar, with average errors of 0.63 m, 0.65 m, and 0.75 m for the WLS, nonlinear, and LS methods, respectively. These localization errors are lower than the average level for different methods, as these four test points are in the center of the test area, and the distances between the blind node and each anchor node are similar, which reflects a stable localization pattern, as discussed in Section 3.1. For test points 2, 3, 5, 8, 9, 12, 14, and 15, the average localization errors are 0.68 m, 0.72 m, and 0.84 m for the WLS, nonlinear, and LS methods, respectively. These errors are higher than the errors in the first group noted above. For test points 1, 4, 13, and 16, the average localization errors are 0.78 m, 0.82 m, and 0.97 m for the WLS, nonlinear, and LS methods, respectively, and these values are much higher than those of the other two groups, as these test points reflect an unstable localization pattern.

Figures 13 and 14 show the average localization error in the experiment with different values of the log-normal channel model parameter. The WLS algorithm exhibits better localization accuracy than the LS algorithm. Additionally, the accuracy of the WLS method is better than that of the nonlinear method for most of the channel model parameters. Considering the high complexity of the nonlinear method, WLS displays good efficiency in localization procedures. The curves in the figures exhibit a valley, which reflects the optimal parameters for the current channel environment that minimizes the localization errors. In the current situation, the approximate values of the parameters are $\eta = 3.7$ and $A = -34$ dBm. In real applications of a WSN, the wireless radio channel model should not be constant, and WLS should exhibit higher robustness to channel model variance, which can help improve the localization accuracy of the localization algorithm.

5.2. Localization Experiment with Error Control. In the second experiment, we deploy 16 sensor nodes in a 4×4 square grid area, as shown in Figure 15. The positions of the nodes are indicated by red dots in the subfigure in Figure 15, and the vertical or horizontal distance between neighboring nodes is 10 meters. To construct a multihop network in a relatively small area, we must modify the traditional sensor network. During the experiments, the transmission power of the sensor nodes is set lower than that in a traditional network, and a constant neighborhood is written into the nodes, which means that node 1 can only be heard by sensors 2, 5, and 6 (sensor 6 can only be heard by sensors 1, 2, 3, 5, 7, 9, 10, and 11). To test the efficiency of the iterative localization algorithm for different anchor setting conditions, we conduct the experiment following an experimental scheme, as shown in Table 5. With different numbers of anchor nodes and different anchor node setups, the experiment is separated into 10 different schemes. In the first five schemes, three sensor nodes are selected as the anchor node. Not all the possible combinations of the anchor nodes are covered in these five schemes, as the layout of the network is symmetrical. In the last five schemes, the same considerations are applied.

During the experiments, the sensor nodes run the localization algorithm introduced in Section 3. Every time the blind node obtains a position estimation, it uploads the position results and the RSS value used for localization to the sink node. Additionally, the data received by the sink node are transmitted to the laptop through a USB cable. When the position of the blind node becomes stable, the experiment is stopped. For each experimental scheme, we run the experiment 50 times; then, we change the radio channel parameters and perform the experiment again. After all the radio parameters are assessed, the nodes are established for the next experimental scheme.

The distance estimation accuracy is important for the proposed localization algorithm. For the given deployments of the sensor network, the real distances between sensor nodes are 10 m, 14 m, 20 m, 22 m, 28 m, 30 m, 32 m, 36 m, and 42 m. During the experiment, the distance estimation results of the sensor nodes are collected, and the estimation errors are shown in Figure 16. The average distance estimation error

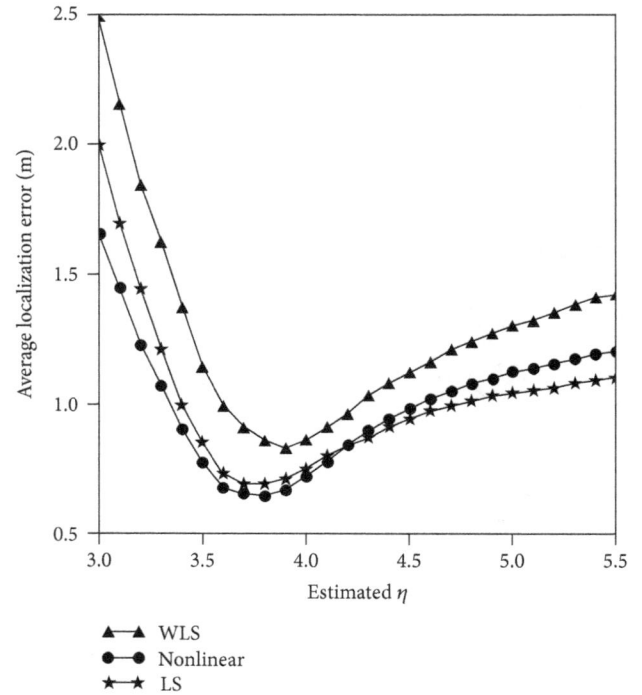

FIGURE 13: Average localization error for different values of η and $A = -34$ dBm.

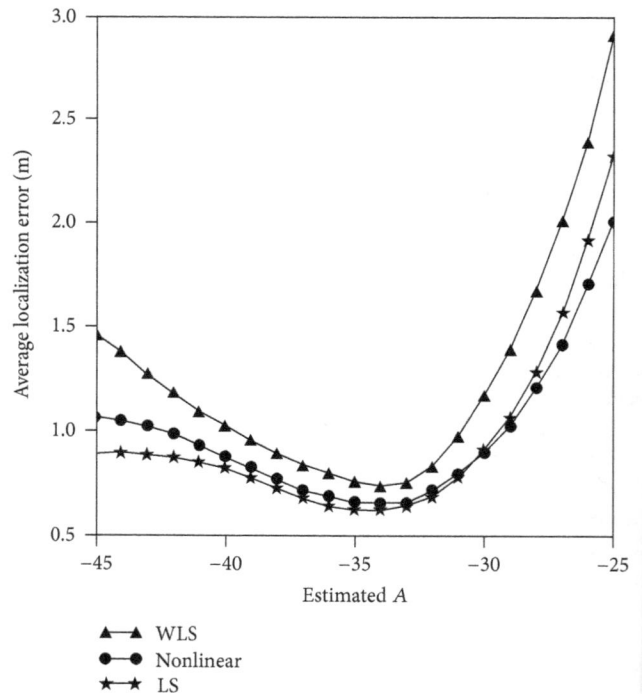

FIGURE 14: Average localization error for different values of A and $\eta = 3.7$.

exhibits little difference as the real distance increases from 10 m to 42 m. However, the variance in the error considerably increases as the real distance increases.

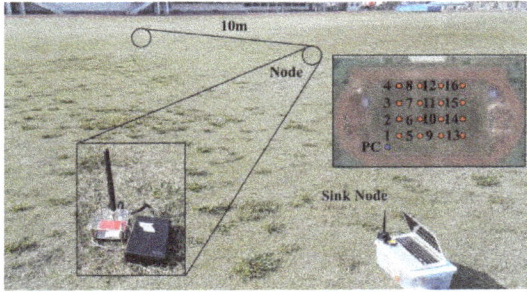

FIGURE 15: Deployment area in the iterative localization experiment.

TABLE 5: Experimental schemes.

Scheme	Anchor ID	Anchor ID	Anchor ID	Anchor ID
T1	1	2	5	×
T2	1	2	3	×
T3	1	4	13	×
T4	1	10	16	×
T5	7	10	11	×
F1	1	2	5	6
F2	1	4	13	16
F3	1	2	3	4
F4	1	7	10	16
F5	6	7	10	11

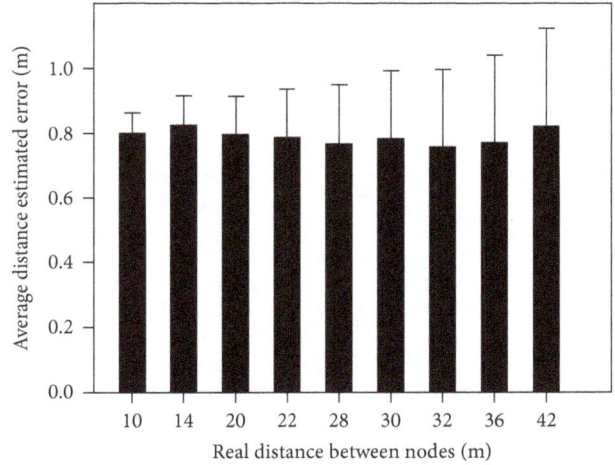

FIGURE 16: Average distance estimation error.

TABLE 6: Localization results for different schemes.

Scheme	Average error	Error variance
T1	0.62	0.38
T2	2.31	1.41
T3	1.43	0.6
T4	1.83	1.24
T5	0.67	0.43
F1	0.61	0.36
F2	0.47	0.35
F3	2.38	1.28
F4	1.71	1.11
F5	0.8	0.41
Total	0.8	0.41

The localization errors of all the schemes listed in Table 5 are shown in Table 6. The results show that the average localization error for all the schemes is 1.23 m and that the variance of the error is 0.8 m. Moreover, a comparison of the results of the schemes with three anchor nodes (series T) with those of the schemes with four anchor nodes (series F) reveals that the average localization errors are similar. This finding suggests that the proposed localization algorithm displays similar performance for sensor networks with different anchor node settings because the error control mechanism of the proposed localization algorithm always tries to find a better localization reference to update the position estimation. The results of series F exhibit slightly less error variance because these schemes have one more anchor node, which can increase the localization stability. The errors of schemes T2 and T4 are higher than those of the other results in the T series, as the layout of the anchor nodes in these two schemes tends to be linear. Although the error control mechanism can reduce the localization error to 2.42 m (T4 1.9 m), the initial localization instability will affect the final results. Comparing schemes T3 and T5, the average error of T5 is lower than that of T3 because the anchor nodes in T3 are set in a corner. The anchor node layout of T3 can cause the blind nodes on the edge of the area to have a higher localization error because an unstable localization pattern exists. Additionally, the proposed algorithm is sensitive to the initial positions of the anchor nodes. However, in real applications of large-scale sensor node deployment, the network is always fully random with more than three anchor nodes; therefore, this situation would rarely occur.

5.3. Computational Complexity. In some applications of WSNs, the computational ability is constrained, which requires the localization to be performed on chip. Computational complexity, which is defined as the number of operations performed by the algorithm, is used to evaluate the possibility of chip implementation.

The proposed localization algorithm has three parts, which are presented in Algorithms 1, 2, and 3. Algorithms 1 and 2 are the initialization phase of the localization algorithm. Additionally, the computational complexity is $O(n^2)$ for a numerical sorting algorithm and a linear LS algorithm. n is the number of anchor nodes used for position estimation. Algorithm 3 is more complex; it involves feature computation, OC-SVM prediction, and WLS position estimation. Thus, the computational complexity is $O(3n^3)$. This computational complexity is not high in real applications, as the value of n is never high (the average number of anchor nodes used by a blind node to perform localization is 9 in our simulation). An anchor node selection scheme can also be

---

Input: The unique ID of current node *ID_c*
      The node Type of current node *Type_c*
Define: Localization Request Packet *LRP*
      Location Packet *LP* = {*ID, Position, Type*}
      Reference List *RL* = [*ID_i, p_i, d_i*]
      Localization Confidence Score *LCS*
      Estimated Position *EP*
      Confidence List *CL* = [*ID, EP, LCS, Type*]
      Localization Decision *LD*
Procedures:
1: send an *LRP*
2: **If** received an *LP*:
3:    parse the *LP* to obtain *ID, position, Type*
4:    obtain the *RSSI*
5:    **if** Type==reference:
6:       obtain distance estimation *d* via Eq. (1)
7:       append *ID, position, d,* to *RL*
8: **If** sizeof (*RL*) > =3:
9:    calculate current geometric features via Eqs. (13) (14) and (15)
10:   calculate the *LCS* of this localization by Eq. (19)
11:   **If** *LCS* > *LCS* in the last *CL*
12:      input the features to one-class SVM classifier to
           obtain a decision result *LD*
13:      **If** *LD*==localization:
14:        estimate the location of current node *ID_c* by
           Eq. (10).
15:        change the type of current node to pseudo-anchor
16:        append [*ID_c, EP, LCS, pseudo-anchor*] to *CL*

ALGORITHM 3: Error control phase for blind nodes.

---

designed into the algorithm to further decrease the computational complexity of the algorithm.

6. Conclusions

This paper presents a novel iterative localization algorithm that mitigates the effects of error propagation with an anchor geometric feature-based error control mechanism. The advantage of this approach is that it can be used when only a few anchor nodes are involved. Traditional methods often require a large number of well-placed anchor nodes to work well; this limitation is overcome by using a combination of DV-hop localization and a novel error control mechanism. First, we analyze the geometric relationships between the blind node and the anchor nodes. Several features are proposed to describe the geometric features. Then, an OC-SVM-based decision machine is trained to select a good localization condition for the blind node. The performance of the algorithm is assessed by simulation experiments as well as real-life outdoor experiments, and the results show that the localization accuracy is improved compared with that of several global localization methods.

Conflicts of Interest

The authors declare that they have no conflicts of interest.

Acknowledgments

This work was supported by the National Natural Science Foundation of China (11402121).

References

[1] M. Obaidat and S. Misra, *Principles of Wireless Sensor Networks*, Cambridge University Press, 2014.

[2] Y. Yang and C. Wang, *Wireless Rechargeable Sensor Networks*, Springer International Publishing, 2015.

[3] M. Demirbas, Xuming Lu, and P. Singla, "An in-network querying framework for wireless sensor networks," *IEEE Transactions on Parallel and Distributed Systems*, vol. 20, no. 8, pp. 1202–1215, 2009.

[4] V. K. Chaurasiya, N. Jain, and G. C. Nandi, "A novel distance estimation approach for 3D localization in wireless sensor network using multi dimensional scaling," *Information Fusion*, vol. 15, pp. 5–18, 2014.

[5] J. K. Lee, Y. Kim, J. H. Lee, and S. C. Kim, "An efficient three-dimensional localization scheme using trilateration in wireless sensor networks," *IEEE Communications Letters*, vol. 18, no. 9, pp. 1591–1594, 2014.

[6] W. D. Wang and Q. X. Zhu, "RSS-based Monte Carlo localisation for mobile sensor networks," *IET Communications*, vol. 2, no. 5, pp. 673–681, 2008.

[7] D. Wang, T. Pei, Y. Zhou, and Y. Wei, "Distributed multi-object localisation by consensus on compressive sampling

received signal strength fingerprints," *IET Communications*, vol. 9, no. 14, pp. 1738–1745, 2015.

[8] F. Yin, C. Fritsche, F. Gustafsson, and A. M. Zoubir, "TOA-based robust wireless geolocation and Cramér-Rao lower bound analysis in harsh LOS/NLOS environments," *IEEE Transactions on Signal Processing*, vol. 61, no. 9, pp. 2243–2255, 2013.

[9] N. Wu, Y. Xiong, H. Wang, and J. Kuang, "A performance limit of TOA-based location-aware wireless networks with ranging outliers," *IEEE Communications Letters*, vol. 19, no. 8, pp. 1414–1417, 2015.

[10] J. Xu, C. L. Law, and M. Ma, "Performance of time-difference-of-arrival ultra wideband indoor localisation," *IET Science, Measurement & Technology*, vol. 5, no. 2, pp. 46–53, 2011.

[11] K. C. Ho, "Bias reduction for an explicit solution of source localization using TDOA," *IEEE Transactions on Signal Processing*, vol. 60, no. 5, pp. 2101–2114, 2012.

[12] Z. Irahhauten, H. Nikookar, and M. Klepper, "A joint ToA/DoA technique for 2D/3D UWB localization in indoor multipath environment," in *2012 IEEE International Conference on Communications (ICC)*, pp. 4499–4503, Ottawa, ON, Canada, June 2012.

[13] H. Chenji and R. Stoleru, "Toward accurate mobile sensor network localization in noisy environments," *IEEE Transactions on Mobile Computing*, vol. 12, no. 6, pp. 1094–1106, 2013.

[14] C. Medina, J. Segura, and Á. de la Torre, "Ultrasound indoor positioning system based on a low-power wireless sensor network providing sub-centimeter accuracy," *Sensors*, vol. 13, no. 3, pp. 3501–3526, 2013.

[15] J. Liu, Y. Zhang, and F. Zhao, "Robust distributed node localization with error management," in *2006 International Symposium on Mobile Ad Hoc Networking and Computing (MobiHoc)*, pp. 250–261, Florence, Italy, May 2006.

[16] M. Erol-Kantarci, H. T. Mouftah, and S. Oktug, "A survey of architectures and localization techniques for underwater acoustic sensor networks," *IEEE Communications Surveys & Tutorials*, vol. 13, no. 3, pp. 487–502, 2011.

[17] H. Suo, J. Wan, L. Huang, and C. Zou, "Issues and challenges of wireless sensor networks localization in emerging applications," in *2012 International Conference on Computer Science and Electronics Engineering*, pp. 447–451, Hangzhou, China, March 2012.

[18] J. Liu and Y. Zhang, "Error control in distributed node self-localization," *EURASIP Journal on Advances in Signal Processing*, vol. 2008, no. 1, Article ID 162587, 2007.

[19] Z. Yang and Y. Liu, "Quality of trilateration: confidence-based iterative localization," *IEEE Transactions on Parallel and Distributed Systems*, vol. 21, no. 5, pp. 631–640, 2010.

[20] X. Wu, S. Tan, and Y. He, "Effective error control of iterative localization for wireless sensor networks," *AEU - International Journal of Electronics and Communications*, vol. 67, no. 5, pp. 397–405, 2013.

[21] Y. Hu, L. Zhang, L. Gao, X. Ma, and E. Ding, "Linear system construction of multilateration based on error propagation estimation," *EURASIP Journal on Wireless Communications and Networking*, vol. 2016, no. 1, 2016.

[22] T. K. Sarkar, Zhong Ji, Kyungjung Kim, A. Medouri, and M. Salazar-Palma, "A survey of various propagation models for mobile communication," *IEEE Antennas and Propagation Magazine*, vol. 45, no. 3, pp. 51–82, 2003.

[23] D. Niculescu and B. Nath, "DV based positioning in ad hoc networks," *Telecommunication Systems*, vol. 22, no. 1–4, pp. 267–280, 2003.

[24] Y. Chen, X. S. Zhou, and T. S. Huang, "One-class SVM for learning in image retrieval," in *Proceedings 2001 International Conference on Image Processing (Cat. No.01CH37205)*, pp. 34–37, Thessaloniki, Greece, 2001.

[25] Y. Guerbai, Y. Chibani, and B. Hadjadji, "The effective use of the one-class SVM classifier for handwritten signature verification based on writer-independent parameters," *Pattern Recognition*, vol. 48, no. 1, pp. 103–113, 2015.

[26] N. Patwari, A. O. Hero, M. Perkins, N. S. Correal, and R. J. O'Dea, "Relative location estimation in wireless sensor networks," *IEEE Transactions on Signal Processing*, vol. 51, no. 8, pp. 2137–2148, 2003.

[27] Y. Shang, W. Rumi, Y. Zhang, and M. Fromherz, "Localization from connectivity in sensor networks," *IEEE Transactions on Parallel and Distributed Systems*, vol. 15, no. 11, pp. 961–974, 2004.

[28] P. Biswas and Y. Ye, "Semidefinite programming for ad hoc wireless sensor network localization," in *Proceedings of the third international symposium on Information processing in sensor networks - IPSN'04*, pp. 46–54, Berkeley, CA, USA, April 2004.

[29] K. Whitehouse and X. Jiang, "Calamari: a localization system for sensor networks," 2004, http://www.cs.virginia.edu/~whitehouse/research/localization/.

3

A Novel Saliency Detection Method for Wild Animal Monitoring Images with WMSN

Wenzhao Feng(ID),[1] Junguo Zhang(ID),[1] Chunhe Hu,[1] Yuan Wang,[1] Qiumin Xiang,[2] and Hao Yan[1]

[1]*School of Technology, Beijing Forestry University, Beijing 100083, China*
[2]*Chongqing Mobile Communications Limited Company, Chongqing 404100, China*

Correspondence should be addressed to Junguo Zhang; zhangjunguo@bjfu.edu.cn

Academic Editor: Francesco Dell'Olio

We proposed a novel saliency detection method based on histogram contrast algorithm and images captured with WMSN (wireless multimedia sensor network) for practical wild animal monitoring purpose. Current studies on wild animal monitoring mainly focus on analyzing images with high resolution, complex background, and nonuniform illumination features. Most current visual saliency detection methods are not capable of completing the processing work. In this algorithm, we firstly smoothed the image texture and reduced the noise with the help of structure extraction method based on image total variation. After that, the saliency target edge information was obtained by Canny operator edge detection method, which will be further improved by position saliency map according to the Hanning window. In order to verify the efficiency of the proposed algorithm, field-captured wild animal images were tested by using our algorithm in terms of visual effect and detection efficiency. Compared with histogram contrast algorithm, the result shows that the rate of average precision, recall and F-measure improved by 18.38%, 19.53%, 19.06%, respectively, when processing the captured animal images.

1. Introduction

Preservation of wild animal is crucial for the balance and stability of the whole ecosystem. However, the phenomenon of excessive hunting and killing of wild animal around the world is serious [1]. Over 300 kinds of terrestrial vertebrates are in an endangered state according to preliminary statistics. Saliency region detection are capable of effectively extracting the wild animal region information. Besides, it can provide an option for scanning and matching important target regions in wild animal detection and recognition [2, 3]. Hence, saliency region detection for wild animal images is becoming more and more significant in animal protection realm, which has become a focus of recent researches.

Traditional wild animal detection and recognition method [4] mainly use the collected images of wild animal as a test while a training set for learning and training purpose. These experimental samples need to treated by several screening and preprocessing since they generally contain complete, clear, and low-noise image features. However, traditional detection algorithms cannot effectively process captured animal images during wild animal monitoring mission due to the character of complex background and nonuniform illumination that exist in original images. Therefore, proposing an appropriate and effective detection method is a crucial prerequisite to solve the existing problem.

At present, visual saliency detection technique [5] can quickly and automatically extract the main image information and remove the redundant background information, which has won wide attention from both domestic and foreign researchers [6–8]. Most visual attention-related saliency detection methods are based on the foundation of biological theory [9–11]. However, these algorithms have low saliency image resolutions, and their computational complexity is high at the same time. Another popular method is based on the basis of model analysis [12–15]. Although it has a good detection efficiency and the detection result coordinates well with the human eye characteristics, it cannot effectively process rich texture information in monitoring scenes. Saliency object detection [16–18] can efficiently separate the salient

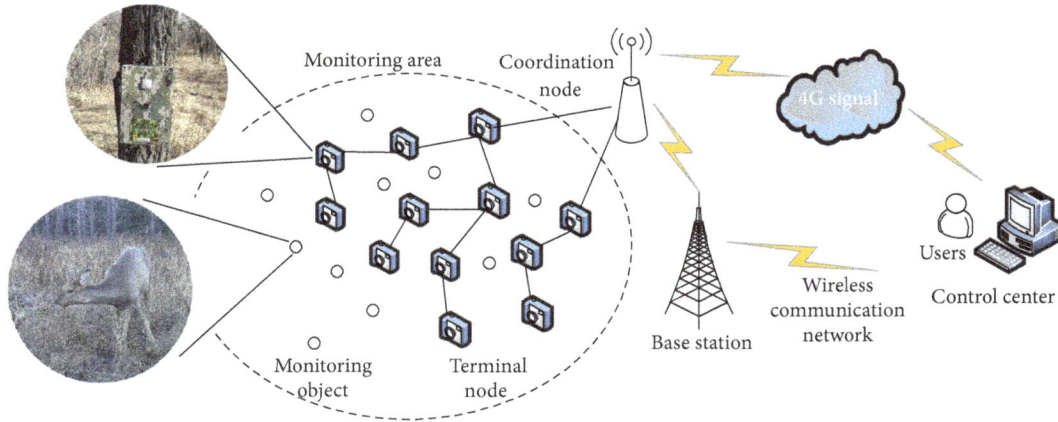

FIGURE 1: Wild animal monitoring system.

objects from image background. Klein et al. developed a salient object detection based on the standard structure with the help of cognitive visual attention models, and Yong et al. presented a framework that models semantic contexts for key-frame extraction based on wild animal images. Nevertheless, most existing algorithms can only process images with simple background and ordinary resolution features. Besides, they are not applicable to field-captured images in practical wildlife monitoring mission.

Therefore, this paper aims at the demands of actual wild animal monitoring and focuses on solving the problem of high resolution, complex background, and nonuniform illumination that exist in wild animal image saliency detection research.

In this paper, a sample library of wild animal monitoring images was established. The dataset contains images captured from Saihan Ula Nature Reserve in Inner Mongolia province using WMSN monitoring system. The WMSN monitoring system was configured and developed independently in the laboratory with wireless remote, real-time, precise, and meticulous modules. The image dataset covers 12 species and 1000 HD wild animal images. We have established a standard ground-truth image library through manual stamp operation. Based on the field monitoring images, we proposed an improved histogram contrast detection method. The correctness and validity of this method are shown by implementing to the wild animal images.

The contribution of the present study included the following: (1) developed the wild animal monitoring system based on WMSN to capture experimental materials in Saihan Ula Nature Reserve in Inner Mongolia province; (2) established actual wild animal monitoring image database with the unique characteristics; (3) a novel saliency detection method was introduced in this paper for wild animal images with high resolution, complex background, and nonuniform illumination.

2. Wild Animal Monitoring Based on WMSN

2.1. Wild Animal Monitoring System. Traditional wild animal monitoring methods include crewed field survey, GPS (global positioning system) collar [19], infrared camera [20], and

TABLE 1: Parameters of WMSN node.

Monitoring node	Function parameters
Camera	OV7725 QVGA 30 fps
Physical pixel	640×480
Support memory card	SD 16G
Controller	STM32 control ship (72 MHz CPU, 512 K SRAM)
Measuring range	310 m
Rate	200 kbps
Trigger mode	Infrared trigger

satellite remote sensing monitoring [21]. However, these methods have defects such as limited monitoring range, data acquisition lag, and unmeasured local microinformation.

Wireless multimedia sensor network (WMSN) is mainly used to capture the wild animal image materials utilizing industrial grade cameras with terminal node equipment embedded. The wild animal monitoring system which achieves remote, real-time, all-weather, and friendly monitoring goals consists of WMSN terminal nodes, coordination nodes, gateway nodes, and data storage center (back-end sever). The detailed configurations are shown in Figure 1. The monitoring node devices developed by our laboratory are based on ZigBee network protocols. Detailed parameters are shown in Table 1.

The monitoring node devices using ZigBee network protocols established a wireless image sensor network in a self-organizing way. When wild animals enter the monitoring view field, the infrared sensor embedded in terminal node will trigger the camera to capture images. Captured images were firstly saved in SD card. Then these images will be transmitted to the coordination node via multihop method. After the coordination nodes successfully receive and converging transmitted image data information from all terminal nodes, the monitoring image information will be transmitted to data center through gateway node utilizing 4G signal by wireless and remote way.

2.2. Fieldwork Material Collection. The WMSN monitoring system for wild animal monitoring was deployed in Saihan

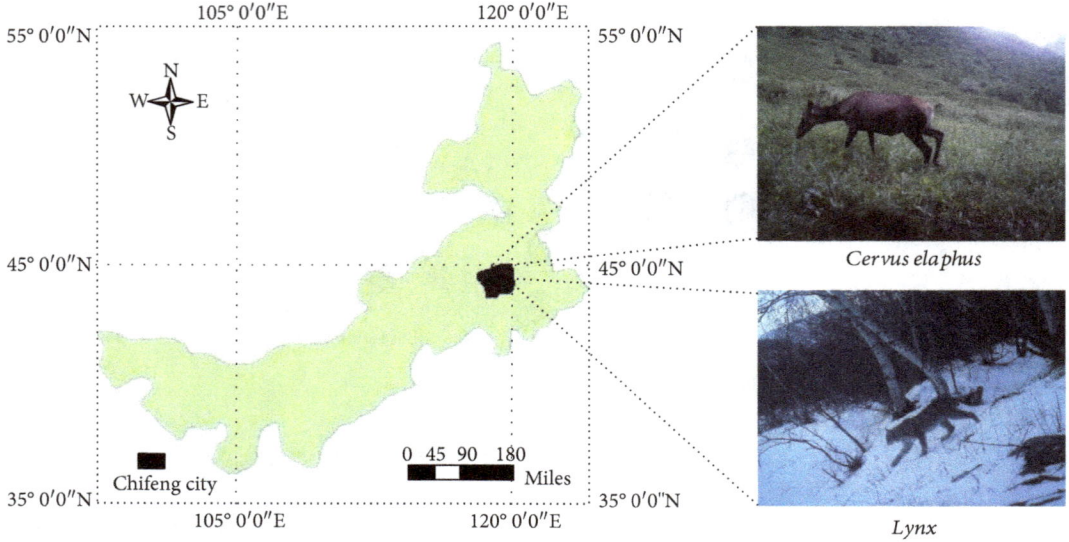

Cervus elaphus

Lynx

FIGURE 2: Wild animal monitoring images in Saihan Ula Nature Reserve.

Ula National Nature Reserve in Inner Mongolia, which is subordinative to the Greater Khingan mountains. The experiment area with temperate semihumid rainy climate has an average altitude of 1000 m above the sea level. Wild animals collected in the experimental area include *Cervus elaphus*, *Lynx*, *Capreolus pygargus*, *Sus scrofa*, and *Naemorhedus goral*. *Cervus elaphus* and *Lynx* are national secondary protected animals (shown in Figure 2). In this paper, 1600 images of more than 12 wild animal species are acquired, and the total image data storage volume is 2.4 G.

By analyzing the captured images, we found that most images have complex background, nonuniform illumination, and different image target ratio features. Those image features will cause effects on the saliency detection work, especially in the regions with large grayscale gradients.

3. Method Analysis

The improved histogram contrast area detection method is proposed to process the wild monitoring images with high resolution and complex background in this section. Due to the particularity of materials, both structure extraction and edge detection are introduced in this paper. As shown in Figure 3, we firstly implemented the image structure extraction method based on the image total variation to extract the structure of input images, which aims to smooth the image texture and reduce image noise. Then the saliency detection method-based histogram contrast is implemented to capture the color saliency information of the image. By quantifying the input image to a small color range, the calculation procedure becomes simple and the computation efficiency can be improved. Finally, the edge detection and the position saliency map are synthesized to obtain the final optimization results.

3.1. Structure Extraction Based on Image Total Variation. During the first step, we have not assumed or manually

determined the type of textures, as the patterns could vary a lot in different examples. We applied image window total variation and internal variation to the structure extraction method.

Firstly, the window total variation $D_x(p)$ and $D_y(p)$ of image samples pixel p in x and y directions are obtained by

$$D_x(p) = \sum_{q \in R(p)} g_{p,q} \cdot |(\partial_x S)_p|,$$
$$D_y(p) = \sum_{q \in R(p)} g_{p,q} \cdot |(\partial_y S)_p|,$$
$$g_{p,q} \propto \exp\left(-\frac{(x_p - x_q)^2 + (y_p - y_q)^2}{2\sigma^2}\right),$$

where $R(p)$ denotes the 19×19 square region centered at pixel p. $g_{p,q}$ denotes the Gaussian filter kernel function in which x and y are the pixel coordinates and σ means scale parameter of the function, controlling the spatial scale of the window.

To help distinguish prominent structures from the texture elements, the window internal variation is calculated to extract the prominent structures from the texture elements according to

$$L_x(p) = \left|\sum_{q \in R(p)} g_{p,q} \cdot (\partial_x S)_p\right|,$$
$$L_y(p) = \left|\sum_{q \in R(p)} g_{p,q} \cdot (\partial_y S)_p\right|.$$

Finally, the optimized model with window total variation and internal variation is established. With the optimization

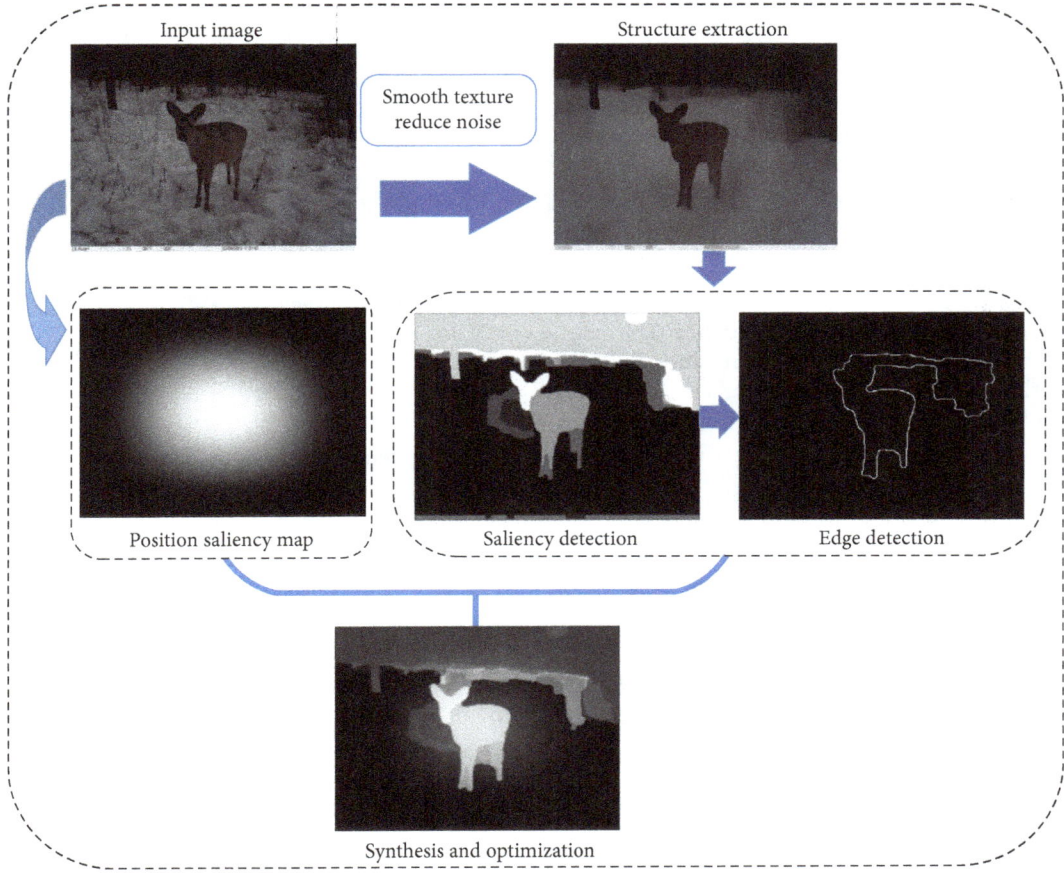

FIGURE 3: Process of wild animal image saliency detection.

result, the contrast between texture and structure of the visually salient areas can be further enhanced.

$$\arg \min_{S} \quad \sum_{p} \left(S_p - I_p\right)^2 + \lambda \cdot \left(\frac{D_x(p)}{L_x(p) + \varepsilon} + \frac{D_y(p)}{L_y(p) + \varepsilon} \right), \quad (3)$$

where S_p denotes extracted structure image while I_p denotes the input image. The term $(S_p - I_p)^2$ makes the input and result not deviate. λ is the smoothness coefficient of the image, and ε is a small positive number to avoid division by zero.

3.2. Histogram Contrast Saliency Detection.

The structure extraction can smooth the texture and reduce the image noise, therefore the extraction result could be used as input of saliency detection. Then the input images are quantified according to the number of quantify channels CN. The main color is arranged into a color matrix by histogram statistics. After, the image pixels are reordered by color value, such that the terms with the same color value are grouped together. The saliency value $S(C_i)$ between different colors is calculated and expressed as shown in (4). The saliency values are the same when the color of the pixels is the same.

$D(c_i, c_j)$ denotes the color distance metric between the pixel c_i and c_j in Lab space.

$$S(C_i) = \sum_{j=1}^{n} f_j D\left(c_l, c_j\right), \quad (4)$$

where c_j is the color value of the P_k in structure extraction image, and $S(P_k)$ represents the saliency value of P_k. n denotes the total color numbers of input image and f_j represents the ratio of the pixels whose color value is c_i to the total pixel numbers in the image.

Color quantization greatly simplifies calculation procedure, but similar colors may be quantized to different values during the process. In order to reduce noisy saliency results caused by such randomness, we replace the saliency value of each color by the weighted average of the saliency value of similar colors.

$$S^{'}(c) = \frac{1}{(m-1)T} \sum_{i=1}^{m} (T - D(c, c_i)) S(c_i), \quad (5)$$

where the equation $T = \sum_{i=1}^{m} D(c, c_i)$ denotes the distance between the color c and its nearest colors. Typically, m is quarter of color numbers n in the images after quantification.

$$\sum_{i=1}^{m}(T - D(c, c_i)) = (m - 1)T. \qquad (6)$$

Then the saliency area is obtained by comparing appearance frequency [22] of the first n kinds color. If the appearance frequency is greater than CR (color retention rate), the color with low frequency is discarded subsequently and replaced with the closest color. The color appearance frequency of the first n kinds color will be increased in accordance with the number of quantify channels until it is greater than CF.

3.3. Edge Detection Based on Canny Operator.
The edge detection process aims to measure the convolution of the Gaussian smoothing filter g and the above saliency detection result $I(i, j)$ to obtain the most optimized approximation operator.

$$S(i, j) = g * I(i, j), \qquad (7)$$

where S denotes convolutional result, $*$ refers to convolution function, and (i, j) is the position of the pixel in saliency result.

Then the partial derivative is obtained by calculating the first-order finite difference of the filter result.

$$P(i, j) \approx \frac{(S(i, j + 1) - S(i, j) + S(i + 1, j + 1) - S(i + 1, j))}{2},$$

$$Q(i, j) \approx \frac{(S(i, j) - S(i + 1, j) + S(i, j + 1) - S(i + 1, j + 1))}{2}. \qquad (8)$$

Among them, $P(i, j)$ represents the gradient partial derivative of image in x direction, and $Q(i, j)$ is the gradient partial derivative in y direction. Therefore, the pixel amplitude matrix and gradient direction matrix are calculated as shown in the following equations:

$$M(i, j) = \sqrt{p(i, j)^2 + Q(i, j)^2}, \qquad (9)$$

$$\theta(i, j) = \arctan\left(\frac{Q(i, j)}{P(i, j)}\right). \qquad (10)$$

Finally, nonmaximal value suppression is completed through seeking amplitude maximum of the matrix along the gradient direction. The pixels with maximum amplitude are considered as the edge pixel. To make the image edge close, this paper selects double appropriate threshold (high threshold and low threshold). As a consequence, the nonedge points that do not satisfy the threshold condition are removed. Then the connected domain is expanded to get the final edge detection result.

3.4. Synthesis and Optimization.
In this section, we optimize the center position weight of the two-dimensional images with Hanning window function established by center-edge and contrast degree theory. The one-dimensional Hanning window function is constructed as follows.

$$w(n) = 0.5\left(1 - \cos\left(2\pi \frac{n}{N}\right)\right). \qquad (11)$$

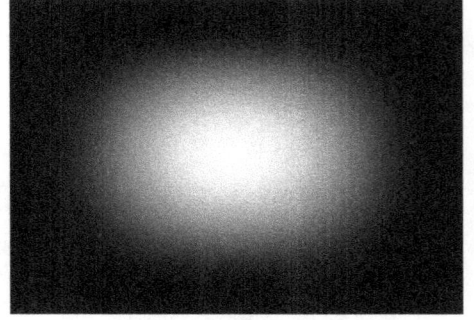

FIGURE 4: Position saliency map.

where N denotes the input data length, $n \in [0, N]$. Consequently, Y in the two-dimensional Hanning window construction function is the scalar product of two one-dimensional function,

$$Y = w_{\text{Row}} \cdot w_{\text{Col}}'. \qquad (12)$$

On the basis of position saliency map, the saliency and edge detection based on structure extraction are introduced to obtain more accurate saliency detection results,

$$S = Q_1 \text{Smap} + Q_2 \text{Emap} + Q_3 \text{Hmap}, \qquad (13)$$

where S denotes the synthetic saliency map and coefficient $Q_1 = Q_2 = Q_3 = 1/3$. Smap refers to saliency detection result of histogram contrast. Emap is the edge detection result by image structure extraction. Hmap is the constructed position saliency map.

3.5. Experimental Parameters Selection.
The color value of a single pixel in RGB image ranges from 0 to 256^3, therefore the number of colors that needs to be processed will reach 10^7 during the color saliency calculation process. In this paper, the color value of each channel (R, G, and B) is quantified to 0–12 (CN) so that the number of colors that need to be calculated will reduce to $12^3 = 1728$. In order to ensure the smoothing effects, the filter function scale parameter is taken as 3 after several single-variable experiments. The number of iterations is 3, and smoothness coefficient λ is taken as 0.015. In addition, the color retention rate CF is 0.95 to obtain high-frequency colors in saliency area. The position saliency map whose size is 300×400 is shown in Figure 4.

4. Comparison and Discussion

In order to verify the effectiveness of the proposed algorithm, we selected the images of actual wild animal monitoring images and public image library as experimental sample and compared the result of our algorithm with other saliency detection algorithms.

Both precision and recall rate [13] are used as objective criteria to evaluate the accuracy of saliency detection. Precision/recall is the ratio of correctly detected salient region to the detected/"ground truth" salient region which means

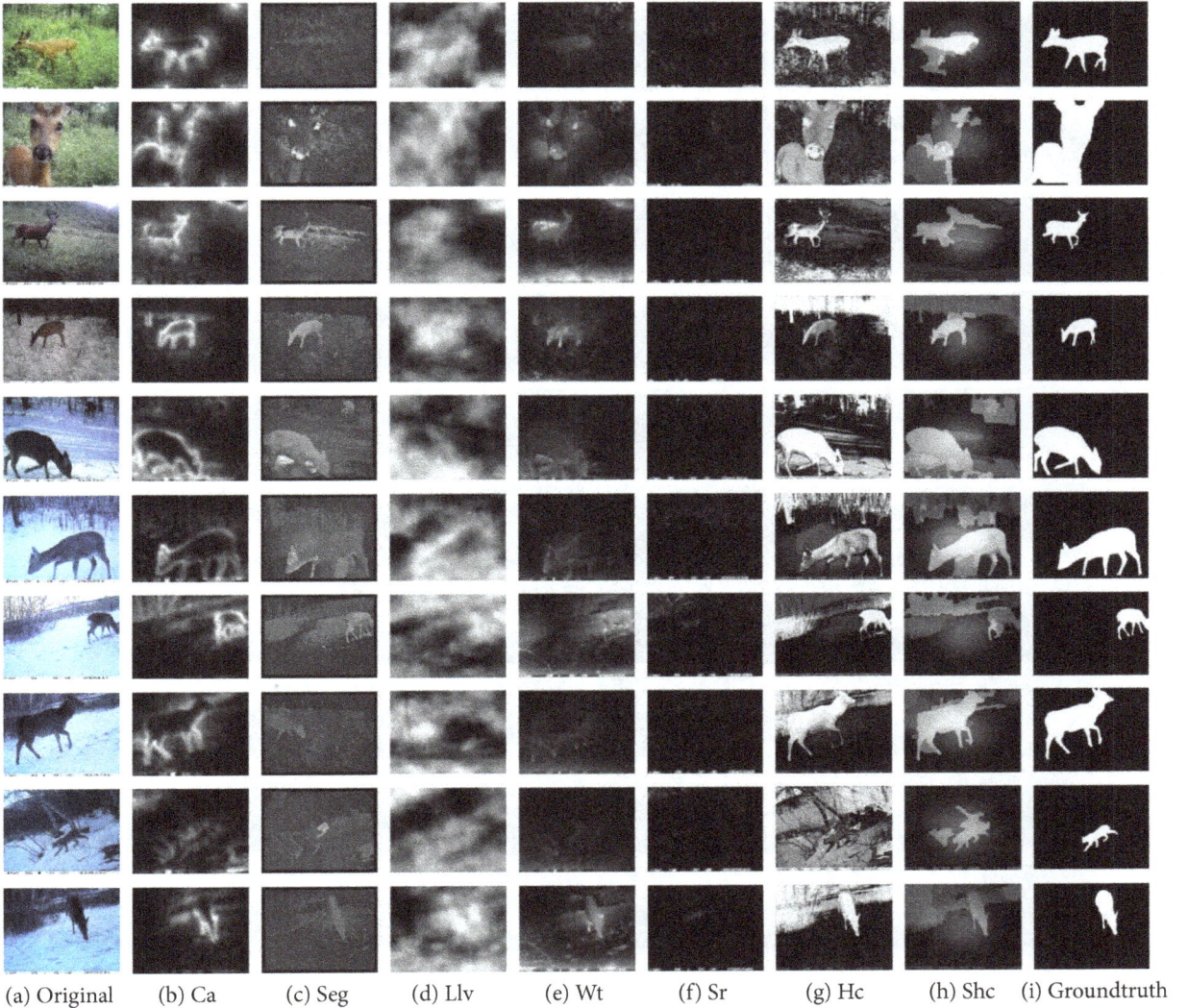

(a) Original   (b) Ca   (c) Seg   (d) Llv   (e) Wt   (f) Sr   (g) Hc   (h) Shc   (i) Groundtruth

FIGURE 5: Visual comparison of saliency detection in wild animal monitoring images.

that the precision rate is the ratio of the correct area and the true saliency area of saliency object detected by the visual saliency model.

$$Pr = \frac{\sum(G_x \cap G_t)}{\sum G_t}. \quad (14)$$

The recall rate is the ratio of the correct area and the detected saliency area that are calculated by the saliency model in saliency map.

$$Re = \frac{\sum(G_x \cap G_t)}{\sum G_x}, \quad (15)$$

where G_t is the true saliency area and x is the corresponding image index. G_x is the calculated saliency area.

4.1. Experiment to Wild Animal Images. The field samples of wild animal monitoring are selected from private image database with different light intensities, capture distances, and backgrounds due to seasonal variations. Six classical

saliency detection algorithms consisting of CA (context aware) [23], SEG (segment) [24], LLV (low level vision) [25], WT (wavelet transform) [26], SR (spectral residual) [27], HC (histogram contrast) [28], and five classical object detection algorithms consisting of GS (global saliency) [29], FD (frequency domain) [30], GP (gestalt principle) [31], MC (multiscale contrast) [32], and DF (dynamic feature) [13] are compared with our proposed algorithm (SHC) to verify the detection effect in this section.

As the results shown in Figures 5 and 6, the detection method proposed in this paper is more accurate than above classical algorithms in detecting the object areas. The algorithm in this paper preserves a better quality of edge information, and its object area is more smooth as the images containing rich color and complex background.

We set the segmentation threshold from 0 to 255, and the average precision and recall rate of all the images of private image database are shown in Figure 7.

The precision and recall rate of our algorithm is higher than other eleven alternatives. We believe that the main

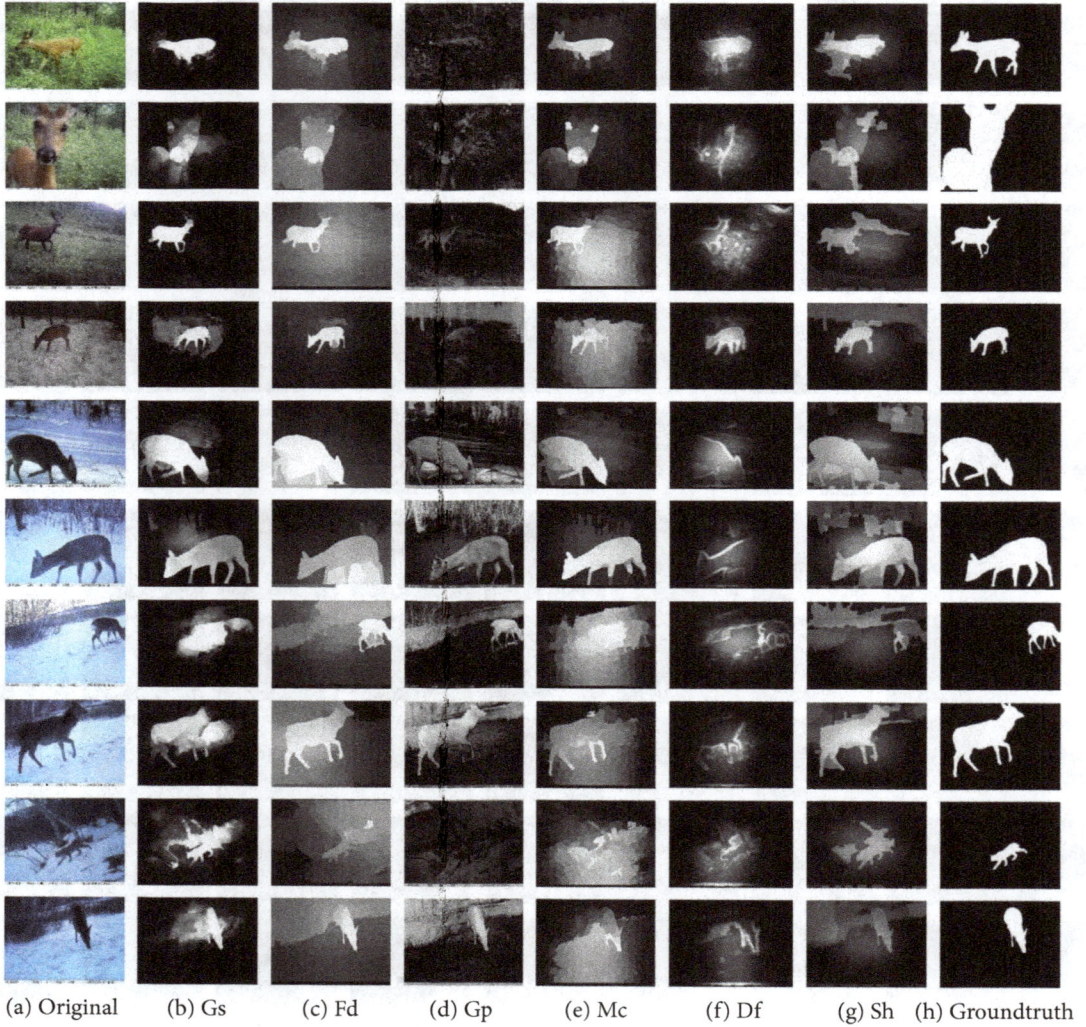

(a) Original   (b) Gs   (c) Fd   (d) Gp   (e) Mc   (f) Df   (g) Sh   (h) Groundtruth

FIGURE 6: Visual comparison of object detection in wild animal monitoring images.

structure extraction utilized in our algorithm successfully suppresses the influence of the texture information on the detection result. The relatively high recall rate tends to have higher precision rate. The edge detection and the position saliency map are further improved both in uniform and smooth aspects.

In addition, as all the pixels in the saliency map are considered to be foreground, where the segmentation threshold is 0, all algorithms tend to have the same precision and recall rate (precision rate is about 0.1, recall rate is 1.0).

According to formula (14) and (15), the evaluation indicators of precision and recall rate are negative correlation. Therefore, we use F-measure (also known as F-score) to evaluate the effectiveness of saliency detection algorithms. F-measure value is the harmonic mean parameter calculated from precision and recall rate by a certain weight that can be obtained by

$$F_\beta = \frac{(1 + \beta^2) \cdot Pr \cdot Re}{\beta^2 \cdot Pr + Re}. \tag{16}$$

The F-measure is an overall performance measurement, among them, β is the weight parameter for controlling the precision and recall rate. The smaller β means less important the precision is, which is set as $\sqrt{3}$ herein.

We also introduce an adaptive threshold that is image saliency dependent, instead of using a constant threshold for each image. The adaptive threshold T_α is used to segment the saliency detection results as follows.

$$T_\alpha = \frac{2}{W \times H} \sum_{x=0}^{W-1} \sum_{y=0}^{H-1} S(x, y), \tag{17}$$

where T_α is the obtained threshold value. W and H are the width and height of the saliency map, respectively. S denotes the saliency map and (x, y) refers to the corresponding coordinate of the saliency map.

After obtaining the segmentation result based on the saliency map, the precision, recall rate, and F-measure of all segmentation results are calculated. Taking their average,

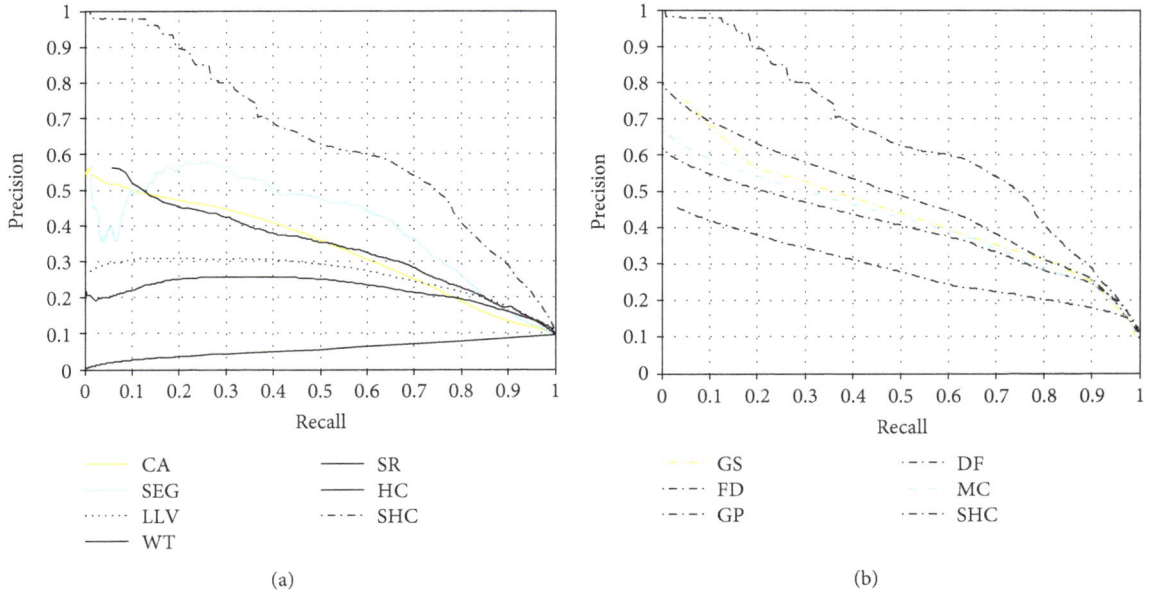

FIGURE 7: Precision-recall rate curves.

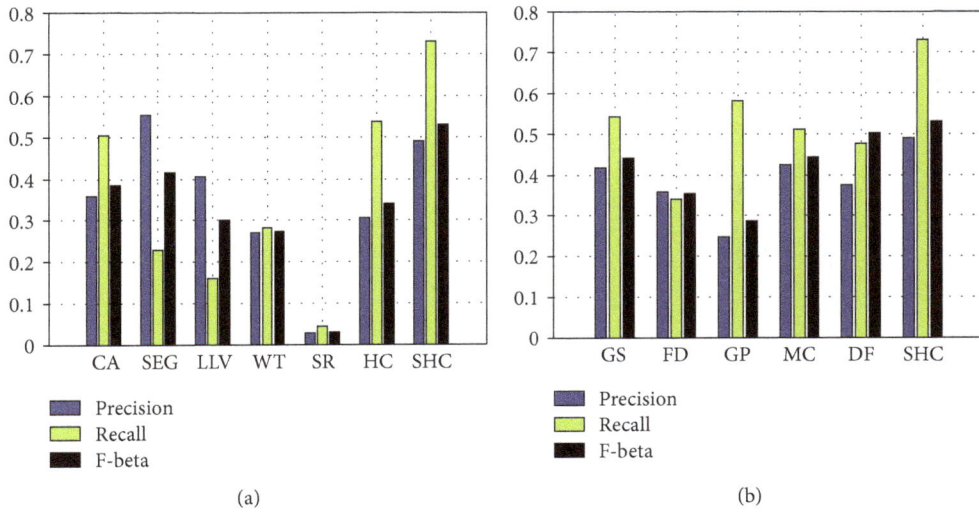

FIGURE 8: Average precision, recall, and F-measure.

TABLE 2: Detailed data with different algorithms in saliency detection.

Method	CA	SEG	LLV	WT	SR	HC	SHC
Average precision rate	0.3585	0.5537	0.4060	0.2696	0.0268	0.3058	0.4895
Average recall rate	0.5045	0.2274	0.1600	0.2807	0.0431	0.5368	0.7321
F-measure	0.3842	0.4159	0.2997	0.2721	0.0294	0.3395	0.5300

TABLE 3: Detailed data with different algorithms in object detection.

Method	GS	FD	GP	MC	DF	SHC
Average precision rate	0.4176	0.3593	0.2479	0.4253	0.3745	0.4895
Average recall rate	0.5423	0.3410	0.5805	0.5122	0.4764	0.7321
F-measure	0.4410	0.3549	0.2857	0.4426	0.5022	0.5300

TABLE 4: Mean time comparison of detection.

Method	CA	SEG	LLV	WT	SR	HC	GS	FD	GP	MC	DF	SHC
Average time(s)	90.1796	17.2373	2.7255	8.998	0.0705	2.7383	5.3849	4.3892	6.7341	10.7568	18.4795	3.9462
Code type	MATLAB	MATLAB	MATLAB	MATLAB	MATLAB	MATLAB	MATLAB	MATLAB	MATLAB	MATLAB	MATLAB	MATLAB

respectively, as the comparison result by using different methods and the detailed results are shown in Figure 8.

As shown in Figure 6, the average performance of the proposed algorithm is more efficient than the other six algorithms. Among them, the average precision rate, recall rate, and F-measure of detected results of our algorithm are 0.4895, 0.7321, and 0.5300 (shown in Tables 2 and 3), which increased by 18.38%, 19.53%, and 19.06%, respectively, while comparing with the HC algorithm. Although the average precision rate of SEG algorithm is higher than our algorithm, its higher computational complexity makes it not efficient.

We have compared the average running time of each algorithm, and the comparison table is shown in Table 4. All experiments are performed using MATLAB (R2014a) on the workstation with Intel (R) Core (TM) i3-2330 and 4 GB RAM.

SR algorithms cost the least calculation time because they used domain transformation and simple filtering to get the saliency map. However, the detection accuracy and quality of the above two algorithms are not satisfactory. Compared with the HC algorithm, the saliency detection result of our algorithm is better despite slightly higher calculation time. Above results show that our algorithm is more suitable for the application of wild animal image saliency detection.

5. Conclusion

In this paper, we proposed a novel saliency detection algorithm based on histogram contrast for wild animal monitoring images, which can be used for dealing with high resolution, rich colors, and high noise. The proposed method consists of four steps, namely, structure extraction, saliency detection, edge detection, and synthesis optimization. Firstly, the structure extraction is required to smooth image texture and reduce image noise. Saliency detection using histogram contrast aims to extract the area of wild animal from images. Then canny operator is further implemented to edge detection to obtain complete saliency target edge information. Finally, Hanning window is applied to make saliency areas prominent. To demonstrate the efficiency and validation of the proposed method, the images from field-captured wild monitoring database are processed. The final result shows that the proposed algorithm has better performance than existing classical algorithms, especially for captured wild animal monitor images. Compared with the classical detection algorithms, the average precision rate, recall rate, and F-measure of detected results obtained by our algorithm are increased, respectively, by 18.38%, 19.53%, and 19.06% for the wild animal images when compared with the HC algorithm.

Conflicts of Interest

The authors declare that they have no conflicts of interest.

Acknowledgments

This study was financially supported by National Natural Science Foundation of China (Grant no. 31670553), Fundamental Research Funds for the Central Universities (Grant no. 2016ZCQ08), and Import Project under China State Forestry Administration (Grant no. 2014-4-05).

References

[1] B. Hori, R. J. Petrell, G. Fernlund, and A. Trites, "Mechanical reliability of devices subdermally implanted into the young of long-lived and endangered wildlife," *Journal of Materials Engineering and Performance*, vol. 21, no. 9, pp. 1924–1931, 2012.

[2] P. Manipoonchelvi and K. Muneeswaran, "Region-based saliency detection," *IET Image Processing*, vol. 8, no. 9, pp. 519–527, 2014.

[3] Y. Xue, R. Shi, and Z. Liu, "Saliency detection using multiple region-based features," *Optical Engineering*, vol. 50, no. 5, article 057008, 2011.

[4] A. G. Villa, A. Salazar, and F. Vargas, "Towards automatic wild animal monitoring: identification of animal species in camera-trap images using very deep convolutional neural networks," *Ecological Informatics*, vol. 41, pp. 24–32, 2017.

[5] K. Duncan and S. Sarkar, "Saliency in images and video: a brief survey," *IET Computer Vision*, vol. 6, no. 6, pp. 514–523, 2012.

[6] M. Zeppelzauer, "Automated detection of elephants in wildlife video," *Eurasip Journal on Image and Video Processing*, vol. 2013, no. 1, 2013.

[7] P. Kamencay, T. Trnovszky, M. Benco, R. Hudec, P. Sykora, and A. Satnik, "Accurate wild animal recognition using PCA, LDA and LBPH," in *Proceedings of the 11th International Conference on ELEKTRO*, pp. 62–67, Strbske Pleso, Slovakia, 2016.

[8] T. N. Vikram, M. Tscherepanow, and B. Wrede, "A saliency map based on sampling an image into random rectangular regions of interest," *Pattern Recognition*, vol. 45, no. 9, pp. 3114–3124, 2012.

[9] L. Itti, C. Koch, and E. Niebur, "A model of saliency-based visual attention for rapid scene analysis," *IEEE Transactions on Pattern Analysis and Machine Intelligence*, vol. 20, no. 11, pp. 1254–1259, 1998.

[10] A. Rahman, D. Houzet, D. Pellerin, S. Marat, and N. Guyader, "Parallel implementation of a spatio-temporal visual saliency model," *Journal of Real-Time Image Processing*, vol. 6, no. 1, pp. 3–14, 2011.

[11] Q. Lin, X. G. Xu, Y. Z. Zhan, and D. A. Liao, "Extracting regions of interest based on visual attention model," in *2011 International Conference on Multimedia Technology*, pp. 313–316, Hangzhou, China, 2011.

[12] J. Harel, C. Koch, and P. Perona, "Graph-based visual saliency," in *Proceedings of the 20th Annual Conference on Neural Information Processing Systems, NIPS 2006*, pp. 545–552, Vancouver, Canada, 2006.

[13] T. Liu, Z. Yuan, J. Sun et al., "Learning to detect a salient object," *IEEE Transactions on Pattern Analysis and Machine Intelligence*, vol. 33, no. 2, pp. 353–367, 2011.

[14] W. Chen, T. Sun, M. Li, H. Jiang, and C. Zhou, "A new image co-segmentation method using saliency detection for surveillance image of coal miners," *Computers & Electrical Engineering*, vol. 40, no. 8, pp. 227–235, 2014.

[15] P. Subudhi and S. Mukhopadhyay, "A fast texture segmentation scheme based on active contours and discrete cosine transform," *Computers & Electrical Engineering*, vol. 62, pp. 105–118, 2017.

[16] D. Klein and S. Frintrop, "Center-surround divergence of feature statistics for salient object detection," in *2011 International Conference on Computer Vision*, pp. 2214–2219, Barcelona, Spain, 2011.

[17] S. P. Yong, J. D. Deng, and M. K. Purvis, "Wildlife video keyframe extraction based on novelty detection in semantic context," *Multimedia Tools and Applications*, vol. 62, no. 2, pp. 359–376, 2013.

[18] L. Yang, G. Yang, Y. Yin, and R. Xiao, "Sliding window-based region of interest extraction for finger vein images," *Sensors*, vol. 13, no. 3, pp. 3799–3815, 2013.

[19] J. M. L. Pérez, M. E. A. de la Varga, J. J. García, and V. R. G. Lacasa, "Monitoring lidia cattle with GPS-GPRS technology; a study on grazing behaviour and spatial distribution," *Veterinaria Mexico*, vol. 4, no. 4, 2017.

[20] A. Fernández-Caballero, M. López, and J. Serrano-Cuerda, "Thermal-infrared pedestrian ROI extraction through thermal and motion information fusion," *Sensors*, vol. 14, no. 4, pp. 6666–6676, 2014.

[21] R. Handcock, D. Swain, G. Bishop-Hurley et al., "Monitoring animal behaviour and environmental interactions using wireless sensor networks, GPS collars and satellite remote sensing," *Sensors*, vol. 9, no. 5, pp. 3586–3603, 2009.

[22] K. Smet, W. R. Ryckaert, M. R. Pointer, G. Deconinck, and P. Hanselaer, "Colour appearance rating of familiar real objects," *Color Research & Application*, vol. 36, no. 3, pp. 192–200, 2011.

[23] S. Goferman, L. Zelnik-Manor, and A. Tal, "Context-aware saliency detection," *IEEE Transactions on Pattern Analysis and Machine Intelligence*, vol. 34, no. 10, pp. 1915–1926, 2012.

[24] E. Rahtu, J. Kannala, M. Salo, and J. Heikkilä, "Segmenting salient objects from images and videos," in *Proceedings of the 11th European Conference on Computer Vision, ECCV 2010*, pp. 366–379, Heraklion, Crete, Greece, 2010.

[25] N. Murray, M. Vanrell, X. Otazu, and C. A. Parraga, "Saliency estimation using a non-parametric low-level vision model," in *Proceedings of the IEEE Computer Society Conference on Computer Vision and Pattern Recognition, CVPR 2011*, pp. 433–440, Providence, RI, USA, 2011.

[26] N. Imamoglu, W. Lin, and Y. Fang, "A saliency detection model using low-level features based on wavelet transform," *IEEE Transactions on Multimedia*, vol. 15, no. 1, pp. 96–105, 2013.

[27] X. D. Hou and L. Q. Zhang, "Saliency detection: a spectral residual approach," in *2007 IEEE Conference on Computer Vision and Pattern Recognition*, pp. 1–8, Minneapolis, MN, USA, 2007.

[28] M.-M. Cheng, N. J. Mitra, X. Huang, P. H. S. Torr, and S.-M. Hu, "Global contrast based salient region detection," *IEEE Transactions on Pattern Analysis and Machine Intelligence*, vol. 37, no. 3, pp. 569–582, 2015.

[29] J. Feng, "Salient object detection for searched web images via global saliency," in *2012 IEEE Conference on Computer Vision and Pattern Recognition*, pp. 3194–3201, Providence, RI, USA, 2012.

[30] R. Arya, N. Singh, and R. K. Agrawal, "A novel hybrid approach for salient object detection using local and global saliency in frequency domain," *Multimedia Tools and Applications*, vol. 75, no. 14, pp. 8267–8287, 2016.

[31] G. Kootstra and D. Kragic, "Fast and bottom-up object detection, segmentation, and evaluation using gestalt principles," *2011 IEEE International Conference on Robotics and Automation*, 2011, pp. 3423–3428, Shanghai, China, 2011.

[32] H. Wang, L. Dai, Y. Cai, X. Sun, and L. Chen, "Salient object detection based on multi-scale contrast," *Neural Networks*, vol. 101, pp. 47–56, 2018.

4

A Robust Data Interpolation based on a Back Propagation Artificial Neural Network Operator for Incomplete Acquisition in Wireless Sensor Networks

Mingshan Xie ◉,[1,2] Mengxing Huang ◉,[1] Yong Bai ◉,[1] Zhuhua Hu ◉,[1] and Yanfang Deng[1]

[1]*State Key Laboratory of Marine Resource Utilization in South China Sea, College of Information Science & Technology, Hainan University, Haikou 570228, China*
[2]*College of Network, Haikou University of Economics, Haikou 571127, China*

Correspondence should be addressed to Yong Bai; bai@hainu.edu.cn

Guest Editor: Aniello Falco

The data space collected by a wireless sensor network (WSN) is the basis of data mining and data visualization. In the process of monitoring physical quantities with large time and space correlations, incomplete acquisition strategy with data interpolation can be adopted to reduce the deployment cost. To improve the performance of data interpolation in such a scenario, we proposed a robust data interpolation based on a back propagation artificial neural network operator. In this paper, a neural network learning operator is proposed based on the strong fault tolerance of artificial neural networks. The learning operator is trained by using the historical data of the data acquisition nodes of WSN and is transferred to estimate the value of physical quantities at the locations where sensors are not deployed. The experimental results show that our proposed method yields smaller interpolation error than the traditional inverse-distance-weighted interpolation (IDWI) method.

1. Introduction

The purpose of a wireless sensor network (WSN) is to obtain the data field or data space of the physical world as accurate and complete as possible through acquisition technology. It is an important part of forecasting, simulation, and prediction to obtain the spatial-temporal distribution information of the monitored object accurately. However, in some scenarios, WSN can take an incomplete acquisition strategy, due to the development cost of the sensing device and the deployment environment factor, energy limitation, equipment aging, and other factors, or because it is not necessary to collect the data in each corner of the monitoring area. The incomplete acquisition strategy is divided into three cases: (a) spatial incomplete acquisition strategy—the actual collected area is smaller than the interested area or the actual acquisition location set is part of the entire acquisition locations in the monitoring area; (b) temporal incomplete acquisition strategy—the actual collection time period is less than the time period in which all devices work. The sleeping schedule is a temporal incomplete acquisition strategy. (c) Incomplete acquisition of attributes—the actual physical quantities collected are less than the interested physical quantities.

Because the constraints of interpolation are relatively small, it is more appropriate to use the interpolation algorithm to complete or refine the entire data space in the case of spatial incomplete acquisition. The interpolation completion algorithm takes advantage of the strong correlation between the data in the data space. At present, data interpolation is the main method to complement the data space of the entire region. In [1], Ding and Song used the linear interpolation theory to evaluate the working status of each node and the whole network coverage case. In [2], Alvear et al. applied interpolation techniques for creating detailed pollution maps.

However, WSN is often affected by many unfavorable factors. For example, it is usually arranged in a harsh environment, the node failure rate is relatively high, it is very difficult to physically replace the failure sensor, and the wireless communication network is susceptible to interference, attenuation, multipath, blind zone, and other unfavorable factors. Data is prone to errors, security is not guaranteed, etc. Therefore, WSN data interpolation technology must be highly fault tolerant to ensure high credibility and robustness of the completed data space [3].

Data interpolation is used to predict and estimate the information at an unknown location by means of using known information. Transfer learning opens up a new path for data interpolation. The goal of transfer learning is to extract useful knowledge from one or more source domain tasks and apply them to new target tasks. It is essentially the transfer and reuse of knowledge. Transfer learning has gradually received the attention of scholars. In [4], the authors are motivated by the idea of transfer learning. They proposed a novel domain correction and adaptive extreme learning machine (DC-AELM) framework with transferring capability to realize the knowledge transfer for interference suppression. To improve the radar emitter signal recognition, Yang et al. use transfer learning to obtain the robust feature against signal noise rate (SNR) variation in [5]. In [6], the authors discuss the relationship between transfer learning and other related machine learning techniques such as domain adaptation, multitask learning, and sample selection bias, as well as covariate shift.

Artificial neural networks have strong robustness. One of the requirements to ensure the accuracy of transfer learning is the robustness of the learning algorithm. Many scholars have combined the neural artificial network with transfer learning. In [7] Pan et al. propose a cascade convolutional neural network (CCNN) framework based on transfer learning for aircraft detection. It achieves high accuracy and efficient detection with relatively few samples. In [8], Park et al. showed that the transfer learning of the ImageNet pretrained deep convolutional neural networks (DCNN) can be extremely useful when there are only a small number of Doppler radar-based spectrogram data.

The research aim of data interpolation of WSN is to complete the data space of the entire monitoring area by using the limited data of the acquisition node to estimate the data at the locations where sensors are deployed. However, data errors of WSN due to various reasons have great impact on the accuracy of data interpolation. Due to the strong robustness of an artificial neural network, an artificial neural network learning operator is generated by using historical measurement data of limited data acquisition nodes in this paper. At the same time, this paper applies the learning property of the artificial neural network to the inverse-distance-weighted interpolation method, which is conducive to improving precision and accuracy of data interpolation. On the basis of analyzing the demand of network models, this paper proposes a robust data interpolation based on a back propagation artificial neural network operator for incomplete acquisition in wireless sensor networks. The detailed steps of the algorithm are discussed in detail, and the algorithm is analyzed based on the MATLAB tool and the measured data provided by the Intel Berkeley

Research laboratory [9]. The experimental results are good evaluations of the fault-tolerant performance and lower error of our proposed method.

The main contributions of this paper are as follows:

(1) Aiming at the data loss and disturbance error, we propose a fault-tolerant complementary algorithm based on the robustness of the artificial neural network

(2) We combine the artificial neural network with the inverse-distance-weighted interpolation algorithm to obtain a novel back propagation artificial neural network operator

(3) We use the inverse tangent function to reconcile the relationships among multiple prediction values

The rest of the paper is organized as follows. In Section 2, we summarize the related work. Section 3 introduces the interpolation model in the condition of data error. Section 4 presents how to construct the learning operator set of data acquisition nodes. Section 5 elaborates how to generate the interpolation by the method based on the back propagation artificial neural network operator. We will show the experimental results of our proposed methods compared with the inverse-distance-weighted interpolation in Section 6. The conclusions are given in Section 7.

2. Related Works

2.1. The Inverse-Distance-Weighted Interpolation Method. The inverse-distance-weighted interpolation (IDWI) method is also called "inverse-distance-weighted averaging" or the "Shepard Method." The interpolation scheme is explicitly expressed as follows:

Given n locations whose the plane coordinates are (x_i, y_i) and the values are z_i, where $i = 1, 2, \ldots, n$, the interpolation function is

$$f(x, y) = \begin{cases} \dfrac{\sum_{i=1}^{n} \left(d_i/\text{dist}_i^p \right)}{\sum_{i=1}^{n} \left(1/\text{dist}_i^p \right)} & \text{if } (x, y) \neq (x_i, y_i), \quad i = 1, 2, \ldots, n, \\ d_i & \text{if } (x, y) = (x_i, y_i), \quad i = 1, 2, \ldots, n, \end{cases}$$

(1)

where $\text{dist}_i = \sqrt{(x - x_i)^2 + (y - y_i)^2}$ is the horizontal distance between (x, y) and (x_i, y_i), where $i = 1, 2, \ldots, n$. p is a constant greater than 0, called the weighted power exponent.

It can easily be seen that the interpolation $f(x, y) = \sum_{i=1}^{n} (d_i/\text{dist}_i^p)/\sum_{i=1}^{n} (1/\text{dist}_i^p)$ of the location (x, y) is the weighted mean of $\text{dist}_1, \text{dist}_2, \ldots, \text{dist}_n$.

The application of inverse-distance-weighted interpolation is more extensive. Because of its simple computation and having less constraints, the interpolation precision is higher. In [10], Kang and Wang use the Shepard family of interpolants to interpolate the density value of any given computational point within a certain circular influence domain of the point. In [11], Hammoudeh et al. use a

Shepard interpolation method to build a continuous map for a new WSN service called the map generation service.

From (1), we can see that the IDWI algorithm is sensitive to the accuracy of the data. However, WSN is usually deployed in a harsh environment, and the probability of data being collected is high. The error tolerance of the interpolation algorithm is required. This paper improves the robustness of interpolation algorithm on the basis of the inverse-distance interpolation algorithm.

2.2. Artificial Neural Network. An artificial neural network (ANN) is an information processing paradigm that is inspired from biological nervous systems, such as how the brain processes information. ANNs, like people, have the ability to learn by example. An ANN is configured for a specific application, such as pattern recognition, function approximation, or data classification, through a learning process. Learning in biological systems involves adjustments to the synoptic connections that exist among neurons. This is true for ANNs as well. They are made up of simple processing units which are linked by weighted connections to form structures that are able to learn relationships between sets of variables. This heuristic method can be useful for nonlinear processes that have unknown functional forms. The feed forward neural networks or the multilayer perceptron (MLP) among different networks is most commonly used in engineering. MLP networks are normally arranged in three layers of neurons; the input layer and output layer represent the input and output variables, respectively, of the model; laid between them is one or more hidden layers that hold the network's ability to learn nonlinear relationships [12].

The natural redundancy of neural networks and the form of the activation function (usually a sigmoid) of neuron responses make them somewhat fault tolerant, particularly with respect to perturbation patterns. Most of the published work on this topic demonstrated this robustness by injecting limited (Gaussian) noise on a software model [13]. Velazco et al. proved the robustness of ANN with respect to bit errors in [13]. Venkitaraman et al. proved that neural network architecture exhibits robustness to the input perturbation: the output feature of the neural network exhibits the Lipschitz continuity in terms of the input perturbation in [14]. Artificial neural networks have strong robustness. The robustness of the algorithm is a requirement to ensure the accuracy of artificial neural network operator transferring. We can see from the literatures [7, 8] that operator transferring can combine well with an artificial neural network. The learning operator in this paper also adopts an artificial neural network algorithm.

3. Problem Formulations

3.1. Data Acquisition Nodes. To assess the entire environmental condition, WSN collects data by deploying a certain number of sensors in the location of the monitoring area; thus, the physical quantity of the monitoring area is discretized and the monitoring physical quantity is digitized.

Definition 1. Interested locations: in the whole monitoring area, they are the central locations of the segment of the monitoring area that we are interested in.

Sensors can be deployed at each interested location to capture data. The data at all interested locations reflects the information status of the entire monitoring area.

We assumed that S is the set of the interested locations, which is a matrix of $1 \times n$.

$$S = \{s_1(x_1, y_1), s_2(x_2, y_2), s_3(x_3, y_3), \ldots, s_n(x_n, y_n)\}, \quad (2)$$

where $s_i(x_i, y_i)$ is the ith interested location in the monitoring area. (x_i, y_i) is the coordinates of the s_i interested location in the monitoring area. Due to the difficulty and limitation of deployment, not all the interested locations can deploy sensors. This paper studies the spatial incomplete collection strategy. We select a subset of S as the data acquisition node. $\|S\|$ represents the potential of a set, that is, the number of elements of S. $\|S\| = n$.

Definition 2. Data acquisition nodes: they are the interested locations where the sensors are actually deployed.

The sensors are deployed in these interested locations, so that these locations become data acquisition nodes. The all-data acquisition nodes in the monitoring area act as the sensing layer of the WSN, and the information is transmitted to the server through the devices of the transport layer.

In our research, when the sensors are not deployed at the interested location, we use zero as a placeholder to replace the data acquisition node. When the interested location becomes the data acquisition node, we use 1 as a placeholder to replace the data acquisition node. Suppose that M is the set of data acquisition nodes.

$$M = \{s_i \mid s_i = 1, s_i \in S, i = 1 \ldots n\}, \quad (3)$$

where $s_i = 1$ represents the ith interested location where the sensors are deployed. $\|M\|$ indicates the potential of the set M. It reflects the total number of elements in the set of data acquisition nodes. This paper investigates the case where multiple types of sensors are deployed at a data acquisition node. The data that the data acquisition node s_i can correctly collect at time t_l is defined as $d_{s_i}(t_l) = [d_{s_i,1}(t_l), d_{s_i,2}(t_l), \ldots, d_{s_i,k}(t_l)]$, which is k dimensional data that is perceived by the ith data acquisition location s_i in S. The physical quantity of temperature, humidity, etc. can be measured at the same time. The data is defined as

$$D_M(t_l) = \left\{ d_{s_i}(t_l) \mid s_i \in M, d_{s_i}(t_l) \in R^k \right\}. \quad (4)$$

If $\|M\| < \|S\|$, then WSN implements incomplete coverage; if $\|M\| = \|S\|$, then WSN implements complete coverage. The data of the interested location where the sensors are not deployed can be assessed by the data of the data acquisition node. The interested location where the sensors are not deployed is indicated as non-data acquisition location.

3.2. Data Acquisition Error. Because the wireless communication network is susceptible to interference, attenuation, multipath, blind zone, and other unfavorable factors, the data error rate is high. Nodes and links in wireless sensor networks are inherently erroneous and unpredictable. The error data which greatly deviated from the ideal truth value is divided into two types: data loss and data disturbance.

(1) Data Loss. These reasons, such as nodes cannot work, links cannot be linked, or data cannot be transmitted, cause the data of the corresponding data acquisition nodes to not reach the sink node.

(2) Data Disturbance. Due to the failure, the local function of the WSN is in an incorrect (or ambiguous) system state. This state may cause a deviation between the data measured by the sensors of the corresponding data acquisition node and the true value, or the signal is disturbed during the transmission, and the data received at the sink node is deviant from the true value. The data that corresponds to the data acquisition node is not the desired result. The collected data oscillate in the data area near the true value. In this paper, we assumed that the data disturbance obeys the Gauss distribution.

The main idea of our method is based on the fundamental assumption that the sensing data of WSN are regarded as a vector indexed by the interested locations and recovered from a subset sensing data. As demonstrated in Figure 1, the data acquisition consists of two stages: the data-sensing stage and the data-recovering stage. At the data-sensing stage, instead of deploying sensors and sensing data at all interested locations, a subset of interested locations which are the shaded ones in the second subfigure is selected to sense physical quantity and deliver the sensing data to the sink node at each data collection round. Some locations are drawn by the fork in the second subfigure because their data is lost or disturbed. The fork represented the data errors. When the hardware and software of the network node failures or the communication links of the network are broken, the set of sensors which encounter data errors is only the subset of M. At the data-recovering stage, the sink node receives these incomplete sensing data over some data collection rounds shown in the third subfigure in which the shaded entries represent the valid sensing data and the white entries are unknown. And then we could use them to recover the complete data by our method.

Here we adopt a mask operator $A()$ to represent the process of collecting data-encountering errors:

$$A(D_M(t_l)) = B(t_l), \qquad (5)$$

where B is the data set that is actually received for interpolation. For the sake of clarity, the operator $A()$ can be specified as a vector product as follows:

$$A(D_M(t_l)) = Q \odot D_M(t_l), \qquad (6)$$

where $\odot$ denotes the product of two vectors, i.e., $b_{s_i}(t_l) = d_{s_i}(t_l) \times q_{s_i}$. Q is a vector of $1 \times \|M\|$. b_{s_i} indicates the data

that is actually received by the ith data acquisition node for interpolation.

$$q_{s_i} = \begin{cases} 1, & \text{if the data is not error,} \\ \begin{cases} 0, & \text{if the data is lost,} \\ \dfrac{\text{No}(d_{s_i}(t_l), \sigma^2)}{d_{s_i}}, & \text{if the data is disturbed,} \end{cases} \end{cases} \qquad (7)$$

where $\text{No}(d_{s_i}(t_l), \sigma^2)$ represents Gauss distribution with a mean of $d_{s_i}(t_l)$ and a variance of σ^2.

In this paper, the error rate of the received data is defined as follows: $\|Q\|_{q_{s_i} \neq 1}/\|M\|$, where $\|Q\|_{q_{s_i} \neq 1}$ is the number of non-1 elements.

3.3. Completion of Data Space with Interpolation. Due to conditional restrictions, there is no way to deploy sensors in $S - M$. The data generated in $S - M$ can be estimated by interpolation based on the data in M. The data space of the entire monitoring area is set to

$$D_S(t_l) = D_M(t_l) \cup \widehat{D}_{S-M}(t_l), \qquad (8)$$

where $D_M(t_l)$ represents the data set collected from the data acquisition node in M at epoch t_l. $D_{S-M}(t_l)$ is the data set collected from non-data acquisition locations $S - M$ at the epoch t_l in the ideal case, if the sensors are deployed in the non-data acquisition location. $\widehat{D}_{S-M}(t_l)$ is the data interpolation set based on the data in M.

The problem definition is how to process the $D_M(t_l)$ data so that $\widehat{D}_{S-M}(t_l)$ is as close as possible to $D_{S-M}(t_l)$, that is, the problem of minimizing the error between $\widehat{D}_{S-M}(t_l)$ and $D_{S-M}(t_l)$. Its mathematical expression is as follows:

$$\begin{aligned} \arg\min \quad & \left\| \widehat{D}_{S-M}(t_l) - D_{S-M}(t_l) \right\| \\ \text{s.t.} \quad & Q \neq 1, \end{aligned} \qquad (9)$$

where $\|\cdot\|$ is the Euclidean norm form used to evaluate the error between $\widehat{D}_{S-M}(t_l)$ and $D_{S-M}(t_l)$; $Q \neq 1$ indicates that Q is not a full 1 matrix.

$$\begin{aligned} \widehat{D}_{S-M}(t_l) = \Big\{ \widehat{d}_{s_j}(t_l) \mid \widehat{d}_{s_j}(t_l) \\ = E\left(\widehat{d}_{s_j}'(t_l), \widehat{d}_{s_j}''(t_l), \dots, \widehat{d}_{s_j}^n(t_l) \right), s_j \in S - M \Big\}, \end{aligned} \qquad (10)$$

where $d_{s_j}'(t_l), d_{s_j}''(t_l), \dots, d_{s_j}^n(t_l)$ represents each value close to $d_{s_j}(t_l)$. Suppose s_i is the closest data acquisition node to s_j. The value received from the data acquisition node nearest the non-data acquisition location is also close to $d_{s_j}(t_l)$.

$$\because \text{dist}_{s_i \longrightarrow s_j} = \min\left(\text{dist}_{M \longrightarrow s_j} \right), \quad s_i \in M, \qquad (11)$$

where $\text{dist}_{M \longrightarrow s_j}$ indicates the distance from each data acquisition node in M to the non-data acquisition location s_j. The

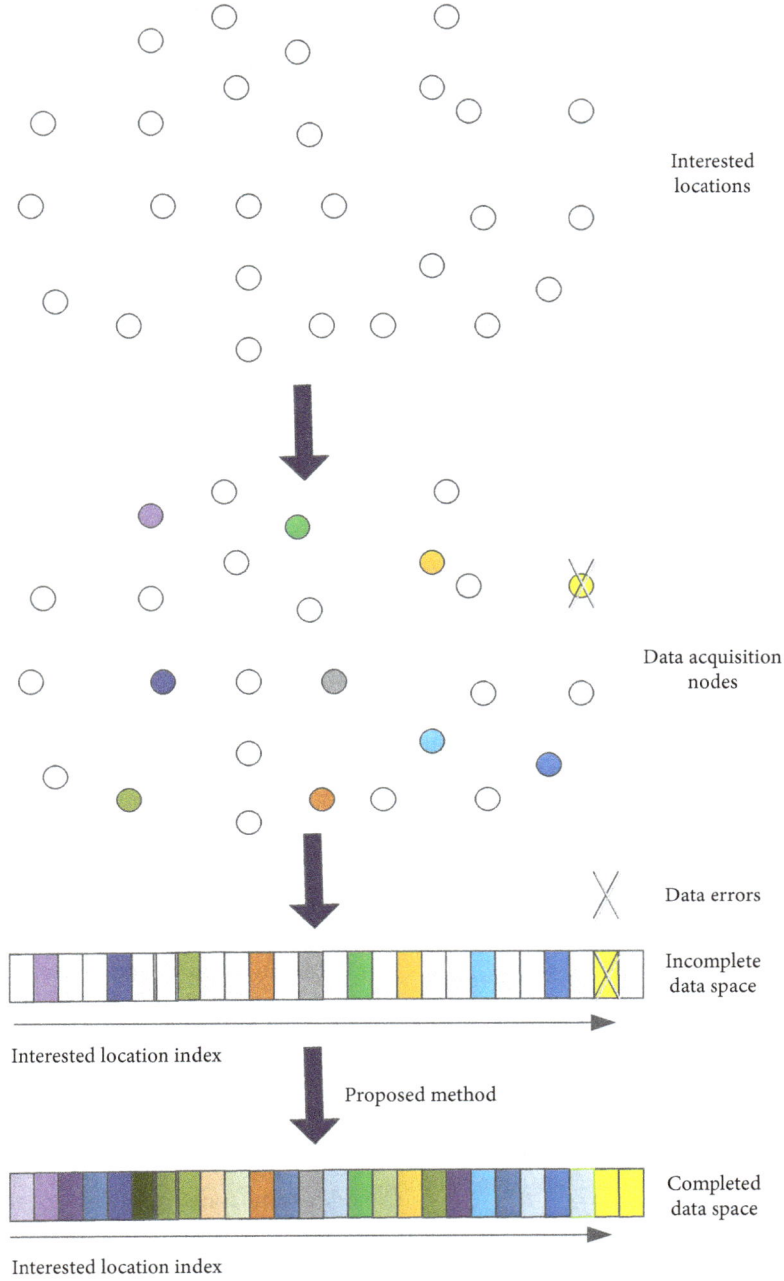

FIGURE 1: Incomplete acquisition process based on our proposed method.

closer the information collected by the node, the greater the correlation [15].

$$\therefore d_{s_j}'(t_l) = d_{s_i}(t_l), \qquad (12)$$

where $d_{s_i}(t_l)$ represents the value of the physical quantity actually collected by the s_i data acquisition node.

Suppose that $\widehat{d}_{s_j}''(t_l)$ represents the assessed value at the jth non-data acquisition location, which is obtained by the back propagation artificial neural network operator. In this

paper, we get the $\widehat{d}_{s_j}''(t_l)$ that is close to $d_{s_j}(t_l)$. The data set of non-data acquisition locations for interpolation is

$$d_{s_j}''(t_l) = \Upsilon_{s_j}\big(b_{s_i}(t_l)\big), \quad s_j \in S - M, s_i \in M, \qquad (13)$$

$$\widehat{D}_{S-M}''(t_l) = \Upsilon_{S-M}(B(t_l)),$$

where $\Upsilon_{S-M}(\cdot)$ represents the learning operator to assess $d_{s_j}(t_l)$. $B(t_l)$ represents the received data set at the epoch t_l. We can use the back propagation artificial neural network operator of the data acquisition node closest to the non-data acquisition location to predict the data of the non-data acquisition location.

4. Learning Operator Set of Data Acquisition Nodes

The mathematical model of the learning operator for the reconstruction is as follows:

$$\arg\min \quad \|\Upsilon(\text{Inp}) - \text{Tar}\|$$
$$\text{s.t.} \quad \text{Inp} \neq \text{Tar}. \tag{14}$$

where Tar represents the learning goals, $\Upsilon(\cdot)$ is the learning operator of Inp, and Inp can be individuals, variables, and even algorithms or functions, sets, and so on. The input of $\Upsilon(\cdot)$ is Inp. Inp and Tar are different. If they do not have differences, there is no need to learn for reconstruction. The purpose of learning is to make $\Upsilon(\text{Inp})$ gradually approach Tar.

Because data of WSN is error-prone, we need fault-tolerant and robust learning operators. The learning operator in this paper uses a back propagation (BP) artificial neural network. We can use data of data acquisition nodes to predict the data of the non-data acquisition location, thus assessing the data space of the entire monitoring area. Because of the strong robustness and adaptability of the artificial neural network, we can use the artificial neural network to interpolate the data of non-data acquisition locations in the case of data error.

The BP artificial neural network is a multilayer (at least 3 levels) feedforward network based on error backpropagation. Because of its characteristics, such as nonlinear mapping, multiple input and multiple output, and self-organizing self-learning, the BP artificial neural network can be more suitable for dealing with the complex problems of nonlinear multiple input and multiple output. The BP artificial neural network model is composed of an input layer, hidden layer (which can be multilayer, but at least one level), and output layer. Each layer is composed of a number of juxtaposed neurons. The neurons in the same layer are not connected to each other, and the neurons in the adjacent layers are connected by means of full interconnection [16].

After constructing the topology of the artificial neural network, it is necessary to learn and train the network in order to make the network intelligent. For the BP artificial neural network, the learning process is accomplished by forward propagation and reverse correction propagation. As each data acquisition node is related to other data acquisition nodes, and each data acquisition node has historical

data, it can be trained through the historical data of data acquisition nodes to generate the artificial neural network learning operator of the data acquisition node.

4.1. Transform Function of the Input Unit. The data of the non-data acquisition location s_j is assessed by using the inverse-distance learning operator of the data acquisition node s_i closest to s_j. Because there is spatial correlation between interested locations in space acquisition, in this paper, we adopt the IDWI algorithm combined with the BP artificial neural network algorithm.

At a certain epoch t_l, the data set of all other data acquisition nodes except the s_i is as follows:

$$B_{M-s_i}(t_l) = \begin{bmatrix} b_{s_o,1}(t_l) & \cdots & b_{s_o,1}(t_l) \\ \vdots & b_{s_q,i}(t_l) & \vdots \\ b_{s_m,k}(t_l) & \cdots & b_{s_m,k}(t_l) \end{bmatrix}, \quad b_{s.} \in M - s_i. \tag{15}$$

In this paper, the inverse-distance weight is used to construct the transform function. The transform function $h_{s_i}(B_{M-s_i}(t_l))$ of the input unit $B_{M-s_i}(t_l)$ of the artificial neural network with data acquisition node s_i is as follows:

$$h_{s_i}\left(B_{M-s_i}(t_l)\right) = \frac{\left(1/\text{dist}^{\gamma}_{s_q \longrightarrow s_i}\right) \times B_{M-s_i}(t_l)}{\sum_{s_q \in M}^{\|M\|-1}\left(1/\text{dist}^{\gamma}_{s_q \longrightarrow s_i}\right)}, \quad s_q \in M - s_i, s_i \in M, \tag{16}$$

where $\text{dist}_{s_q \longrightarrow s_i}$ represents the distance between the data acquisition node s_i and the non-data acquisition location s_q. γ represents the weighted power exponent of the distance reciprocal. $\sum_{s_q \in M}^{\|M\|-1}(1/\text{dist}^{\gamma}_{s_q \longrightarrow s_i})$ represents the sum of the weighted reciprocal of the distance from the data acquisition node s_i to the rest of data acquisition nodes.

The artificial neural network requires a training set. We take the historical data of the period T of all data acquisition nodes as the training set. $T = [t_0, t_1, \cdots, t_l, \cdots, t_m]$. In practical engineering, it is feasible for us to get the data from data acquisition nodes in a period to learn.

In this paper, data collected from all data acquisition nodes are used as the training set $B_{M-s_i}(T)$. $B_{M-s_i}(T)$ is a three-order tensor, as shown in the following Figure 2. $B_{M-s_i}(T) \in R^{k \times \|M-s_i\| \times t_m}$. The elements of the training set are indexed by physical quantities, acquisition node, and epoch. The $B_{M-s_i}(T)$ matrix is obtained by

$$B_{M-s_i}(T) = \begin{bmatrix} \begin{vmatrix} b_{s_o,1}(t_0) & \cdots & b_{s_o,1}(t_0) \\ \vdots & b_{s_q,i}(t_l) & \vdots \\ b_{s_m,k}(t_0) & \cdots & b_{s_m,k}(t_0) \end{vmatrix} \begin{vmatrix} b_{s_o,1}(t_1) & \cdots & b_{s_o,1}(t_1) \\ \vdots & b_{s_q,i}(t_1) & \vdots \\ b_{s_m,k}(t_1) & \cdots & b_{s_m,k}(t_1) \end{vmatrix} \cdots \begin{vmatrix} b_{s_o,1}(t_l) & \cdots & b_{s_o,1}(t_l) \\ \vdots & b_{s_q,i}(t_l) & \vdots \\ b_{s_m,k}(t_l) & \cdots & b_{s_m,k}(t_l) \end{vmatrix} \cdots \begin{vmatrix} b_{s_o,1}(t_m) & \cdots & b_{s_o,1}(t_m) \\ \vdots & b_{s_q,i}(t_l) & \vdots \\ b_{s_m,k}(t_m) & \cdots & b_{s_m,k}(t_m) \end{vmatrix} \end{bmatrix}, \quad b_{s.} \in M - s_i. \tag{17}$$

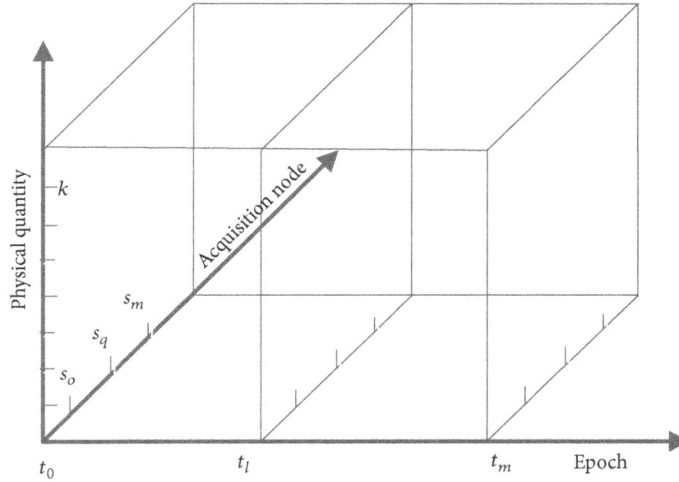

FIGURE 2: The model of $B_{M-s_i}(T)$.

In actual engineering, $B_{M-s_i}(T)$ has a lot of noise. It is necessary to clean it. If it is not cleaned, it will affect the estimation accuracy of the artificial neural network and the precision of learning. Data cleaning is the process of reexamining and verifying data to remove duplicate information, correct existing errors, and provide data consistency. We can use various filter algorithms to get $\bar{B}_{M-s_i}(T)$.

4.2. Artificial Neural Network Learning Operator. The sensing data of the real network include several physical quantities, such as temperature, humidity, and illumination. Usually, a variety of sensors are deployed at an acquisition node, and multiple physical quantities are collected at the same time. Physical quantities at the same acquisition node have the same temporal and spatial variation trend, but in the process of recovery using the artificial neural network, each quantity has a mutual promotion effect. The distance between data acquisition nodes and non-data acquisition locations is very easy to obtain. Based on the inverse-distance interpolation algorithm and BP artificial neural network algorithm, we propose a multidimensional inverse-distance BP artificial neural network learning operator.

The BP artificial neural network, which is the most widely used one-way propagating multilayer feedforward network, is characterized by continuous adjustment of the network connection weight, so that any nonlinear function can be approximated with arbitrary accuracy. The BP artificial neural network is self-learning and adaptive and has robustness and generalization. The training of the BP artificial neural network is the study of "supervisor supervision," The training process of the BP artificial neural network is shown in Figure 3.

For the input information, it is first transmitted to the node of the hidden layer through the weighted threshold summation, and then the output information of the hidden node is transmitted to the output node through the weighted threshold summation after the operation of the transfer function of each element. Finally, the output result is given. The purpose of network learning is to obtain the right weights and thresholds. The training process consists of two parts: forward and backward propagation.

The BP artificial neural network is a multilayer feedforward network trained by the error backpropagation algorithm. The BP artificial neural network structure is a multilayer network structure, which has not only input nodes and output nodes but also one or more hidden nodes. As demonstrated in Figure 3, according to the prediction error, the v and w are continuously adjusted, and finally the BP artificial neural network learning operator for reconstructing the data of data acquisition node s_i can be determined, as shown in the part $\Upsilon_{s_i}(\cdot)$ of Figure 3.

The input layer of the BP artificial neural network learning operator is the data collected at a certain time at all acquisition nodes except the s_i acquisition node, which is the vector $h_{s_i}(\bar{B}_{M-s_i}(t_l))$. The input of the jth neuron in the hidden layer is calculated by

$$\Sigma_j = \sum_{q=1}^{\|M\|-1} \sum_{p=1}^{k} v_{q,j,p} h_{s_i}\left(\bar{B}_{M-s_i}(t_l)\right) - \theta_j, \qquad (18)$$

where $v_{q,j,p}$ represents the weight between the (q,p)th input neurons and the jth neurons in the hidden layer. The (q,p)th input neurons are $h_{s_i}(b_{s_q,k}(t_l))$, that is, the k-th dimension data of the s_q acquisition node at time t_l. θ_j is the threshold of the jth neuron in the hidden layer. The output of neurons in the hidden layer is calculated by

$$ho_j = \frac{1}{1+e^{-\Sigma_j}}. \qquad (19)$$

Similarly, the output of each neuron in the output layer is set to o_p.

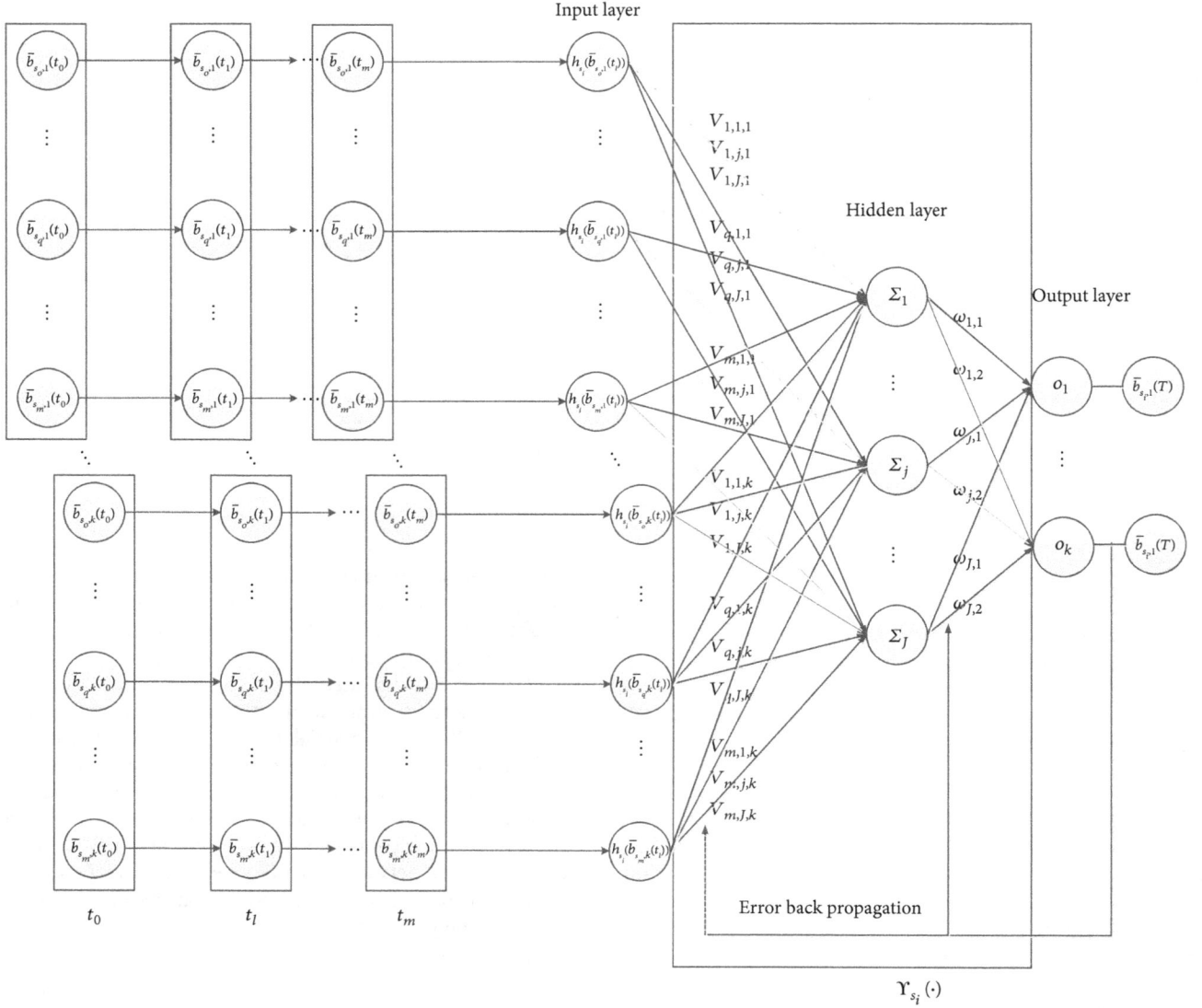

FIGURE 3: The training process of the BP artificial neural network.

The sum of squared errors of all neurons in the output layer is the objective function value optimized by the BP artificial neural network algorithm, which is calculated by

$$E = \frac{1}{(\|M\| - 1) \times k \times k} \sum_{q=1}^{\|M\|-1} \sum_{p=1}^{k} \sum_{p=1}^{k} \left(\bar{b}_{s_i,p} - o_p\right)^2, \quad (20)$$

where $\bar{b}_{s_i,p}$ represents the expected output value of neuron o_p in the output layer, corresponding to the value of the pth physical quantity sensed by the data acquisition node s_i. According to the gradient descent method, the error of each neuron in the output layer is obtained by

$$E_p = o_p \times \left(1 - o_p\right) \times \left(\bar{b}_{s_i,p} - o_p\right). \quad (21)$$

The weights and thresholds of the output layer can be adjusted by

$$\Delta\omega = \partial \times E_p \times ho_j, \quad (22)$$

$$\Delta\theta_p = \partial \times E_p, \quad (23)$$

where $\partial \in [0, 1]$ is the learning rate, which reflects the speed of training and learning. Similarly, the weight and threshold of the hidden layer can be obtained. If the desired results cannot be obtained from the output layer, it needs to constantly adjust the weights and thresholds, gradually reducing the error. The BP neural network has strong self-learning ability and can quickly obtain the optimal solution.

The mathematical model of Figure 3 is as follows:

$$f_{o_p}\left(\sum_{j=1}^{J}\sum_{p=1}^{k}\omega_j f_{\Sigma_j}\left(\sum_{q=1}^{\|M\|-1}\sum_{p=1}^{k}v_{q,j,p}h_{s_i}\left(\bar{B}_{M-s_i}(T)\right)\right)\right) \longrightarrow \bar{b}_{s_i,p}(T),$$

(24)

where $\bar{B}_{s_q}(T)$ is the training data, that is, the historical data of acquisition nodes except s_i in the period T. $h_{s_i}(\cdot)$ is the transform function of the input layer. $v_{q,j}$ is the weight of the input layer. $f_{\Sigma_j}(\cdot)$ is the transfer function of the input layer to the hidden layer. ω is the weight of the hidden layer to the output layer. $f_{o_p}(\cdot)$ is the transfer function of the hidden layer to the input layer. $\bar{b}_{s_i}(T)$ is the "supervisor supervision" that is the historical data of acquisition node s_i in the period T.

From (14) and (24), the following formula can be obtained:

$$\Upsilon_{s_i}(\cdot)=f_{o_p}\left(\sum_{j=1}^{J}\omega_j f_{\Sigma_j}\left(\sum_{q=1}^{\|M\|-1}\sum_{p=1}^{k}v_{q,j,p}(\cdot)\right)\right)|\bar{b}_{s_i}(T),$$

(25)

where $\Upsilon_{s_i}(\cdot)$ represents the inverse-distance BP artificial neural network learning operator of acquisition node s_i. $|\bar{b}_{s_i}(T)$ represents that this neural network operator $\Upsilon_{s_i}(\cdot)$ is a learning operator trained by the data of s_i as tutor information.

The data collected by each data acquisition node in the monitoring area is related to each other. The data of a data acquisition node can be learned by inputting the data of other data acquisition nodes into the learning operator. The learning operator set of data acquisition nodes in the whole monitoring area is as follows:

$$\Upsilon(\cdot)=\left\{\Upsilon_{s_i}(\cdot)\mid s_i\in M\right\}.$$

(26)

5. Interpolation at Non-Data Acquisition Locations

We transfer the BP artificial neural network operators from data acquisition nodes to non-data acquisition locations. We can use the learning operator of the data acquisition node closest to the non-data acquisition location to estimate the data of the non-data acquisition location. There are four ways to implement the learning operator transferring, including sample transferring, feature transferring, model transferring, and relationship transferring. In this paper, model transferring (also called parameter transferring) is used to better combine with the BP artificial neural network; that is, we can use the pretrained BP artificial neural network to interpolate. The BP artificial neural network learning operator has strong robustness and adaptability. In the interested area, if the physical quantity collected is highly correlated, and there is no situation in which the physical quantity changes

drastically between the various interested locations, the learning operator can be transferred.

Since the construction of the artificial neural network requires historical data for training, and there is no historical data at the non-data acquisition location deployed without sensors, it is very difficult to construct the artificial neural networks of the non-data acquisition location. However, we can predict the physical quantities of non-data acquisition locations by using the learning operator from the nearest data acquisition node.

5.1. Transform Function Corresponding to Nonacquisition Location s_j. First, we use the IDWI method to construct the transform function. The transform function of the artificial neural network input unit s_q of the acquisition node s_i is defined as $h_{s_i}(B_{M-s_i}(t_l))$:

$$h_{s_j}\left(B_{M-s_i}(t_l)\right)=\frac{B_{M-s_i}(t_l)/\text{dist}_{s_q\rightarrow s_j}^{\gamma}}{\sum_{s_q\in M-s_i}^{\|M\|-1}\left(1/\text{dist}_{s_q\rightarrow s_j}^{\gamma}\right)},$$

(27)

where $\text{dist}_{s_q\rightarrow s_j}$ represents the distance between s_q and s_j. γ represents the weighted exponentiation of the distance reciprocal. $\sum_{s_q\in M}^{\|M\|-1}(1/\text{dist}_{s_q\rightarrow s_j}^{\gamma})$ represents the sum of the weighted reciprocal of the distance from the non-data acquisition location s_j to the rest of the acquisition nodes. The physical data collected from the remaining $\|M\|-1$ data acquisition nodes except s_i are as follows:

$$B_{M-s_i}=\left\{b_{s_q}(t_l)\mid s_q\in M-s_i\right\}.$$

(28)

5.2. Learning Operator Transferring. Since the data of the data acquisition node closest to the non-data acquisition location is important to the non-data acquisition location, and its data is most correlated with the data of the non-data acquisition location, we estimate the data of the non-data acquisition location with data from the nearest data acquisition node and its learning operator.

Since the data we are targeting is spatially correlated, the smaller the distance between the two interested locations is, the smaller the difference of the collected data is. Conversely, the greater the difference of the collected data is. We can use the data of the data acquisition node close to the non-data acquisition location to assess the data at the non-data acquisition location.

Because the BP artificial neural network learning operator has strong robustness and adaptability, we can transfer the inverse BP artificial neural network learning operator of s_i to s_j in order to estimate the data at s_j in this paper. The transform function $h(\cdot)$ does not change with time, and it is a function of the distance between the sampling locations. $h_{s_i}(B_{M-s_i}(t_l))\longrightarrow h_{s_j}(B_{M-s_i}(t_l))$, the number of input parameters is constant, while the input parameters vary with time. So the change of the transform function will not affect the trend of input parameters. The change of the transform function will not affect the accuracy of prediction, and the operator transferring can be implemented.

$$\hat{a}''_{s_j}(t_l) = \Upsilon_{s_i}\left(h_{s_j}\left(B_{M-s_i}(t_l)\right)\right)$$
$$= f_{s_i}\left(\sum_{j=1}^{J}\omega_j f_{z_j}\left(\sum_{q=1}^{\|M\|}v_{q,j}h_{s_j}\left(B_{M-s_i}(t_l)\right)\right)\right). \quad (29)$$

The assessment network based on the BP artificial neural network operator is shown in Figure 4.

In this paper, our proposed method is improved on the basis of the BP artificial neural network. According to spatial correlation, the physical quantities of the monitored and interested location close to the data acquisition nodes s_q can be approximated by learning the operator of s_q.

Suppose d_{s_j} represents the estimated value of the physical quantity of the interested location s_j. Due to conditional restrictions, no sensor is deployed at the interested location s_j. We choose the physical quantity of the data acquisition node s_i nearest to the non-data acquisition location s_j for estimation. We can use Algorithm 1 to achieve the determination of s_i.

5.3. Assessment at the Non-Data Acquisition Location. s_i is the nearest data acquisition node to s_j. We can use the learning operator of the data acquisition node s_i to estimate d_{s_j}. This paper is an improvement on the inverse-distance interpolation method. Because s_i is closest to s_j, the correlation of their data is the largest. The data collected actually by s_i have the greatest impact on the predictive value at s_j.

$$d_{s_j} = E\left(\widehat{d'}_{s_j}(t_l), \widehat{d''}_{s_j}(t_l)\right)$$
$$= \alpha \times \widehat{d'}_{s_j}(t_l) + \beta \times \widehat{d''}_{s_j}(t_l), \quad s_i \in M, s_j \in S-M, \quad (30)$$

where α and β denote the weight of $\widehat{d'}_{s_j}(t_l)$ and $\widehat{d''}_{s_j}(t_l)$, respectively, to estimate physical quantities at s_j. $\alpha + \beta = 1$.

$\widehat{d'}_{s_j}(t_l)$ is the value of the actual measurement, so its credibility is higher. The closer s_i is to s_j, the greater the correlation between physical quantity at s_i and the data collected by s_i based on spatial correlation.

We assume that $\mathrm{dist}_{s_i \longrightarrow s_j}$ represents the distance between s_i and s_j. The influence weight of data collected by s_i on the assessment of s_j decreases with increasing $\mathrm{dist}_{s_i \longrightarrow s_j}$. Conversely, the greater the distance $\mathrm{dist}_{s_i \longrightarrow s_j}$, the smaller the impact. The change field of $\mathrm{dist}_{s_i \longrightarrow s_j}$ is $[0, +\infty)$. We find that the inverse tangent function is an increasing function on $[0, +\infty)$. The function curve is shown in Figure 5.

We use the boundedness and monotonically increasing characteristics of the arctangent function curve and are

inspired by the idea of IDWI. The formula for calculating β is as follows:

$$\beta = \frac{\arctan\left(\mathrm{dist}_{s_i \longrightarrow s_j}\right)}{\pi/2}, \quad \mathrm{dist}_{s_i \longrightarrow s_j} \in [0, +\infty). \quad (31)$$

We limit the value of β to the interval $[0, 1]$. When $\mathrm{dist}_{s_i \longrightarrow s_j}$ is close to 0, β is close to 0, and $1 - \beta$ is close to 1, then the data measured by the data acquisition point is closer to the value at the non-data acquisition location. When $\mathrm{dist}_{s_i \longrightarrow s_j} = 0$, it means that the sensors have deployed in s_j, and we do not need other values calculated by the prediction algorithm and directly use the actual measured data.

$$\hat{d}_{s_j} = \frac{\arctan\left(\mathrm{dist}_{s_i \longrightarrow s_j}\right)}{\pi/2} \times \widehat{d''}_{s_j}(t_l)$$
$$+ \left(1 - \frac{\arctan\left(\mathrm{dist}_{s_i \longrightarrow s_j}\right)}{\pi/2}\right) \times \widehat{d'}_{s_j}(t_l), \quad s_i \in M, s_j \in S-M. \quad (32)$$

If the interested location of the interpolation is still far from the nearest data acquisition node, then this algorithm will cause a large error. Since we are using the sensor placement based on the iterative dividing four subregions, the sensors of the data acquisition node are omnidirectional throughout the space, not only concentrated in a certain domain. The error is not too large.

6. Experiments and Evaluation

6.1. Parameter Setup. The data set we used is the measured data provided by the Intel Berkeley Research lab [9]. The data is collected from 54 data acquisition nodes in the Intel Berkeley Research lab between February 28th and April 5th, 2004. In this case, the epoch is a monotonically increasing sequence number from each mote. Two readings from the same epoch number were produced from different motes at the same time. There are some missing epochs in this data set. Mote IDs range from 1–54. In this experiment, we selected the data of these 9 motes as a set of interested locations, because these 9 points have the same epoch.

$$S = \{s_7(22.5, 8), s_{18}(5.5, 10), s_{19}(3.5, 13), s_{21}(4.5, 18),$$
$$s_{22}(1.5, 23), s_{31}(15.5, 28), s_{46}(34.5, 16), s_{47}(39.5, 14),$$
$$s_{48}(35.5, 10)\}. \quad (33)$$

When the environment of the monitoring area is not very complicated, the physical quantity of acquisition is a very spatial correlation, and it is feasible to use the placement based on the iterative dividing four subregions to deploy sensors. In this experiment, we use the method of the iterative dividing four subregions to select the interested locations from S as data acquisition nodes. The closest $s_i(x_i, y_i)$ $(s_i(x_i, y_i) \in S)$ to the deployment location

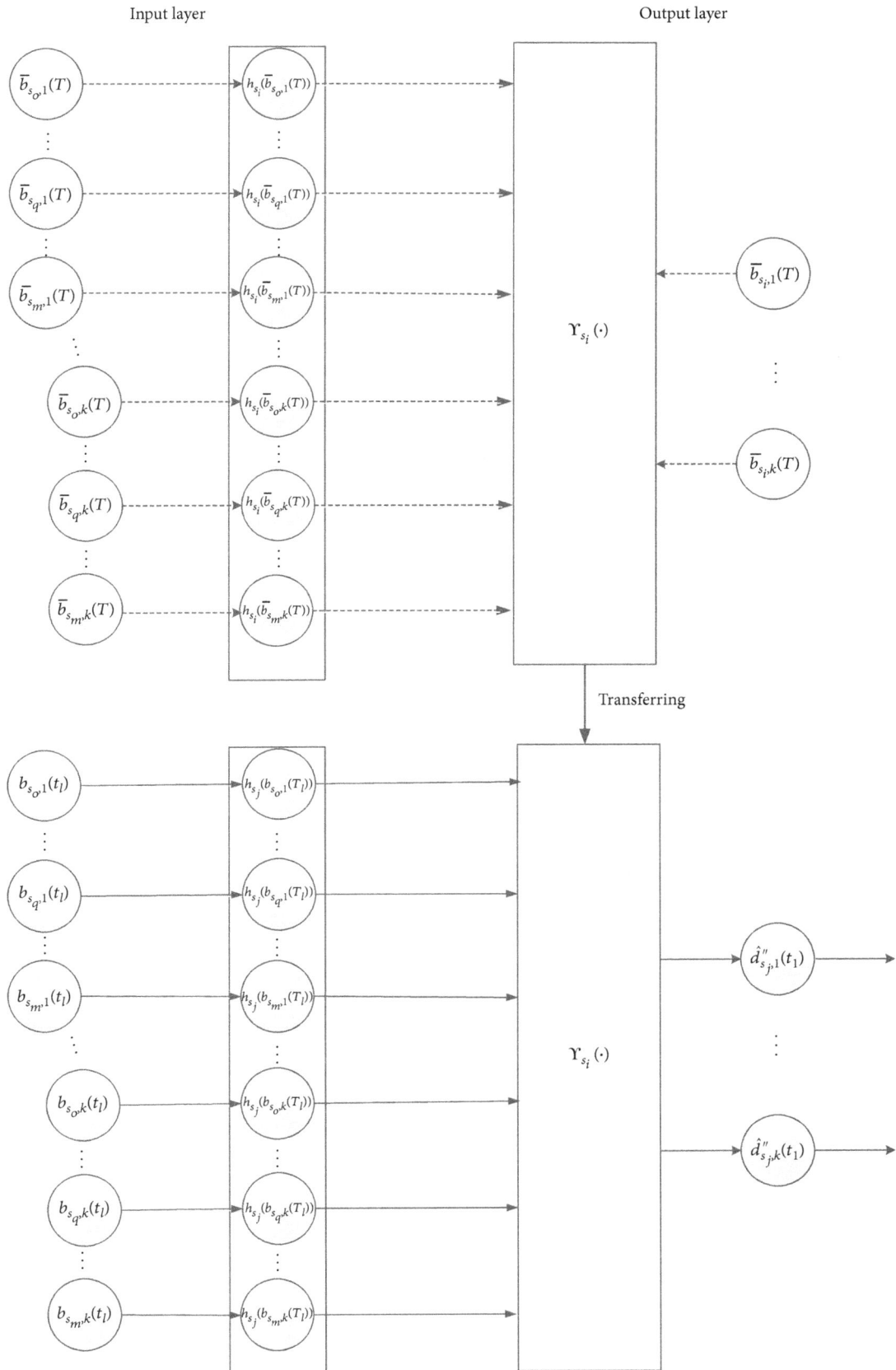

FIGURE 4: The assessment network based on the BP artificial neural network operator.

generated by the method of the iterative dividing four subregions is selected in M.

In this experiment, we set the dimension of the collected data to 2, that is, $k = 2$. We take two physical quantities: temperature and humidity. Data acquisition nodes need at least two, i.e., $\|M\| \geq 2$. Because one data acquisition node should be the closest node from the interested location for interpolation, and $s_i \notin \varnothing$, the data of other acquisition nodes should be used as the training set of the artificial neural network. $M - s_i \not\subset \varnothing$, $\|M - s_i\| \geq 1$.

The epochs we chose are: $T = \{33,35,42, 44,51,56,61,62, 70,73,86,88,89,91,95,99,102,107,112,113,123,124,127\}$. These data whose epoch is an element in T are used as a training set epoch. Unfortunately, the actual sampling sensing data are always corrupted and some values go missing. We need real clean sensing data to train the BP artificial neural networks to improve the interpolation precision.

In order to not lose the trend of data change in the process of data filtering, we simply use the mean processing with a filter window on the temporal direction of the measurements as the clean sensing data. If $|d_{s_i}(t - 1) - d_{s_i}(t)| \gg \delta \times |\bar{d}_{M-s_i}(t - 1) - \bar{d}_{M-s_i}(t)|$, then $d_{s_i}(t)$ can be replaced by the value calculated

$$d_{s_i}(t) = \frac{d_{s_i}(t - 1) + d_{s_i}(t + 1)}{2}, \qquad (34)$$

where δ is the adjusting coefficient. In this experiment, we take $\delta = 10$.

The filtered result of the temperature measurements is shown in Figure 6(a), and the actual sensing data is shown in Figure 6(b). The filtered result of the humidity measurements is shown in Figure 7(a), and the actual sensing data is shown in Figure 7(b).

When comparing the test results, we need that the epoch of the data supplied by each mote is the same. Because in the actual application, the data space of the monitoring area we built is the data space at a certain moment. We assumed that $t_l = 138$. In this experiment, the time we selected was the 138th epoch for interpolation. We compare the actual collected value at the 138th epoch with the interpolation calculated by algorithms.

To evaluate the accuracy of the reconstructed measurement, we choose the mean relative error (MRE). It reflects the precision of the estimated data relative to the measured data. The formula for the calculation of MRE is as follows [17]:

$$\text{MRE} = \frac{1}{n} \sum_{i=1}^{n} \left| \frac{d(x_i) - \hat{d}(x_i)}{d(x_i)} \right|, \qquad (35)$$

where $d(x_i)$ is the actual acquisition value of the ith sensor. Correspondingly, $\hat{d}(x_i)$ is the assessed value. n is the total number of data acquisition nodes.

6.2. Results of the Experiment. This experiment is in a small range, the physical quantity of the collected data is highly correlated, and there is no situation in which the physical

Input: M, $s_j(x_j, y_j)$
Output: s_i
Initialize mindist
For each s_i in M
    $\text{dist}_{s_j \longrightarrow s_q} = \sqrt[2]{(x_j - x_q)^2 + (y_j - y_q)^2}$
    if mindist > $\text{dist}_{s_j \longrightarrow s_q}$
       then $i = q$, mindist = $\text{dist}_{s_j \longrightarrow s_q}$
    end if
End for

ALGORITHM 1: Culling s_i from M.

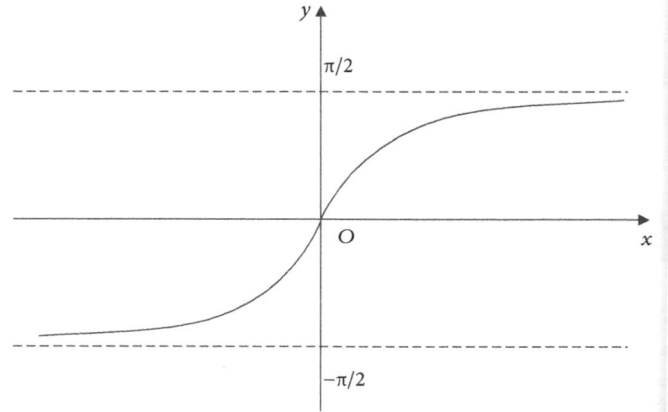

FIGURE 5: Inverse tangent function curve.

quantity changes drastically between the various interested locations, so the learning operator can be transferred.

In the case of data loss, we compare our method with the inverse-distance-weighted interpolation algorithm in terms of interpolation accuracy. Then $q_{s_i} = 0$. Since the data acquisition nodes for data loss are random, we conducted 20 tests for a more accurate comparison. Obviously, it is necessary that the number of acquisition node points for data loss be less than or equal to the total number of acquisition nodes. The results are shown in Figures 8 and 9.

As can be seen from Figures 8 and 9, as the proportion of lost data in all collected data decreases, the error of interpolation is gradually reduced. The curve of the proposed algorithm is relatively flat, indicating that it is less affected by data loss.

In the case of data loss, especially when the number of data acquisition nodes is relatively small, the interpolation error of our algorithm is much smaller than that of the inverse-distance-weighted interpolation.

In the case of data disturbance, we compare our method with the inverse-distance-weighted interpolation algorithm in terms of interpolation accuracy. Then $q_{s_i} = \text{No}(d_{s_i}(t_l), \sigma^2)/d_{s_i}$. For the interpolation of temperature, we use the mean temperature of all acquisition nodes as the mean value of the Gauss distribution. We set the parameters

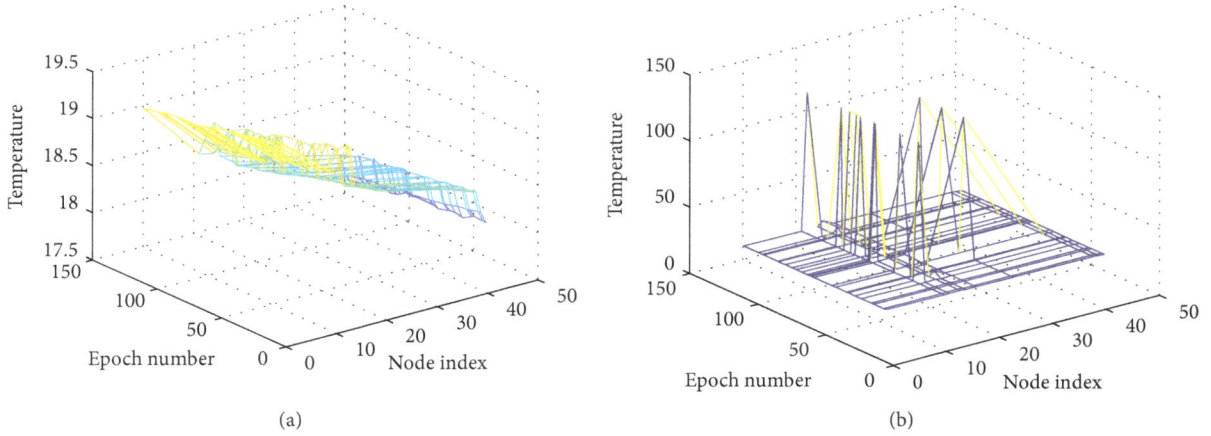

FIGURE 6: The temperature sensing data from the Intel Berkeley Research lab: (a) the filtered sensing data; (b) the actual sensing data.

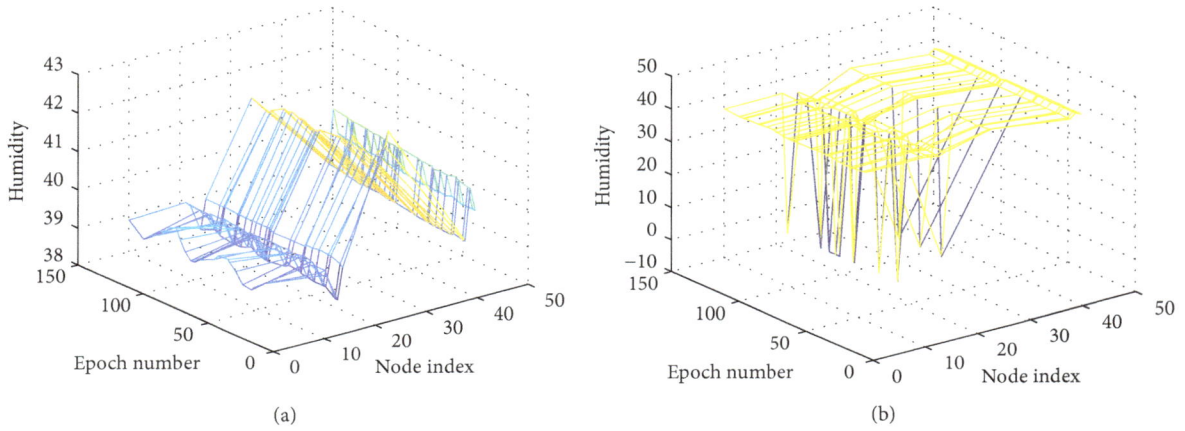

FIGURE 7: The humidity sensing data from the Intel Berkeley Research lab: (a) the filtered sensing data; (b) the actual sensing data.

as $d_{s_i}(t_l) = 18.47895$ and $\sigma^2 = 18$; for the interpolation of humidity, we use the mean humidity of all acquisition nodes as the mean value of the Gauss distribution. We set the parameters as $d_{s_i}(t_l) = 39.44691$ and $\sigma^2 = 39$. Again, we did 20 tests. The results are shown in Figures 10 and 11.

It can be seen from Figures 10 and 11 that the variation curve of the proposed algorithm is relatively flat, while the curve of the inverse-distance-weighted interpolation algorithm fluctuates greatly. The interpolation error of the IDWI algorithm is not only affected by data disturbance but also affected by the deployment location of the data acquisition nodes. When 7 acquisition nodes are deployed, the error of the IDWI algorithm is the smallest. Because sensor placement based on the iterative dividing four subregions is near uniform deployment when 7 acquisition nodes are deployed, the interpolation error is small.

In the case where there are not many acquisition nodes, the density of the acquisition nodes where data disturbance occurs is large, so the error of interpolation is more prominent. The number of sensors that can be deployed increases, and the error of interpolation is also reduced.

As we can see from Figures 8–11, our algorithm is insensitive to errors and strong in robustness when the data is wrong, while the inverse-distance interpolation method has a great influence on the interpolation accuracy of the error data. In particular, when the error rate is high, the relative error of our algorithm is much lower than that of the inverse-distance interpolation algorithm. When the error rate is 50%, our algorithm has a relative error of 0.1 and the relative error of the inverse-distance interpolation algorithm is 3.5, as shown in Figure 9(a).

7. Conclusions

In this paper, we proposed a robust data interpolation based on a back propagation artificial neural network operator for incomplete acquisition in a wireless sensor network. Under the incomplete collection strategy of WSN, the effect of complete acquisition can be approximately obtained by interpolation. In the case of limited data acquisition nodes, the data of the acquisition nodes are used to train to obtain the learning operator. Then, the learning operator of the acquisition node

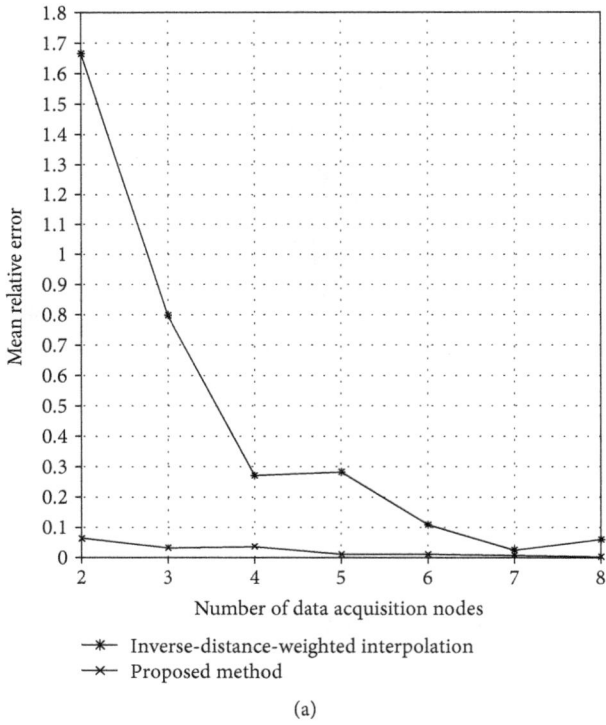

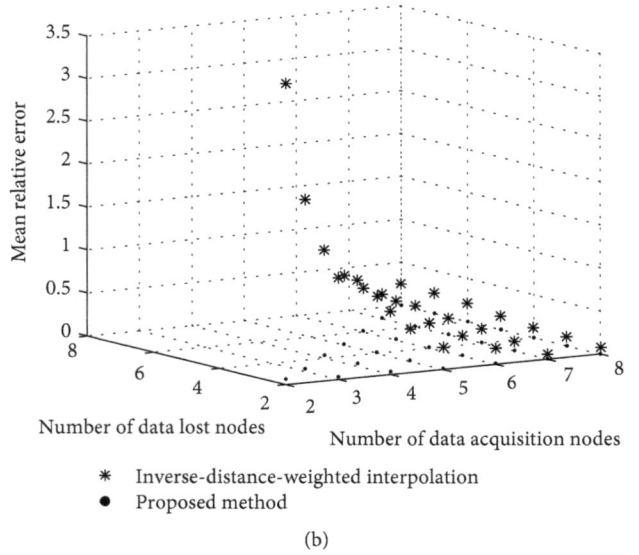

FIGURE 8: The temperature sensing data from the Intel Berkeley Research lab: (a) data loss occurred at 1 data acquisition node; (b) data loss occurred at data acquisition nodes increased from 2 to 8.

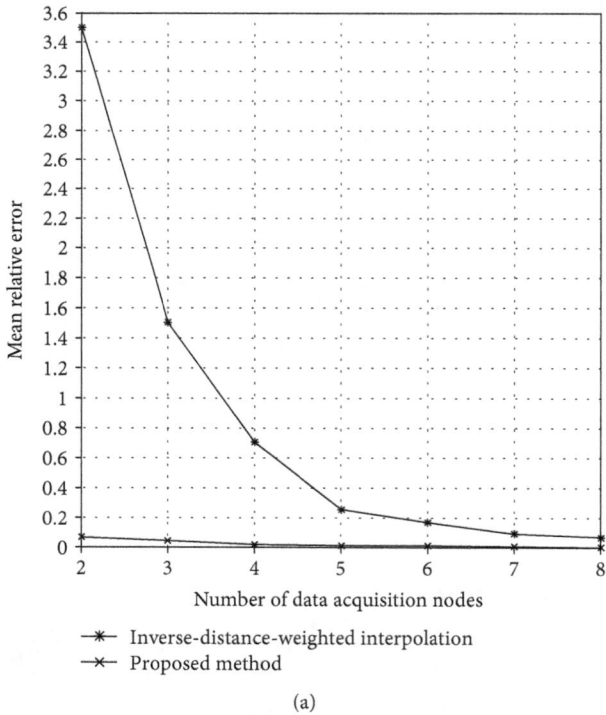

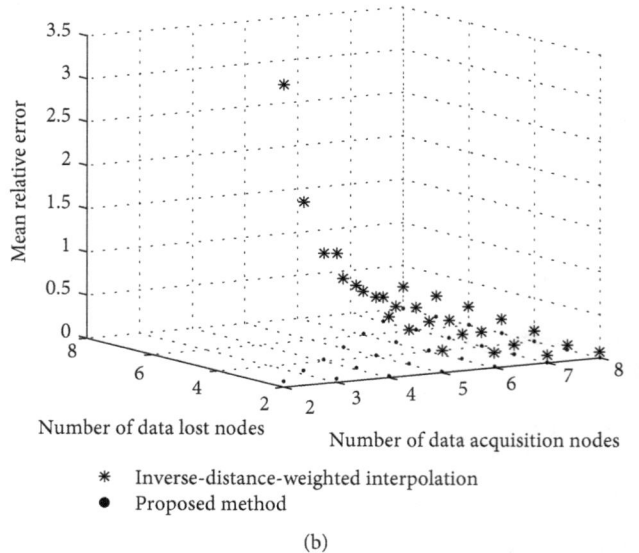

FIGURE 9: The humidity sensing data from the Intel Berkeley Research lab: (a) data loss occurred at 1 data acquisition node; (b) data loss occurred at data acquisition nodes increased from 2 to 8.

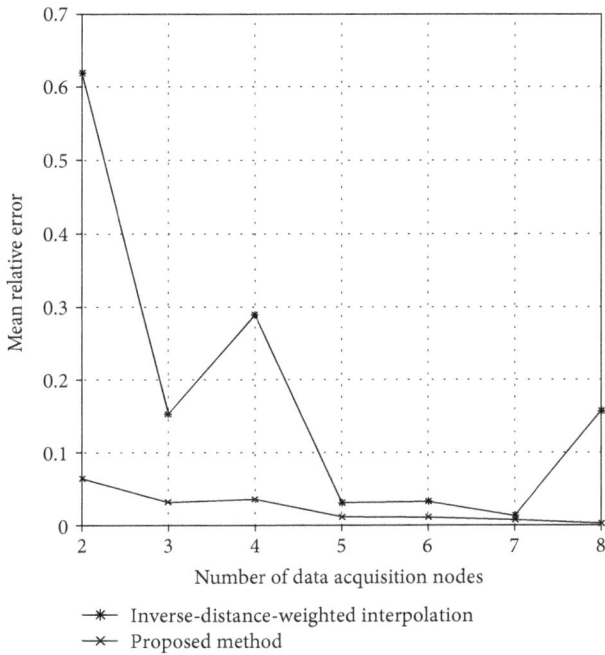

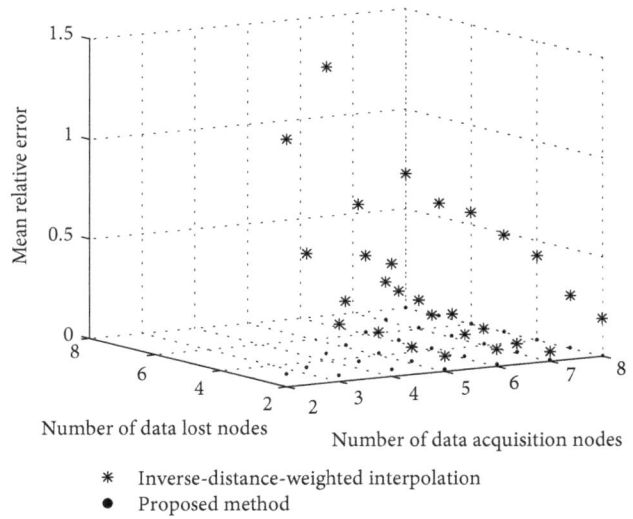

(a)   (b)

FIGURE 10: The temperature sensing data from the Intel Berkeley Research lab: (a) data disturbance occurred at 1 data acquisition node; (b) data disturbance occurred at data acquisition nodes increased from 2 to 8.

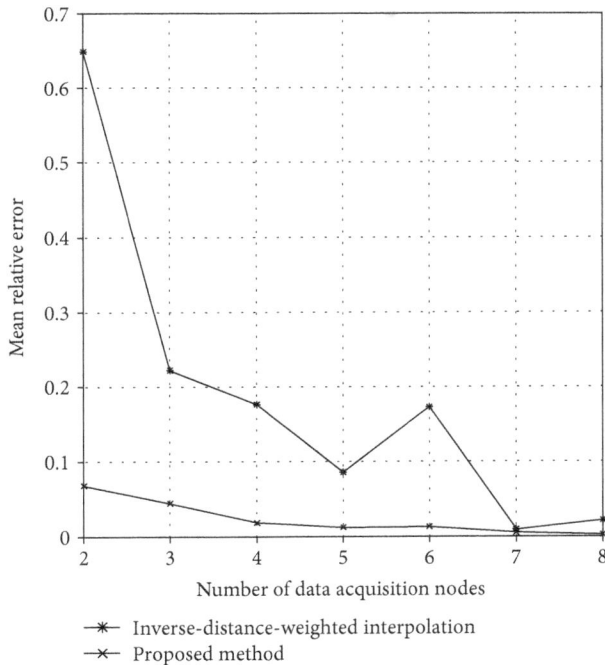

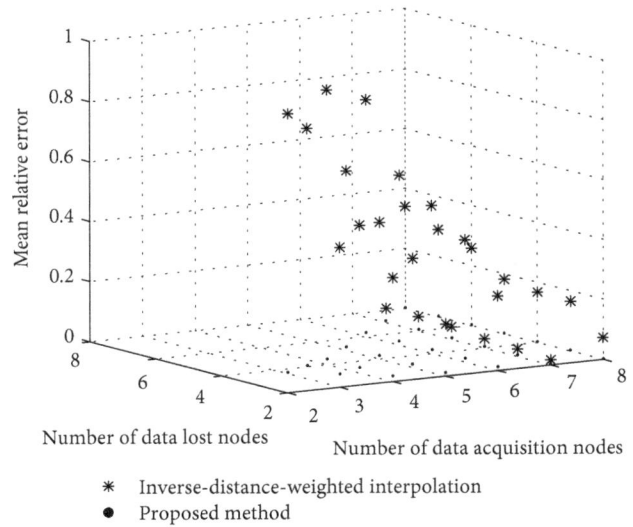

(a)   (b)

FIGURE 11: The humidity sensing data from the Intel Berkeley Research lab: (a) data disturbance occurred at 1 data acquisition node; (b) data disturbance occurred at data acquisition nodes increased from 2 to 8.

closest to the non-data acquisition location is transferred to the non-data acquisition location for interpolation. Considering that the data collected by WSN is prone to error, we

analyzed the reasons for the error. In order to improve the fault tolerance and robustness of the interpolation algorithm, we proposed a BP artificial neural network learning operator. From the experiments, we demonstrated that our algorithm has strong robustness, and it has a lower error in the case of data errors collected by the WSN. This method has strong potential for practical data visualization, data analysis, WSN monitoring, etc.

Conflicts of Interest

The authors declare that there is no conflict of interests regarding the publication of this paper.

Acknowledgments

This work was financially supported by the National Natural Science Foundation of China (Grant Nos. 61561017 and 61462022), the Innovative Research Projects for Graduate Students of Hainan Higher Education Institutions (Grant No. Hyb2017-06), the Hainan Province Major Science & Technology Project (Grant No. ZDKJ2016015), the project of the Natural Science Foundation of Hainan Province in China (Grant No.617033), the open project of the State Key Laboratory of Marine Resource Utilization in South China Sea (Grant No. 2016013B), and the Key R&D Project of Hainan Province (No. ZDYF2018015).

References

[1] F. Ding and A. Song, "Development and coverage evaluation of ZigBee-based wireless network applications," *Journal of Sensors*, vol. 2016, Article ID 2943974, 9 pages, 2016.

[2] O. Alvear, W. Zamora, C. Calafate, J. C. Cano, and P. Manzoni, "An architecture offering mobile pollution sensing with high spatial resolution," *Journal of Sensors*, vol. 2016, Article ID 1458147, 13 pages, 2016.

[3] W. Xiao, *The Key Technology of Adaptive Data Fault Tolerance in Wireless Sensor Networks*, National Defense University of Science and Technology, Hunan, China, 2010.

[4] Z. Liang, F. Tian, C. Zhang, H. Sun, A. Song, and T. Liu, "Improving the robustness of prediction model by transfer learning for interference suppression of electronic nose," *IEEE Sensors Journal*, vol. 18, no. 3, pp. 1111–1121, 2017.

[5] Z. Yang, W. Qiu, H. Sun, and A. Nallanathan, "Robust radar emitter recognition based on the three-dimensional distribution feature and transfer learning," *Sensors*, vol. 16, no. 3, pp. 289–303, 2016.

[6] S. J. Pan and Q. Yang, "A survey on transfer learning," *IEEE Transactions on Knowledge and Data Engineering*, vol. 22, no. 10, pp. 1345–1359, 2010.

[7] B. Pan, J. Tai, Q. Zheng, and S. Zhao, "Cascade convolutional neural network based on transfer-learning for aircraft detection on high-resolution remote sensing images," *Journal of Sensors*, vol. 2017, Article ID 1796728, 14 pages, 2017.

[8] J. Park, R. Javier, T. Moon, and Y. Kim, "Micro-Doppler based classification of human aquatic activities via transfer learning of convolutional neural networks," *Sensors*, vol. 16, no. 12, pp. 1990–2000, 2016.

[9] http://db.csail.mit.edu/labdata/labdata.html.

[10] Z. Kang and Y. Wang, "Structural topology optimization based on non-local Shepard interpolation of density field," *Computer Methods in Applied Mechanics and Engineering*, vol. 200, no. 49-52, pp. 3515–3525, 2011.

[11] M. Hammoudeh, R. Newman, C. Dennett, and S. Mount, "Interpolation techniques for building a continuous map from discrete wireless sensor network data," *Wireless Communications and Mobile Computing*, vol. 13, no. 9, pp. 809–827, 2013.

[12] M. Saberi, A. Azadeh, A. Nourmohammadzadeh, and P. Pazhoheshfar, "Comparing performance and robustness of SVM and ANN for fault diagnosis in a centrifugal pump," in *19th International Congress on Modelling and Simulation*, pp. 433–439, Perth, Australia, 2011.

[13] R. Velazco, P. Cheynet, J. D. Muller, R. Ecoffet, and S. Buchner, "Artificial neural network robustness for on-board satellite image processing: results of upset simulations and ground tests," *IEEE Transactions on Nuclear Science*, vol. 44, no. 6, pp. 2337–2344, 1997.

[14] A. Venkitaraman, A. M. Javid, and S. Chatterjee, "R3Net: random weights, rectifier linear units and robustness for artificial neural network," 2018, https://arxiv.org/abs/1803.04186.

[15] C. Zhang, X. Zhang, O. Li et al., "WSN data collection algorithm based on compressed sensing under unreliable links," *Journal of Communication*, vol. 37, no. 9, pp. 131–141, 2016.

[16] H. L. Chen and W. Peng, "Improved BP artificial neural network in traffic accident prediction," *Journal of East China Normal University (Natural Science)*, vol. 2017, no. 2, pp. 61–68, 2017.

[17] G. Y. Lu and D. W. Wong, "An adaptive inverse-distance weighting spatial interpolation technique," *Computers & Geosciences*, vol. 34, no. 9, pp. 1044–1055, 2008.

5

Fault Detection Modelling and Analysis in a Wireless Sensor Network

Shuang Jia ⓘ,[1] Lin Ma ⓘ,[1] and Danyang Qin ⓘ[2]

[1]*School of Electronics and Information Engineering, Harbin Institute of Technology, Harbin 150080, China*
[2]*Department of Communication Engineering, Heilongjiang University, Harbin 150080, China*

Correspondence should be addressed to Lin Ma; malin@hit.edu.cn

Guest Editor: Ioana Fagarasan

For the serious impacts of network failure caused by the unbalanced energy consumption of sensor nodes, hardware failure, and attacker intrusion on data transmission, a low-energy-consumption distributed fault detection mechanism in a wireless sensor network (LEFD) is proposed in this paper. The time correlation information of nodes is used to detect fault nodes in LEFD firstly, and then the spatial correlation information is adopted to detect the remaining fault nodes, so as to check the states of nodes comprehensively and improve the efficiency of data transmission. In addition, the nodes do not need to exchange information with their neighbor nodes in the detection process since LEFD uses the data sensed by the node itself to detect some types of faults, thus reducing the energy consumption of nodes effectively. Performance analysis and simulation results show that the proposed detection mechanism can improve the transmission performance and reduce the energy consumption of the network effectively.

1. Introduction

A wireless sensor network (WSN) consists of a large number of sensor nodes deployed in a specific area in a self-organized manner. There is no central control node in the network, and the end-to-end information transmission can be achieved by the intermediate nodes in a multihop forwarding way [1]. WSN with the flexible, distributed, and dynamic characteristics has a wide range of applications, such as battlefield, disaster relief, exploration, environmental threat detection, and other fields [2]. The sensor nodes, however, often suffer from various attacks and other external damage since they are usually deployed in severe environments. In addition, sensor nodes have the low manufacturing cost and limited resources and radio coverage. All the factors will cause failure to the nodes and thus will reduce the accuracy of monitoring data. Therefore, the network node fault detection is very important for ensuring the accuracy of monitoring results.

The fault detection algorithms of sensor nodes in WSN can be divided into centralized fault detection and distributed fault detection according to different data processing methods [3]. The centralized fault detection algorithms usually require all the information being collected by a particular node and then determine the states of the other nodes. These algorithms can lead to many problems easily, such as single-node failure, information loss, and much energy consumption [4]. The distributed fault detection algorithms require each node to possess the ability to detect faults and adopt the data collected by itself or the surrounding nodes to determine their own faults [5]. At present, some problems existing in the fault detection algorithms are as follows [6]:

(1) The network energy consumption is sacrificed for higher detection accuracy and lower false-positive ratio. In the existing detection algorithms, the sensor node needs to communicate with its neighbor nodes during fault detection, which will lead to higher energy consumption

(2) The types of fault nodes are not fully considered. Therefore, the detection performance of these

algorithms will decline rapidly if the types of fault nodes increase

(3) The ability of sensor nodes to collect data is not fully utilized, and only the spatial correlation of the sensor network is used to achieve fault detection so that the complexity of algorithms increases significantly

In order to solve the above problems, a low-energy-consumption distributed fault detection mechanism in a wireless sensor network is proposed in this paper. LEFD adopts the time correlation features of the data collected by sensor nodes to detect certain types of fault nodes and then removes them from the network. LEFD may reduce the time and energy consumption in the communication between neighbor nodes. Then, LEFD adopts the spatial correlation properties of WSN to detect the remaining fault nodes that are not detected during the initial detection phase. If the measured value of a node is the same or similar to that of the neighbor which is in the normal state, the node can be considered a normal node. Otherwise, the node is considered a faulty node. The algorithm also considers the transient faults in the sensor readings and corrects the fault data using the data collected in a short period of time when transient faults occur, which avoids mistaking the normal node as a faulty node.

2. Related Work

A distributed Bayesian algorithm for detecting and correcting node faults in WSN (BAFD) is proposed [7]. In BAFD, a sensor node exchanges information with its neighbor nodes to obtain the statistical probability of the event, and the failure ratio of the node is used to identify events and fault nodes. A fault detection scheme is proposed in [8], where each node detects any suspicious behavior using time correlation in its own reading, and the suspected node is required to communicate with the confident neighbor node to find the fault node. The algorithm has high detection accuracy and low communication overhead but does not take into account the impact of transient failure. The authors in [9] propose a distributed byzantine fault detection method based on hypothesis testing, the Neyman-Pearson test method is used to predict the fault states of each sensor node and adjacent sensor nodes, and then the final state of the node is determined by voting in this mechanism. A distributed fault detection method based on metric correlation is proposed in [10]. The algorithm detects the fault nodes through the internal metric correlation of sensor nodes. The computational complexity of the algorithm is low, but it does not consider the influence of transient faults. A distributed localized fault sensor detection algorithm for wireless sensor networks (DLFS) is proposed in [11], and the mutual test results between nodes and neighboring nodes are utilized to determine the states of nodes. DLFS has high detection accuracy and low computational complexity, but the algorithm requires at least two communications between adjacent nodes, thus resulting in much energy consumption. The authors in [12] propose a fault detection mechanism based on the hidden Markov random field. The HMRF model is used to characterize the correlation between the measured value and the actual value of the sensor node, and then the parameters of the HMRF model are obtained through the variable error estimation method to determine the state of the node. The method has high detection accuracy and low false-positive ratio, but it also causes high computational complexity and much energy consumption. In [13], a fully distributed fault detection algorithm is proposed. In this algorithm, the nodes first collect the measurements of their neighborhoods and process them to determine whether they contain an exception value and broadcast the results. Then, nodes determine their own operational state autonomously. Therefore, the computational complexity of the algorithm is low, while the algorithm needs to communicate with its neighbor nodes several times, thus leading to much energy consumption. The authors in [14] propose a novel method for detecting a sensor that generates fault data in a distributed manner. The algorithm detects the fault nodes in the cluster locally through the cluster head and uses the trust concept to identify the type of data failure, which may reduce the influence of a fault node on sensor probability. However, this method leads to the uneven energy consumption of sensor nodes.

It can be seen that the existing detection algorithm does not fully consider the types of the faulty nodes, resulting in poor detection performance. In addition, the multiple communications between adjacent nodes generate much energy consumption.

Aimed at improving the detection accuracy and reducing the energy consumption of the existing detection algorithm, this paper presents a low-energy-consumption distributed fault detection mechanism in a wireless sensor network. The main contributions of this paper are summarized as follows:

(1) A low-energy-consumption distributed fault detection mechanism is proposed, which uses the time correlation information of sensor nodes to detect the fault nodes in the initial detection stage. During the detection of the remaining nodes, the other nodes do not need to communicate with the detected fault nodes, thus reducing the communication traffic and network energy consumption

(2) All kinds of fault nodes are fully considered to ensure the performance of the detection algorithm. In addition, the proposed algorithm also considers the nodes that may have transient faults in the sensor readings. For transient faults, the fault values will be promptly corrected by LEFD to avoid mistaking the normal node as a faulty node, thus improving the utilization of nodes and reducing false-positive ratio

3. System Model

3.1. Network Model. WSN is a special wireless communication system, which does not depend on any fixed communication facilities; it can be deployed in a complex environment for data communication rapidly, and the architecture of WSN is shown in Figure 1. Each node plays the role of a

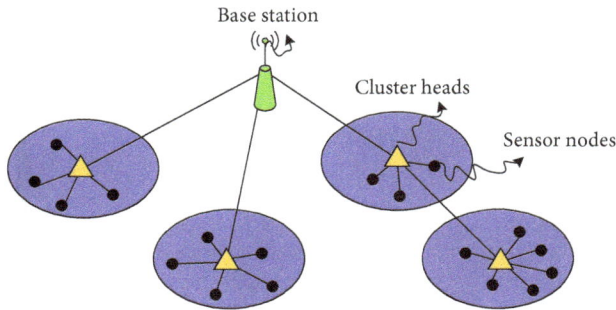

FIGURE 1: Network model.

TABLE 1: Notations and their definitions.

Notation	Definition
n_i	The i-th node
$\text{Num}(N(n_i))$	Number of neighbors of n_i
S_i	Initial states of sensor nodes
v_i^r	Each row in matrix $\mathbf{Q}$
E_c	Total energy consumption
$N(n_i)$	All neighbor nodes of n_i
d_i^t	Measurement data of node n_i at t
FS_i	Final states of sensor nodes
k	Number of data collected by node n_i
$R_{\max}$	Communication radius
DA	Detection accuracy
FPR	False-positive ratio

router and endpoint and has access service and wireless backbone interface [15]. There is no absolute domination of the nodes in WSN, and each node is equal and independent. The data between nodes are transmitted to the destination node by intermediate nodes; that is, data transmission is carried out by multihop forwarding, which can guarantee the flexibility of network topology.

Assume that the number of nodes randomly deployed in a specific area is N. These sensor nodes have the same communication radius $R_{\max}$. A node stores at least k segments of data that has been collected before executing the fault detection algorithm. n_i represents the i-th node in WSN. The node in the node $n_i's$ communication radius is called the neighbor node of n_i. $N(n_i)$ denotes all neighbor nodes of n_i, and $\text{Num}(N(n_i))$ represents the number of neighbor nodes of n_i. d_i^t denotes the measurement data of the node n_i at t. It is assumed that the k segments of the data have been collected in the sensor and stored in the memory before t, that is, $d_i^t, d_i^{t-1}, \ldots, d_i^{t-k-1}$. Node n_i and node $N(n_i)$ are in the same or similar environment, which means that the neighbor nodes of n_i are also in the same event area if the node n_i is in the event area and the neighbor nodes of n_i are also in the same normal area if the node n_i is in the normal area. The remaining parameters are shown in Table 1.

3.2. Fault Model. Nodes could still receive, send, collect, and process data if the network is partially faulty, but the data collected by nodes is usually wrong. According to the abnormal data collected by nodes, the fault of sensor nodes can be divided into the following specific types [16]:

(1) Fixed fault: a sensor with this fault collects data with the same reading, and the data is not affected by the environment

(2) Random fault: node readings are random and uncertain

(3) Offset fault: the node readings deviate from normal values, and the readings may change if the environment changes

(4) Transient fault: transient fault may occur in a short time due to the hardware characteristics and the impact of the environment on the data collecting

process, resulting in data anomalies occurring one or more times

In order to improve the utilization of the sensor nodes, this paper considers that the nodes with transient faults are normal nodes because the readings of these nodes are available at most of the time.

4. Proposed Fault Detection Model

4.1. Detection Principle. The data collected by sensor nodes in a short time is temporally relevant, which means that the collected data is the same or similar in a short time, and the change is not so great [17]. LEFD can detect some types of fault nodes based on this feature, such as random faults and transient faults. The value of the collected data in a short time is unstable when these faults occur. However, this paper will consider that the nodes with transient faults are normal nodes to improve the utilization of nodes, so only the collected data generated when the fault occurs is corrected; the normal node will not be mistaken for a fault node. The matrix $\mathbf{Q}$ is established to determine whether there are transient faults or random faults based on the difference of the data collected by nodes. The faulty data will be replaced by the collected normal data at other times for transient faults; therefore, the false-positive ratio can be effectively reduced. However, it is not enough to use only the time correlation information. For example, the node's reading still satisfies the time correlation feature when fixed faults or offset faults occur, and this type of fault nodes cannot be detected just by using the time correlation feature, so the neighbor nodes are necessary. The node fails if the collected data of most neighbor nodes is not similar to the node's collected data; that is, the sensor nodes have spatial correlation property, which means that most sensor nodes have the same or similar readings in smaller areas.

The differences between LEFD and the existing algorithms can be summarized from the above analysis. Firstly, LEFD uses time correlation information to detect certain types of fault nodes and corrects some values as needed, and then the spatial correlation property of nodes is adopted to detect the remaining fault nodes. However, the existing algorithms do not use time correlation information, or only the spatial correlation information is adopted to detect fault nodes, so there always are undetected fault nodes in the network. Secondly, the existing algorithms do not consider the transient faults of nodes so that the normal node is mistaken as a faulty node, thus reducing the utilization ratio of nodes.

4.2. Detection Method. The latest k segments of data can be obtained after the node n_i collects the data at time t. The matrix **Q** is established according to

$$\mathbf{Q}_{m \times n} = \begin{cases} 0, & \text{if } (|d_i^m - d_i^n| \le \beta_1), \\ 1, & \text{otherwise,} \\ & m, n = \{0, 1, \dots k - 1\}. \end{cases} \tag{1}$$

For each row in matrix **Q**, v_i^r is calculated as

$$v_i^r = \begin{cases} 0, & \text{if } \left(\sum_{j=t-k+1}^{t} \mathbf{Q}_{ij} < \dfrac{k}{2}\right), \\ 1, & \text{otherwise,} \\ & t - k + 1 \le r \le t. \end{cases} \tag{2}$$

At time t, the value of v_i^r is corrected by v_i^t:

$$v_i^t = \begin{cases} 0, & \text{if } \left(\sum_{r=t-k+1}^{t} v_i^r < \dfrac{k}{2}\right), \\ 1, & \text{otherwise.} \end{cases} \tag{3}$$

Any measured value at other times when $v_i^r = 0$ can be considered its value at time t.

Equation (4) is used to determine the initial states of sensor nodes:

$$S_i = \begin{cases} 0, & \text{if } \left(\sum_{r=t-k+1}^{t} v_i^r < \dfrac{k}{2}\right), \\ 1, & \text{otherwise.} \end{cases} \tag{4}$$

For the node n_i with state 0, the neighbor reading whose initial fault condition is 0 is obtained. Then, the final state of the node is determined according to (5) and (6):

$$v_{ij}^t = \begin{cases} 0, & \text{if } \left(|d_i^t - d_j^t| \le \beta_2\right), \\ 1, & \text{otherwise.} \end{cases} \tag{5}$$

$$FS_i = \begin{cases} 0, & \text{if } \left(\sum_{n_j \in N(n_i), S_i=0} v_{ij}^t < \dfrac{\text{Num}(N(n_i), S = 0)}{2}\right), \\ 1, & \text{otherwise,} \end{cases} \tag{6}$$

where $\text{Num}(N(n_i), T = 0)$ denotes the number of neighbor nodes of n_i and the number of nodes whose state may be normal. β represents the node failure thresholds. $FS_i = 0$ represents that the node n_i is a normal node. Otherwise, the node n_i is a faulty node.

For example, assuming that $k = 5$ and $\text{Num}(N(n_i)) = 5$, the k segments of data collected by the node n_i at time t and before t are $\{d_i^t, d_i^{t-1}, d_i^{t-2}, d_i^{t-3}, d_i^{t-4}\}$, and their corresponding values are $\{60.12, 30.23, 31.54, 10.68, 30.87\}$. The neighbor nodes' reading of node n_i is $\{d_1^t, d_2^t, d_3^t, d_4^t, d_5^t\}$ at t, whose values are $\{70.22, 31.35, 65.79, 30.84, 31.10\}$, and $\beta = 2$. Then, the matrix **Q** is established according to (1):

$$\mathbf{Q}_{5 \times 5} = \begin{pmatrix} 0 & 1 & 0 & 0 & 1 \\ 1 & 0 & 1 & 1 & 1 \\ 0 & 1 & 0 & 0 & 1 \\ 0 & 1 & 0 & 0 & 1 \\ 1 & 1 & 1 & 1 & 0 \end{pmatrix}. \tag{7}$$

There are $v_i^r = \{0, 1, 0, 0, 1\}$ and $t - 4 \le r \le t (t \ge 4)$ according to (2). According to (3), the value of v_i^t is corrected to 0; then, there is $v_i^r = \{0, 1, 0, 0, 0\}$. The measured value at r is used to update d_i^t when there is $v_i^r = 0$, namely, $d_i^t = 30.23$. According to (4), the initial fault state of node n_i is considered $S = 0$, which represents that the node n_i is a normal node. Then, the algorithm obtains the neighbor nodes' data of node n_i at time t. Equation (6) shows $FS_i = 0$, and the final state of n_i is normal.

As can be seen from this example, LEFD is a very effective detection method for transient faults and random faults. The detailed description of LEFD is shown in Algorithm 1.

The algorithm adopts the historical data sensed by nodes to determine the initial state of nodes. The node may be a normal node if the collected data is stable in a short time (almost no change). Otherwise, the node may be a faulty node. In other words, only the sensor node's own data can be used to identify some of the fault nodes. After determining the initial states of nodes, LEFD further determines that the initial states of their neighbor nodes are normal for the nodes with a normal initial state. A node will be determined as a normal node if its measured value is similar to that of most of its neighbor nodes. In the whole algorithm implementation process, the fault nodes that are identified in the initial detection process are no longer able to communicate with other normal nodes, and the algorithm adopts the data from nodes whose initial state is normal. This method not only consumes less energy but also reduces the error detection ratio. In addition, LEFD also considers the transient faults of nodes. The algorithm will correct the false readings when

```
(1) Begin
(2) for each node n_i in WSN (i = 1, 2, ..., N)
/* The following method is adopted to establish Q* /
(3)         for each k times before time t (including time t)
(4)             if |d_i^m - d_i^n| ≤ β_1
(5)                 Q_mn = 0
(6)             else
(7)                 Q_mn = 1
(8)             end if
(9)         end for
/* Generate test v_i^r */
(10)        if ∑_{j=t-k+1}^{t} Q_ij < k/2
(11)            v_i^r = 0
(12)        else
(13)            v_i^r = 1
(14)        end if
/* Correct v_i^t */
(15)        if ∑_{r=t-k+1}^{t} v_i^r < k/2 and v_i^t = 1
(16)            v_i^t = 0, d_i^t = d_i^{t-x} //((t - x) ∈ [t - k + 1, t] and v_i^{t-x} = 0)
(17)        else if ∑_{r=t-k+1}^{t} v_i^r ≥ k/2 and v_i^t = 0
(18)            v_i^t = 1, d_i^t = d_i^{t-x} //((t - x) ∈ [t - k + 1, t] and v_i^{t-x} = 1)
(19)        end if
/* Generate a state value S_i = 0 based on the value of v_i^r */
(20)        if ∑_{r=t-k+1}^{t} v_i^r < k/2
(21)            S_i = 0
(22)        else
(23)            S_i = 1
(24)        end if
/* For the nodes with status value S_i = 0, test each member of their neighbors to generate test v_ij^t {0, 1} by adopting the following
way*/
(25)        if |d_i^t - d_j^t| ≤ β_2
(26)            v_ij^t = 0
(27)        else
(28)            v_ij^t = 1
(29)        end if
/* Make the final decision of the nodes' state FS_i */
(29)        if ∑_{n_j ∈ N(n_i) and S_i=0} v_ij^t < Num(N(n_i) and S = 0)/2
(30)            FS_i = 0
(31)        else
(32)            FS_i = 1
(33)        end if
(34) end for
Process end
```

ALGORITHM 1: The pseudocode of low-energy-consumption distributed fault detection algorithm.

transient faults occur, which means that the algorithm adopts the reading at other times instead of the reading at this time to further improve the fault tolerance ability of sensor nodes to transient faults.

5. Simulation Experiment and Performance Analysis

5.1. *Performance Indicators.* The two indicators are usually adopted to evaluate the effect of fault node identification, namely, detection accuracy and false-positive ratio.

Detection accuracy (DA) refers to the ratio between the number of fault nodes that have been correctly identified and the total number of actual fault nodes:

$$DA = \frac{|F \cap A|}{|A|}, \qquad (8)$$

where F represents the set of fault nodes detected by the algorithm and A represents the set of actual fault nodes.

The false-positive ratio (FPR) refers to the ratio between the number of normal nodes which are identified as fault nodes and the total number of normal nodes:

$$FPR = \frac{|F - A|}{N - |A|}. \qquad (9)$$

Most of the energy consumption is caused by communication between nodes [18]. Thus, the total number of communications between nodes can be adopted to represent the total network. When the communication radius of node is R_{max}, it is assumed that the average energy consumption when the node n_i communicates with its neighbor node once is $elec_i$:

$$E_c = \sum_{i=1}^{N} elec_i, \qquad (10)$$

where E_c denotes the total energy consumption.

5.2. Parameter Settings. The performance of LEFD was analyzed using NS2 in this study [15]. In order to maintain the generality, it is assumed that the position of each node is known and all nodes have the same communication radius R_{max}, and the reading of the nodes in the normal region is subject to the distribution of $N(\mu, \sigma)$ ($\mu = 35, \sigma = 1$). At least 5 segments of data ($k = 5$) are stored in each sensor node. The value of k should not be chosen too high because the sensor nodes have limited storage capacity. The data may take up too much storage space if the value of k is too high. The node failure threshold β is 5, and the basic idea of the node failure threshold selection is to determine the node failure threshold according to the allowable deviation of the sensor node. The key step is designing an observer. The output of the observer and the output of the sensor node constitute a redundant signal, and then the two signals are compared to obtain the sensor residual sequence. The allowable error of the sensor node is selected as the node failure threshold [19]. Since the fixed fault is similar to the offset fault, the two types of faults are also regarded as offset faults. The results were obtained from the mean of 100 experiments. All of the simulation parameters are shown in Table 2.

5.3. Experimental Results and Performance Analysis. Figures 2(a) and 2(b) show the performance comparison results of different algorithms in terms of DA and FPR when only offset faults occur. It can be seen that the DA of DLFS is much higher when the sensor fault probability is less than 30% as shown in Figure 2(a), and the DA of LEFD proposed in this paper is similar to that of BAFD; the DA of DLFS is rapidly reduced compared with that of LEFD and BAFD when the sensor fault probability is higher than 30%. However, DLFS has low FPR, and the FPR of LEFD is between the FDR of DLFS and BAFD. Based on all the above factors, the performance of the LEFD algorithm is between the performances of DLFS and BAFD when only offset faults occur.

Figures 3(a) and 3(b) show the performance comparison results of different algorithms in terms of DA and FPR when

TABLE 2: Simulation parameters.

Parameter	Value
The type of nodes	Normal node, fault node
The number of nodes	1024
Network range	32m × 32m
The reading range of offset fault	61–70
The reading range of random fault	1–100
The reading range of transient fault	1–100
R_{max}	2
$elec_i$	10^{-5} J
k	5

only random faults occur. It can be seen that the sensor fault probability has a little effect on the DA, and the FPR increases with the increasing sensor fault probability as shown in Figure 3. However, LEFD also has good performance and always maintains high DA and low FPR even in the case of high sensor fault probability for random faults, since the LEFD algorithm first checks whether the nodes' reading is stable in a short time. The node may be faulty if its reading is unstable, because the data of the random fault sensor is random and unstable. Since the range of random faults is from 1 to 100, DLFS and BAFD are effective for this fault but are less efficient than LEFD. The DA of DLFS and BAFD are also more than 93%, and the DA may increase (such as that of BAFD) when the sensor fault probability increases. But the FPR of DLFS and BAFD also increase when the sensor fault probability increases. However, the FPR of LEFD is almost zero.

Figure 4 shows the relationship between the sensor fault probability and the FPR if only transient faults occur when $k = 5$. It can be seen that the FPR of all algorithms increase with the increasing sensor fault probability as shown in Figure 4. The FPR of LEFD is very low because it has a good ability to handle transient faults. For example, the FPR of LEFD is still less than 5% when the sensor fault probability is 50%, since LEFD determines whether the collected data is correct according to the k segments of data. LEFD will replace the current data with the data collected at another time to avoid the impact of transient faults if data is wrong. This is why LEFD has a good fault tolerance performance for transient faults. However, both DLFS and BAFD do not consider transient faults, so the FPR will continue to increase as the sensor fault probability increases.

Figures 5(a) and 5(b) show the relationship between the sensor fault probability and the DA/FPR when the offset faults, random faults, and transient faults occur randomly, respectively. The DA of DLFS and LEFD is almost the same as shown in Figure 5(a). The DA of DLFS and LEFD are higher than that of BAFD when the sensor fault probability is less than 35%. However, the DA of DLFS will decline rapidly when the sensor fault probability is greater than 35%. The FPR of LEFD is the lowest of the three algorithms. In short, the performance of the LEFD algorithm achieves our expectations in the event of mixed faults.

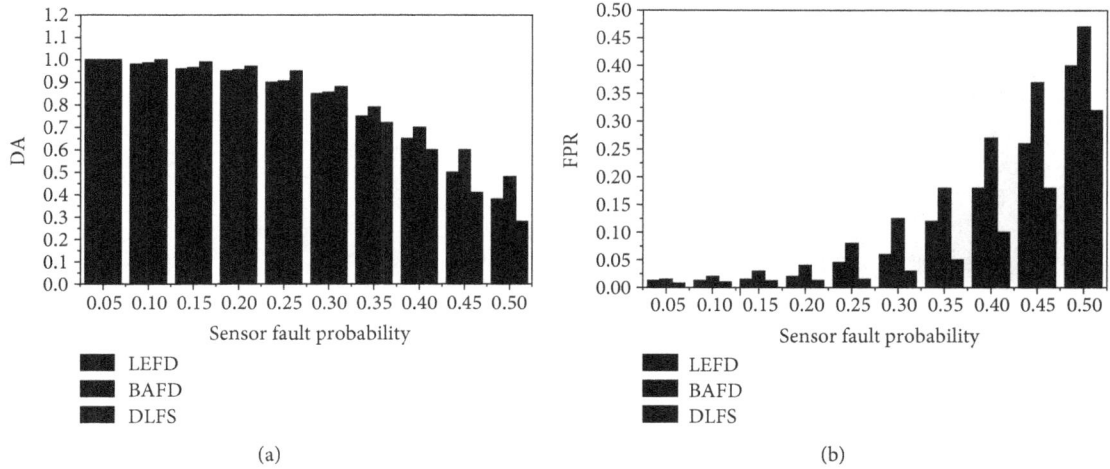

FIGURE 2: (a) Detection accuracy under offset faults. (b) False-positive ratio under offset faults.

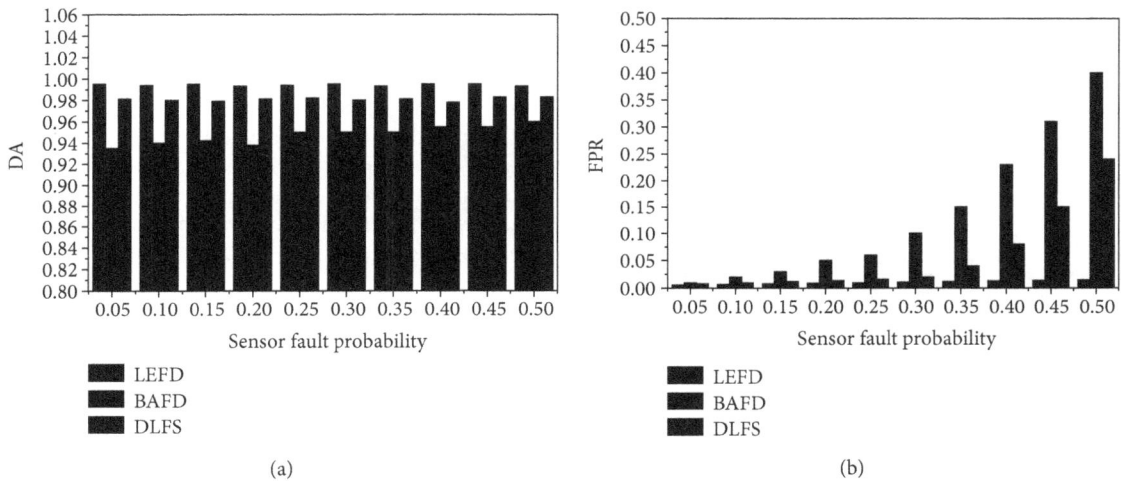

FIGURE 3: (a) Detection accuracy under random faults, (b) False positive ratio under random faults.

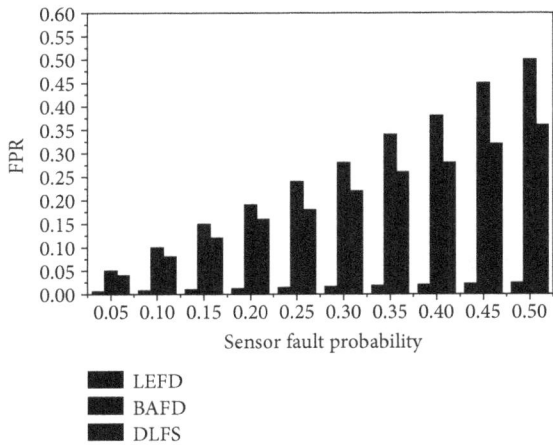

FIGURE 4: The relationship between the false-positive ratio and the sensor fault probability under transient faults.

Figure 6 shows the relationship between the energy consumption (EC) of DLFS, BAFD, and LEFD and the sensor fault probability when the transient fault ratio and the offset fault ratio are 1 : 1, the communication radius R_{max} is 2, and the nontransient faults occur. It can be seen that DLFS has much higher energy consumption as the sensor fault probability increases under the same conditions as shown in Figure 6, since each node needs to communicate with its neighbor nodes at least twice (the first communication is to exchange the initial data set, the second communication is to exchange the initial state of each node). However, nodes that have not yet determined the final state need to make the third communication. As a result, the energy consumption of DLFS is always relatively high. Each node only needs to communicate with its neighbor nodes once for BAFD, so its network energy consumption is moderate and does not change with the sensor fault probability. LEFD first adopts the time correlation information for initial fault detection, and each node does not need to communicate with its neighbor nodes in this process. Only nodes that have been detected

(a)   (b)

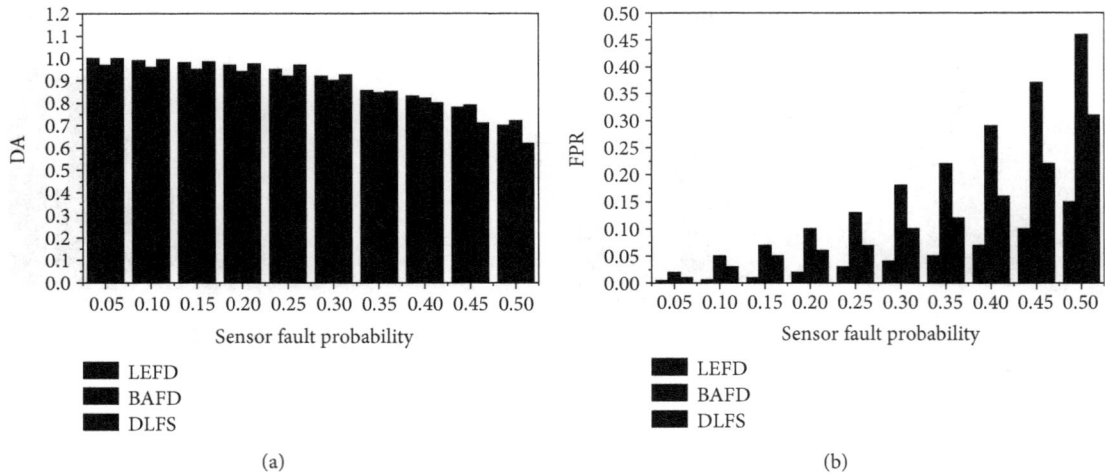

FIGURE 5: (a) The relationship between the detection accuracy and the sensor fault probability under mixed faults. (b) The relationship between the false-positive ratio and the sensor fault probability under mixed faults.

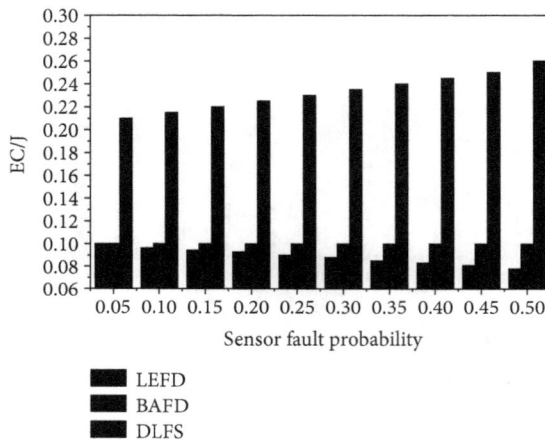

FIGURE 6: Energy consumption of different algorithms.

to have a normal state need to communicate with the neighbor nodes and consume additional energy. Therefore, most of the nodes will be detected as fault nodes by the LEFD algorithm in the case of high sensor fault probability, and the energy consumption of the network also decreases. In summary, the energy consumption of the LEFD algorithm is low.

6. Conclusions

WSN is an important component of modern mobile communication systems. However, network performance is seriously affected due to the breakage of data link and frequent changes of network topology. Therefore, a low-energy-consumption distributed fault detection mechanism in WSN is proposed in this paper. LEFD adopts the data sequence collected by the sensor node itself to detect a particular type of fault and then further uses the neighbor data to determine the states of nodes, thus reducing the communication traffic and network energy consumption. In addition, LEFD also considers the nodes that may have transient faults. For transient faults,

the fault values will be promptly corrected by LEFD to avoid mistaking the normal node as a faulty node, thus reducing false-positive ratio. The simulation results show that LEFD has high detection accuracy, low false-positive ratio, and less energy consumption for various faults. Future research will study a fault tolerance method for WSN, which may provide a new way for the effective transmission of data and ubiquitous routing.

Conflicts of Interest

The authors declare that they have no conflicts of interest.

Acknowledgments

This work is supported by the National Natural Science Foundation of China (61571162, 61771186), Heilongjiang Province Natural Science Foundation (F2016019), Ministry of Education-China Mobile Research Foundation (MCM20170106), and University Nursing Program for Young Scholars with Creative Talents in Heilongjiang Province (UNPYSCT-2017125).

References

[1] Z. Chen, M. He, W. Liang, and K. Chen, "Trust-aware and low energy consumption security topology protocol of wireless sensor network," Journal of Sensors, vol. 2015, Article ID 716468, 10 pages, 2015.

[2] M. P. Fanti, A. M. Mangini, and W. Ukovich, "Fault detection by labeled petri nets in centralized and distributed approaches," IEEE Transactions on Automation Science and Engineering, vol. 10, no. 2, pp. 392–404, 2013.

[3] M. Saihi, B. Boumedyen, A. Zouinkhi, and M. N. Abdlkrim, "A

real time centralized fault detection application in wireless sensor networks," in *International Conference on Automation, Control, Engineering and Computer Science (ACECS'14)*, pp. 95–101, Sousse, Tunisia, March 2014.

[4] D. Kirby, N. Fischer, A. Kalra, and D. Haes, "Case study: centralized ground fault detection system for LADWP ungrounded distribution system," in *The 41st Annual Western Protective Relay Conference*, pp. 1–10, Spokane, Washington, October 2014.

[5] C. Lo, J. P. Lynch, and M. Liu, "Distributed reference-free fault detection method for autonomous wireless sensor networks," *IEEE Sensors Journal*, vol. 13, no. 5, pp. 2009–2019, 2013.

[6] R. Rafeh, "Proposing a distributed fault detection algorithm for wireless sensor networks," *Soft Computing Journal*, vol. 2, no. 2, pp. 26–35, 2014.

[7] B. Krishnamachari and S. Iyengar, "Distributed Bayesian algorithms for fault-tolerant event region detection in wireless sensor networks," *IEEE Transactions on Computers*, vol. 53, no. 3, pp. 241–250, 2004.

[8] K. P. Sharma and T. P. Sharma, "rDFD: reactive distributed fault detection in wireless sensor networks," *Wireless Networks*, vol. 23, no. 4, pp. 1145–1160, 2017.

[9] M. Panda and P. M. Khilar, "Distributed byzantine fault detection technique in wireless sensor networks based on hypothesis testing," *Computers and Electrical Engineering*, vol. 48, pp. 270–285, 2015.

[10] Q. Liu, Y. Yang, and X. Qiu, "A metric-correlation-based distributed fault detection approach in wireless sensor networks," in *Proceedings of the 2015 IEEE Symposium on Computers and Communication (ISCC)*, pp. 526–531, Washington, DC, USA, July 2015.

[11] J. Chen, S. Kher, and A. Somani, "Distributed fault detection of wireless sensor networks," in *Proceedings of the Workshop on Dependability Issues in Wireless ad hoc Networks and Sensor Networks (DIWANS'06)*, pp. 65–71, Los Angeles, Fla, USA, September 2006.

[12] J. Gao, J. Wang, and X. Zhang, "HMRF-based distributed fault detection for wireless sensor networks," in *2012 IEEE Global Communications Conference (GLOBECOM)*, pp. 640–644, Anaheim, CA, USA, December 2012.

[13] W. Li, F. Bassi, D. Dardari, and M. Kieffer, "Low-complexity distributed fault detection for wireless sensor networks," in *IEEE International Conference on Communications*, pp. 6712–6718, Kuala Lumpur, Malaysia, May 2016.

[14] Z. Taghikhaki and M. Sharifi, "A trust-based distributed data fault detection algorithm for wireless sensor networks," in *2008 11th International Conference on Computer and Information Technology*, pp. 1–6, Khulna, Bangladesh, December 2008.

[15] N. Marchang and R. Datta, "Light-weight trust-based routing protocol for mobile ad hoc networks," *IET Information Security*, vol. 6, no. 2, pp. 77–83, 2012.

[16] Q.-J. Wang and J.-L. Cheng, "Research on fault mode and diagnosis of methane sensor," *Journal of China University of Mining and Technology*, vol. 18, no. 3, pp. 386–388, 2008.

[17] P. Jiang, "A new method for node fault detection in wireless sensor networks," *Sensors*, vol. 9, no. 2, pp. 1282–1294, 2009.

[18] A. Cornejo, S. Viqar, and J. L. Welch, "Reliable neighbor discovery for mobile ad hoc networks," *Ad Hoc Networks*, vol. 12, no. 6, pp. 259–277, 2014.

[19] X. H. Wang, "Research on threshold selection principle based on sensor fault detection," *Mechanical Engineering & Automation*, vol. 4, no. 2, pp. 105-106, 2009.

6

A Two-Level Sound Classification Platform for Environmental Monitoring

Stelios A. Mitilineos ⓘD, Stelios M. Potirakis ⓘD, Nicolas-Alexander Tatlas ⓘD, and Maria Rangoussi

Department of Electrical and Electronics Engineering, University of West Attica, Campus 2, 250 Thivon and P. Ralli, Aigaleo, 122 44 Athens, Greece

Correspondence should be addressed to Stelios A. Mitilineos; smitil@gmail.com

Academic Editor: Eduard Llobet

STORM is an ongoing European research project that aims at developing an integrated platform for monitoring, protecting, and managing cultural heritage sites through technical and organizational innovation. Part of the scheduled preventive actions for the protection of cultural heritage is the development of wireless acoustic sensor networks (WASNs) that will be used for assessing the impact of human-generated activities as well as for monitoring potentially hazardous environmental phenomena. Collected sound samples will be forwarded to a central server where they will be automatically classified in a hierarchical manner; anthropogenic and environmental activity will be monitored, and stakeholders will be alarmed in the case of potential malevolent behavior or natural phenomena like excess rainfall, fire, gale, high tides, and waves. Herein, we present an integrated platform that includes sound sample denoising using wavelets, feature extraction from sound samples, Gaussian mixture modeling of these features, and a powerful two-layer neural network for automatic classification. We contribute to previous work by extending the proposed classification platform to perform low-level classification too, i.e., classify sounds to further subclasses that include airplane, car, and pistol sounds for the anthropogenic sound class; bird, dog, and snake sounds for the biophysical sound class; and fire, waterfall, and gale for the geophysical sound class. Classification results exhibit outstanding classification accuracy in both high-level and low-level classification thus demonstrating the feasibility of the proposed approach.

1. Introduction

European countries display one of the richest cultural legacies in the world. With millions of tourists drawn each year to landmark cultural heritage sites, the economic and financial impact of European cultural heritage is considered to be a priority for policymakers but also for the people of Europe [1–4]. Therefore, the conservation of European cultural heritage is critical in order to preserve the European identity but also because cultural heritage may boost economic impact. Alas, heritage sites are exposed to both anthropogenic activity (noise, vandalism, and pollution) and environmental phenomena or natural hazards that may compromise their value. Therefore, preventive measures need to be taken in order to mitigate the negative effects of anthropogenic activity and climate change and preserve cultural heritage artefacts and sites.

In this context, many European institutions have carried out substantial work on preventive strategies aimed at protecting the EU cultural buildings and sites. One of the first related projects, "Carta del Rischio" ("Risk Map"), was carried out in Italy in the early 1990s and completed a long and complex survey of territorial-based environmental and human-caused risks in order to develop the first ever risk map for cultural heritage across Italy [5]. Thereinafter, more countries followed Italy's example and created similar works. An example is the HAR Programme ("Heritage at Risk"), produced by the Historic England organization [6], which resulted in two surveys at 1998 [former "Monuments At Risk Survey (MARS) 1998"] and 2008, helping to establish priorities for action and monument management. The "Carta de Risco do Património Arquitectónico" produced in Portugal by the Direcção-Geral dos Monumentos Nacionais is a similar project but one that is specifically targeted to architectural monuments. The EMERIC programme in Greece was a subproject of the CRINNO project for innovative actions in the island of Crete and included an activity for the tectonic and seismic risk assessment of the historical centers of the main cities of Crete [7]. Finally, the PROHI-TECH project investigated a series of available and novel technologies for the protection of historical buildings against earthquakes and other threats in the Mediterranean and Balkan areas [8].

Our work is part of a larger project ("STORM") that aims in developing a complete platform of technical and managerial resources for cultural heritage sites' safeguarding [4]. STORM will build upon previous work and combine upgraded legacy sensor systems with novel sensing technologies, like wireless acoustic sensor networks (WASNs), in order to provide a novel framework over which to determine how vulnerable structures are affected by risks associated to climatic conditions and anthropogenic activity. Part of the protective mechanisms is the implementation of a WASN platform that will monitor and continuously store sound samples originating from acoustic nodes' surroundings. The collected samples may correspond to environmental sounds regarding natural phenomena or human actions. The WASN-captured data will be transmitted to a central server to populate sound maps and create a database with history of the occurred events, thus alarming stakeholders in cases of potentially malevolent or hazardous events. Herein, we designate as high-level sound classification the act of classifying a sound sample to three classes, namely, *anthropogenic*, *biophysical* (other than human), and *geophysical* sound classes [9, 10]. We further contribute to previous work by extending the proposed classification platform to perform low-level classification too, that is, classify sounds to further subclasses that include airplane, car, and pistol sounds for the anthropogenic sound class; bird, dog, and snake sounds for the biophysical sound class; and fire, waterfall, and gale for the geophysical sound class. We examine two different approaches regarding second-level classification. The first approach is to directly classify each sound sample to one of the nine available classes. The other approach is to first classify each sound sample at high level and then perform classification in a smaller set of available classes depending on the high-level classification result. We report results that indicate the latter method to perform better. Furthermore, we present an integrated platform that includes sound sample denoising using wavelets, feature extraction from sound samples, and Gaussian mixture modeling of these features, as well as the proposed two-layer neural network classifier for the automated classification of incoming sound samples. Numerical results exhibit satisfactory classification accuracy in both high-level and low-level classification levels, thus demonstrating the feasibility of the proposed approach.

The rest of the paper is organized as follows: First, a short literature review of recent sound classification approaches is given in Section 2, providing a justification of the classification approach followed thereinafter. The proposed classification platform is described in Section 3 along with the necessary definition of the employed signal processing tools.

The simulation results are presented in Section 4, and the conclusions are summarized in Section 5.

2. Sound Classification Approaches

In the literature, sound classification is performed using carefully selected sound features that feed a classifier tool like a neural network. The selection of sound features directly affects the performance of the classification procedure and is a demanding task since recorded sounds are typically nonstationary signals while there is also a superimposed background "noise" that originates from natural ambient sounds. Furthermore, sound events are overlapping in space and time and signals originating from neighboring sensors are typically highly correlated [11]. A variety of sound features have been proposed in the literature in order to perform environmental sound monitoring. These features may be either related to the time-domain representation of the signal, for example, zero-crossing rate (ZCR), linear prediction coefficients (LPC), audio signal energy function, and volume, or to the frequency-domain representation, for example, pitch, bandwidth, fundamental frequency, spectral peak track, brightness, mel-frequency cepstral coefficients (MFCC), and short Fourier transform coefficients. There are also many statistical features like the variance, skewness, kurtosis, median, and mean value, as well as various complexity measures (entropies, information) of the signal [12–17]. Other spectral features used in the literature include the 4 Hz modulation energy, percentage of low frames, spectral centroid, spectral roll-off point, spectral frequency, mean frequency, and high and low energy slopes [18–21]. Furthermore, automatic identification is necessary in order to monitor large areas of interest while keeping the operating costs low. Straightforwardly, researchers start up by deploying a network of wireless microphone sensors (WASN) over a large area that capture and transmit environmental sound data samples to a central server. These samples are partitioned into frames and are being processed in order to identify the sound source that created them [11, 17, 22]. Most often, *soundmaps* are created in order to visualize the audio content of large areas, as for example in [23].

Although spectral features are useful in audio classification, they do not provide any information about the temporal evolution of the signal. Therefore, spectral features alone are not enough to represent environmental audio signals that are highly nonstationary in nature. Time-frequency (TF) features have been introduced in order to capture the temporal variation of the spectra of such signals. TF features are effective for revealing nonstationary signal aspects such as trends, discontinuities, and repeating patterns. The usual approach is to extract spectral features for each frame, allowing a certain percentage of overlap between adjacent frames, to produce one of the well-known TF representations like spectrograms, scalograms, or different representations belonging to Cohen's class. However, this approach results in huge feature spaces. Different solutions for the reduction of the resulting data have been proposed. For example, spectrum flux, defined as the average variation value of spectrum between two adjacent frames, can be used.

An effective proposed solution is to use Gaussian mixture models (GMMs) to estimate the probability distribution function of spectral features over all frames. On the other hand, Ghoraani and Krishnan proposed to construct a so-called time-frequency matrix (TFM) of audio signals using a matching pursuit time-frequency distribution technique [18]. Chu et al. also used matching pursuit but with Gabor atom signal representation in order to obtain effective TF features [13]. We consider that the last two approaches impose a prohibitive processing cost (at the embedded level) for our distributed sensor nodes and propose to replace matching pursuit and time-frequency features with probability distribution fitting of temporally varied frequency features (in essence a 1D GMM as it will be explained in detail in the following subsections). After careful consideration and overview of the available literature, we chose to use the following features in our platform: zero-crossing rate, pitch, bandwidth, MFCCs, spectrogram coefficients, and a variety of statistical features, namely, different complexity measures (Shannon, Tsallis, wavelet, and permutation entropies). In order to capture the temporal variation of spectral features, we calculated the GMM of each one of them. Our goals for this selection of features were to keep a high level of performance and robustness together with ease of implementation and low complexity level. Numerical results that are presented in Section 4 of this work justify this approach.

The sound classification task is based on the assumption that every sound source exhibits a specific pattern of distributing its energy over frequency and time. A successful sound classifier should be able to categorize sounds that belong to nonconvex classes of the feature space. Sound classifiers broadly fall under two varieties: *discriminative* and *nondiscriminative*. Examples of the former include the k-means classifier, the polynomial classifier, the multilayer perceptron (neural network), and the support vector machines; such classifiers try to designate a boundary among training data input and match its test input to a specific data class. On the other hand, nondiscriminative classifiers like the hidden Markov model (HMM) attempt to model the underlying distribution of the training data [15, 24, 25]. For the proposed classification platform of the STORM project, we selected to use a generic discriminative classifier, that is, a neural network (NN), since NNs are well-known classifiers that have been extensively used for signal and audio classification purposes. Even though the ANN training needs a high processing power, we assume this to be made available by a central processing server while GMM modeling is adopted at a sensor node level to keep transmission data volume, from sensor node to server, to a minimum.

As long as the specific classes of sounds to be identified are concerned, the available literature is specifically oriented to sound classification for the purposes of environmental monitoring. For example, often we are not interested in identifying a specific bird species or subspecies but rather identify whether or not birds are present at a specific point of an area. The *first level of identification hierarchy* consists in identifying the general sound type, for example, whether it originates from human (anthropogenic) or animal (biogenerated other than human) activity or whether it is an ambient natural sound (e.g., waterfall and fire). This categorization is very popular in the respective literature [26–28]. The *second level of identification hierarchy* consists in further identifying for each sound type a more focused sound origin; for example, whether an animal activity is actually a bird, a snake, or, say, a dog. Deeper levels of identification hierarchy can also be defined where for each next hierarchy step an even narrower and more specific sound class is defined. In this context, a first approach could be to directly detect the second-level sound class. However, we propose to use a two-step approach, where the sound is firstly broadly classified as anthropogenic, animal, or natural ambient, and then it is further classified in a more detailed manner. Numerical results (see Section 4) indicate that this approach delivers much higher performance compared to direct second-level (one-step) sound classification.

Finally, it is worthwhile noting that wavelet analysis has also been used in environmental monitoring for audio signal analysis or signal denoising [25]; herein, we selected to employ wavelet for signal denoising where necessary.

In the following sections, we discuss the proposed sound classification platform approach and present numerical results that demonstrate its applicability in terms of high achieved performance and robust sound classification results.

3. Classification Platform Overview

An overview of the proposed classification platform, together with a discussion on its main components and their interconnectivity, is included in the subsections below.

3.1. General Presentation of the Proposed Platform. The functionality of the proposed platform is illustrated by the flowchart depicted in Figure 1. Every time a sound signal is fed to the platform, there is a decision as to whether the signal will be subject to denoising via wavelet analysis (decomposition and reconstruction) or not. Afterwards, the selected signal features are calculated either for the reconstructed or for the original signal. The features list includes the zero-crossing rate, the pitch, the bandwidth, the MFCCs, the spectrogram coefficients, and a variety of complexity measures including the Shannon, Tsallis, wavelet, and permutation entropies. The zero-crossing rate, the pitch, the bandwidth, and the entropies are scalar features and therefore very efficient in terms of computational cost during classification. On the contrary, the MFCCs and the spectrogram coefficients are TF and thus multidimensional features. More specifically, the signal sample is partitioned into frames and the MFCCs and spectrogram coefficients are calculated for each particular frame. Thereupon, if for example we employ 13 MFCCs and 16 spectrogram coefficients (a popular choice in the literature) for each frame, and a signal is split into 2048 frames (also, a not uncommon case), then we need 2048 vectors of 29 dimensions each (i.e., 59,392 elements); this is a huge number of elements to be used as classifier input. The approach adopted herein as a solution to this problem is to perform a statistical fit of spectral features to a sum of Gaussian probability distribution functions (PDFs) that are fully characterized by only their mean and standard deviation

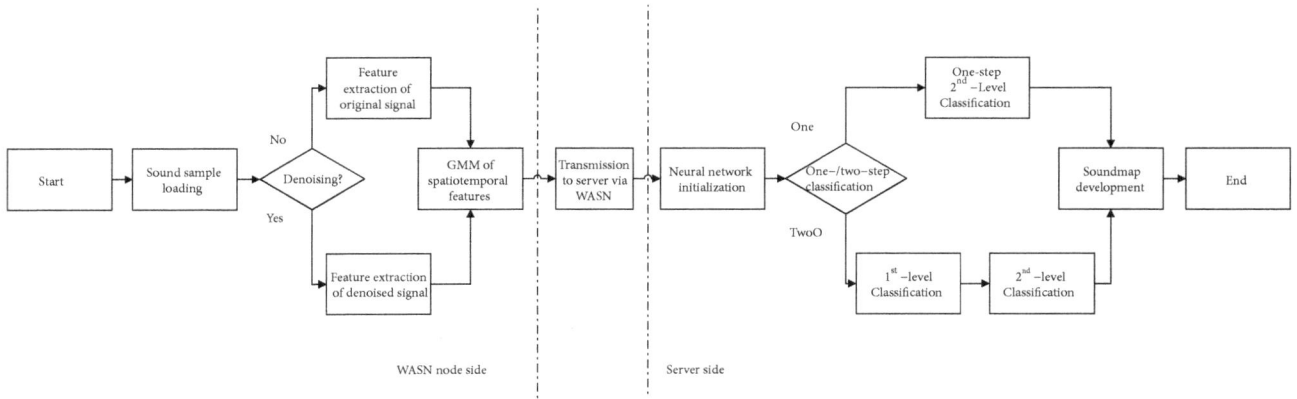

FIGURE 1: Proposed sound classification platform functionality.

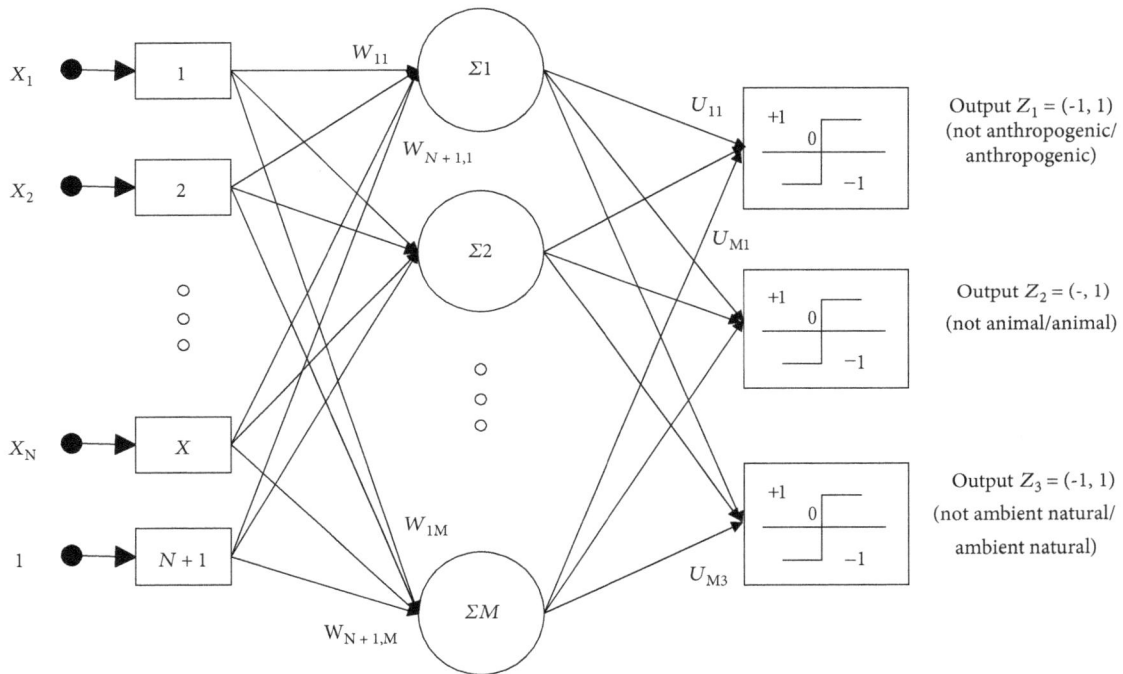

FIGURE 2: Neural network classifier for high-level sound classification.

values; this results to a dramatic reduction of the feature space dimensionality.

Then, a decision is made as to whether sound classification is going to be implemented as a two-step or one-step procedure. In the former case, the sound is first classified to a broad sound category or high-level hierarchy (anthropogenic, ambient natural sound, or animal-originated sound) and then further classified to a more specific class. In the latter case, the sound is directly classified to the more specific class in a one-step procedure. In particular, for the numerical results presented herein, we have used samples from the following sound subclasses: (i) anthropogenic: samples of airplanes', cars', and pistols' sounds; (ii) ambient natural sounds: samples of waterfall, gale, and fire sounds; and (iii) animal-originated sounds: samples of crows', dogs barking, and snakes rattling sounds.

As an example of neural network implementation, consider the high-level classification case. All calculated features are fed to a properly configured NN, as shown in Figure 2. We consider only feedforward artificial neural networks (FANNs) with the training function being an error backpropagation variant. The input layer of the network is used for data entry and weighting. The weights that multiply each data entry are subject to the network's training that is performed off-line and prior to classification. The weighted input features are then forwarded to an intermediate layer of neurons. The middle layer's number of nodes is tuned around the empirical rule-of-thumb value of one and a half times the number of input layer nodes. These neurons sum up all the weighted features and, essentially, configure all possible convex classes of data in the feature space. The output of the intermediate layer is then forwarded to three output neurons. These neurons at the output layer are essentially combining convex classes in order to configure nonconvex classes to classify the input data. Each output is taking a value of "+1" or "−1" that corresponds to "true" or "false" state, respectively.

It is worthwhile noting that 2nd-level classification is implemented in a straightforward manner similar to the procedure described in Figure 2. More specifically, in the second step there are three NNs, namely, the "anthropogenic sounds NN," the "animal sounds NN," and the "geophysical sounds NN"; each one of them is activated only when the respective output of the NN in Figure 2 is true and is fed by the same inputs as the NN of Figure 2. The anthropogenic sounds NN has three output classes, namely, airplanes, cars, and pistols; the animal sounds NN has the birds, dogs, and snakes output classes; finally, the geophysical sounds NN has the gale, rain, and waterfall output classes. This way, at the first step the general class is designated while at the second step the specific subclass of the input sound is identified. On the other hand, as long as one-step second-level classification is considered, there is only one NN with input and intermediate layer similar to that of Figure 2; however, in the specific case there are nine output nodes (i.e., nodes for each one of the subclasses of airplanes, cars, pistols, birds, dogs, snakes, gale, rain, and waterfall) and classification to subclasses is performed by one NN only. Both these cases are not presented using a figure for the sake of brevity.

Finally, postclassification performance results are derived in order to evaluate the applicability of our approach. For each feature's combination mentioned, above we implemented one hundred NNs in order to capture the statistical behavior of node weight assignment during training. For each sound sample and network implementation, we store the confusion matrix and the percentage of correct classifications. The performance of the network is thus evaluated by performance metrics of independent (total correct classifications) as well as dependent (percentage of correct classifications given that a particular sound type is loaded) random variable results.

3.2. Definition of Features

3.2.1. Zero-Crossing Rate. For discrete time signals, a zero-crossing is said to occur if successive samples have different signs [14]. The zero-crossing rate (ZCR) is defined as

$$\text{ZCR} = \frac{1}{2(N-1)} \sum_{n=2}^{N} |\text{sgn}(s(n)) - \text{sgn}(s(n-1))|, \quad (1)$$

where n is the discrete time index, N is the total number of time slots, $s(n)$ is the signal value at time index n, and $\text{sgn}(x)$ is the sign function given by

$$\text{sgn}(x) = \left\{ \begin{array}{l} 1, s(x) \geq 0 \\ -1, s(x) < 0 \end{array} \right\}. \quad (2)$$

3.2.2. Pitch. Pitch is a perceptual feature of the audio signal that depends on the fundamental frequency of the audio waveform. Pitch information can be extracted by using either temporal or frequency analysis. The temporal analysis method is based on the computation of the autocorrelation function or the average magnitude difference function, while with the frequency analysis method the pitch is determined from the periodic structure in the magnitude spectrum of the Fourier transform of an audio frame. The autocorrelation function of a signal is given by

$$R_n(i) = \sum_{n=1}^{N-i} s(n) \cdot s(n+i), \quad (3)$$

where i is the shift. A simple way to calculate the pitch is to estimate $i_{\max}$ that maximizes $R_n(i)$ (i.e., $|R_n(i_{\max})| \geq |R_n(i)|$); the pitch will then be equal to $1/i_{\max} \cdot dt$ where dt is the length of each time slot.

3.2.3. Spectrogram Coefficients. The spectrogram coefficients of a discrete time signal $s(n)$ are essentially the components of the discrete Fourier transform of the signal; the spectrum of such a signal is given by

$$F(f) = \text{DFT}(s(n)), \quad (4)$$

where $\text{DFT}(s(n))$ is the discrete Fourier transform of the signal $s(n)$. The spectrogram is the evolution of the spectrogram coefficients over time and is a TF feature of the signal.

3.2.4. MFCCs. Mel-frequency cepstral coefficients (MFCCs) are very popular in speech/speaker feature extraction and aim at representing the hearing properties of the human ear by using a nonlinear scale of frequencies (i.e., the "mel-frequency" in mel units versus the conventional frequency i.e. measured in Hz) More specifically, the output of the human ear (i.e., output to the auditory processing cells of the human brain) is the convolution of the excitation signal (i.e., the sound under investigation) and the vocal tract filter. The mel transform essentially transforms the spectral coefficients of the sound signal to the mel-frequency domain; then, the cepstral coefficients (as opposed to the spectral coefficients) of the mel-frequency signal components are calculated.

An example of mel transformation is given by

$$M(f) = \left\{ \begin{array}{ll} f, & 0 \leq f \leq 1 \, \text{kHz} \\ 1127 \cdot \ln\left(1 + \dfrac{f}{700}\right), & 1 \, \text{kHz} < f \end{array} \right\}, \quad (5)$$

while an example of a cepstral function calculation formula is given by (the cepstral function is the real cepstrum as opposed to the spectral function i.e. the real spectrum of the signal)

$$c(\tau) = \text{IDTFT}(S(M)), \quad (6)$$

where $\text{IDTFT}(S(M))$ is the inverse discrete-time Fourier transform of the cepstrum magnitude; S(M) is the cepstrum magnitude of the discrete time signal $s(n)$ (i.e., in the mel-frequency domain).

MFCCs, like the spectrogram coefficients, are calculated for each particular frame, and their evolution over time is in essence a TF feature.

3.2.5. Complexity Measures.

There is a wide variety of signal complexity metrics, including different kinds of entropic/information measures. We focus on different entropies that have recently been investigated as to their "insensitiveness" to specific signal compression schemes. As it has recently been shown [17], the precision of specific entropic/information metrics remains reasonably unchanged by certain compression schemes in the sense that the numerical values obtained for these metrics when applied to a compressed signal are very close to the corresponding ones that are obtained when applied to the specific signal in its unprocessed form. In the present paper, we used the Shannon, Tsallis, wavelet, and permutation entropies, the basic formulae of which are presented in the following.

Let $s_k = s(t_k)$ be a discrete measured variable, with $t_k = kT$, $k = 1, 2, \ldots, K$, and T being the sampling period. One can then define a set of N disjoint but adjacent intervals (bins) spanning the observed range of values of the time-series $\{s_k\}$, denoted as $\{x_n\}$, $n = 1, 2, \ldots, N$. Let also $P = \{p(x_1), p(x_2), \ldots, p(x_N)\}$ be a finite discrete probability distribution, with $\sum_{n=1}^{N} p(x_n) = 1$, which describes the probabilities for the samples of the time-series to belong to each one of these N bins; the probability for a sample of the time-series to belong to the nth bin can be denoted as $p(x_n)$. The informational content of the normalized probability distribution P is given by Shannon's information measure as [29]

$$H_{sh} = -K_{sh} \sum_{i=1}^{N} p(x_i) \log[p(x_i)], \qquad (7)$$

where K_{sh} is a positive constant (it merely amounts to a choice of a unit of measure; however, it is usually set equal to 1). The choice of a logarithmic base corresponds to the choice of a unit for measuring information [29]. H_{sh} has been forwarded by Shannon as a measure of information, choice, and uncertainty. The decrease in Shannon entropy is attributed to an increase in the information content and order and, equivalently, to a decrease in complexity. Shannon entropy is recognized as a basic tool for the description of the information, behavior, and complexity of physical, sociological, economic, and so on, systems and their observables, like time-series of measurable quantities that characterize them.

Another statistical representation of a time-series results from the probabilistic analysis of its spectrum. In this approach, instead of analyzing a time-series in terms of the probability of occurrence of its amplitude values, as in the case of Shannon entropy, a time-series is analyzed in terms of the distribution of its energy to frequencies or scales. The Shannon-like total wavelet entropy, or wavelet energy entropy, is defined in this context by [30]

$$H_{WT} = -\sum_{j<1} p_j \ln p_j, \qquad (8)$$

where $p_j = E_j / E_{tot}$ expresses the probability distribution of the energy at different scales of the wavelet spectrum of a signal as it results after the application of the continuous wavelet transform (CWT) on it; it holds that $\sum_j p_j = 1$ and the distribution $\{p_j\}$ can be considered as a time-scale density [30]. Note that the energy at resolution j is $E_j = \sum_k |C_{j,k}|^2$, while the total energy is $E_{tot} = \sum_{j=-N}^{-1} \sum_k |C_{j,k}|^2$, and the signal is considered to be expanded as $y(t) = \sum_{j=-N}^{-1} \sum_k C_{j,k} \psi_{j,k}(t)$, where $j = -1, -2, \ldots, -N$ is the number of resolution levels, corresponding to octave scales [30]. Like the other entropies, wavelet entropy decreases as a result of complexity decrease and order increase.

Symbolic dynamics refers to the mapping of the observables of a complex system (the real values of a time-series) to a sequence of symbols attempting to access useful information ([31] and references therein). The entropic analysis within the context of symbolic dynamics examines the probabilities of appearance of these symbols rather than the probabilities of appearance of the actual real values of the original time-series, thus providing a different kind of statistical representation of the system under analysis. Recently, a new form of symbolic mapping and a corresponding complexity metric has been proposed in the form of permutation entropy (PE) [32]. According to this approach, a time-series is first embedded to a m-dimensional space by building vectors $\mathbf{Y}_k$, each of which contains m values of the original time-series such that every two neighboring vector elements have a time distance equal to L in the original time-series. For every vector $\mathbf{Y}_k$, its m real values are then arranged in an increasing order. This way, each vector $\mathbf{Y}_k$ is uniquely mapped onto a new vector $\pi = [j_1, j_2, \ldots, j_m]$, where π is one of $m!$ possible permutations of the vector of indices of $\mathbf{Y}_k$'s elements $[1, 2, \ldots, m]$. If each of the $m!$ permutations is considered as a symbol, then the procedure allows the mapping of the original continuous time-series to a symbolic sequence [33]. The relative frequency of appearance of each possible permutation π in the time-series, as obtained during the sorting process of all vectors $\mathbf{Y}_k$, is denoted as [34]

$$p(\pi) = \frac{\text{the number of } \pi \text{ permutations found}}{K - (m-1)L}, \qquad (9)$$

while PE is defined according to the Shannon entropy way as

$$H_m = -\sum p(\pi) \ln p(\pi), \qquad (10)$$

where the sum runs over all $m!$ permutations of order m [33]. PE is a measure of regularity in the time-series. When the time-series is so irregular that all $m!$ possible permutations appear with the same probability $p(\pi) = 1/m!$ (completely random), then H_m reaches the maximum value $\ln(m!)$. On the other hand, with increasing regularity, that is, reduced complexity, H_m decreases. For convenience, we usually employ the normalized permutation entropy, by normalizing H_m by $\ln(m!)$ to handle entropy values in the interval $[0, 1]$.

Long-range spatial interactions or long-range memory effects may be observed in a vast variety of complex systems influencing their behavior. A very interesting class of such systems is formed by those characterized by nonextensive statistics. These systems share a very subtle property: they violate the main hypothesis of Boltzmann-Gibbs (B-G) statistics, that is, ergodicity. Inspired by multifractal concepts, Tsallis [35, 36] has proposed a generalization of the B-G statistical mechanics that covers systems that violate ergodicity, that is, systems of the microscopic configurations which cannot be considered as (nearly) independent. This generalization is based on nonadditive entropies, S_q, characterized by an index q which leads to a nonextensive statistics [36] as in

$$S_q = k \frac{1}{q-1} \left\{ 1 - \sum_{i=1}^{N} [p(x_i)]^q \right\}, \qquad (11)$$

where $p(x_i)$ are the probabilities associated with the value bins x_i, as was previously defined for the Shannon entropy case, N is their total number, q is a real number, and k is Boltzmann's constant. The value of q is a measure of the nonextensivity of the system. Notice that in the limit where $q \rightarrow 1$, nonextensive statistics converges to the standard, extensive, B-G statistics [35]. Note that the parameter q itself is not a measure of the complexity of the system but measures the degree of nonextensivity of the system. The value of q represents the strength of the long-range correlations governing the dynamics of the system [37]. The cases $q > 1$ and $q < 1$ correspond to subextensivity or superextensivity, respectively. On the other hand, the time variations of the Tsallis entropy, S_q, for a given q quantify the dynamic changes of the complexity of the system. Lower S_q values characterize signals with lower complexity.

3.3. *Wavelet Denoising.* Denoising refers to processing a noisy signal aiming at the reduction of unwanted noise in such a way that this reduction is as high as possible while at the same time the useful signal is distorted as less as possible. One way to reduce the noise contaminating a signal is to decompose the noisy signal into a number of decomposition levels using the discrete wavelet transform (DWT) and an appropriate orthogonal wavelet basis [38] and then to reconstruct it using only the components that correlate to the useful signal. This is possible by (hard or soft) thresholding that reduces those components' coefficients that correspond to noise [39].

Both the decomposition and the reconstruction processes were performed using Mallat's fast algorithm [38]. According to this algorithm, a hierarchical multiresolution analysis of the signal is performed by using a set of consecutive low- and high-pass filters followed by a decimation; the outputs of these filters are usually referred to as the approximation coefficients and the detail coefficients, respectively. At each level of decomposition, the output of the low-pass filter of the previous level of decomposition (or the original signal for the first level) is fed to a new pair of low- and high-pass filters, the frequency band of which is the half of those of the previous level. As such, the output of each filter can be decimated (downsampled) by a factor of 2. Using this hierarchical approach results in a good time resolution at high frequencies (low scales) and good frequency resolution at low frequencies (high scales).

3.4. *Definition of GMM.* A mixture model is used in statistics in order to represent the presence of data subpopulations within an overall population without the need to identify such subpopulations explicitly. We are using Gaussian mixture models in order to statistically fit MFCC and spectrogram coefficient evolution over time to a PDF. A Gaussian mixture model (GMM) essentially dictates that the empirical PDFs of these coefficients are the weighted sum of Gaussian PDFs of different mean values and standard deviations. In the proposed platform, the user selects the number of Gaussian PDFs to configure the GMM, and the expectation maximization algorithm is used in order to calculate their parameters.

The PDF of a GMM is defined by

$$p(x) = \sum_{i=1}^{G} w_i g(m_i, \sigma_i), \qquad (12)$$

where G is the total number of Gaussian PDFs participating in the GMM, w_i is the weight of the ith Gaussian PDF, and $g(m_i, \sigma_i)$ is a Gaussian PDF of mean m_i and standard deviation σ_i.

It is worthwhile noting that, after multiple statistical fits of the empirical data to GMMs, we decided to abandon the GMM in favor of a simple Gaussian fit since the latter performs comparably with the former with respect to classification while being much less computationally intensive.

4. Numerical Results

In this section, we present numerical results on the performance of the proposed classification platform for several test configurations. We first examine the fundamental first-level classification and assess the performance of the proposed sound features in order to focus on the most high-performing and robust among them. Then, we present results on second-level classification and compare one-step versus two-step implementations.

For the numerical results presented herein, we have used feedforward artificial neural networks with one intermediate hidden layer. The neural network training and performance metric functions are a scaled conjugate backpropagation variant and a mean-square error function, respectively, while the output threshold function is a sigmoid function. The intermediate hidden neuron layer has a varying size according to the number of features used for classification. In the case where scalar features only are used, the number of features is 12 and the number of hidden nodes is 18, while in the case of MFCCs and spectrogram coefficients, the respective figures take a value of 26 features and 42 nodes in the one hand and 64 features and 85 nodes in the other hand, respectively. Furthermore, combinations of features were also used. For the scalar features and MFCC combination, there were 38 features and 40 nodes; for the scalar features and spectrogram coefficients, there were 76 features and 75 nodes; for the MFCCs and spectrogram coefficient combination, there were 90 features and 65 nodes; and, finally, for the scalar features plus MFCCs plus spectrogram coefficient combination there were 102 features and 140 nodes. It is worthwhile

TABLE 1: Summary of classification performance of neural network with selected features: one-level classification.

	Scalar features	MFCCs	Spectrogram coefficients	Scalar feature + MFCCs	Scalar features + spectrogram coefficients	MFCCs + spectrogram coefficients	Scalar features + MFCCs + spectrogram coefficients
Average correct classifications	98.00%	91.06%	88.27%	97.42%	97.57%	91.82%	96.85%
Standard deviation of correct classifications	8.79%	14.30%	16.20%	9.90%	8.91%	16.61%	9.76%
Number of features/hidden layer nodes	12/18	26/42	64/85	38/40	76/75	90/65	102/140

TABLE 2: Confusion matrix of classification results using scalar features only.

	Geophysical samples to:	Animal samples to:	Anthropogenic samples to:	Correctly classified samples
Geophysical class	4446	123	92	4446
Animal class	0	5062	0	5062
Anthropogenic class	54	15	4408	4408
Total samples	4500	5200	4500	13,916/14,200 = 98.00%

noting that the number of hidden layer nodes was optimized after an exhaustive series of trial runs for varying numbers of hidden nodes.

4.1. Neural Network Performance with Selected Features Input: First-Level Classification. Numerical results obtained using three different classes, namely, anthropogenic, geophysical, and animal sounds, are presented in this subsection. Each of these classes was populated with sounds of three different subclasses. More specifically, we used airplane, car, and pistol sounds for the anthropogenic class; gale, waterfall, and fire sounds for the geophysical class; and dog, snake, and crow sounds for the animal sounds. For each subclass, we used a number of 15 different sample recordings of the specific sound type that were extracted from the "505 Digital Sound Effects" audio database [40].

From the list of available features, there exists a set of features (zero-crossing rate, pitch, and entropies) that are scalar and computationally light. Therefore, we group these features together under the label of "scalar features." On the other hand, as long as the MFCCs are concerned, we selected to use a set of 13 time-varying MFCCs for each sound sample and a simple GMM model for each one of them; this results to a set of 26 scalar feature inputs to be fed to the ANN (13 MFCCs; one mean and one standard deviation parameter for each one of them, thereupon a total of 26 scalar parameters). Similarly, we used 13 spectrogram coefficients with simple GMM modeling resulting to another 26 scalar feature inputs. Thereupon, it makes sense to partition the proposed features to three subsets (ZCR-pitch-entropies/MFCCs with GMM modeling/spectrogram coefficients with GMM modeling) and compare their performance. The results for the average and standard deviation of correct classifications are tabulated in Table 1. In the case where only the scalar features are used, the achieved accuracy is 98%; this figure is satisfactorily high and directly comparable (higher) to the respective figures of using scalar features in combination with either MFCCs, spectrograms, or both (see Table 1). Also, the respective standard deviation of accuracy is 8.69%, which is the smallest in Table 1.

Furthermore, confusion matrices are an information-rich and concise way to demonstrate the performance of a classification technique. Confusion matrices demonstrate the performance of a classification platform by illustrating the correct and incorrect classification results of all input samples while also illustrating the nature of the latter by depicting the type of incorrect classifications. Table 2 depicts a confusion matrix that demonstrates the performance of the proposed platform using only scalar features; it demonstrates that the overall classification accuracy is exceptional with a small number of classification errors that originate mostly from mistakenly classifying anthropogenic sound samples to the geophysical class.

4.2. Second-Level Classification of Sounds into Specific Subclasses: One-Step versus Two-Step Implementation. In this subsection, we demonstrate the performance of the proposed platform in the more demanding problem of second-level classification, that is, classifying sounds into more specific subclasses. Table 3 lists the performance of *one-step second-level* classification accuracy achieved by the proposed platform in various combinations of the aforementioned features. One-step second-level classification means that the classification of a sound sample into a specific subclass is performed directly by the neural network, that is, the network has 9 outputs and is directly fed with the signal features. The performance metrics include the average correct classifications and the respective standard deviation.

On the other hand, *two-step second-level classification* means that the classification into specific subclasses is performed in two steps. First, a sound is classified to a generic class using a 3-output neural network, as indicated in Figure 2. Then, according to the result of this first-level classification, the sound is fed to one of three subsequent neural networks each of which is optimized for classifying sounds of either anthropogenic, animal, or natural origin. The accuracy

TABLE 3: Summary of classification performance of neural network with selected features: second-level classification, one-step Implementation.

	Scalar features	MFCCs	Spectrogram coefficients	Scalar features + MFCCs	Scalar features + spectrogram coefficients	MFCCs + spectrogram coefficients	Scalar features + MFCCs + spectrogram coefficients
Average correct classifications	80.07%	79.55%	66.99%	85.34%	85.67%	75.77%	85.98%
Standard deviation of correct classifications	26.45%	23.77%	26.00%	23.38%	21.69%	23.42%	22.74%
Number of features/ hidden layer nodes	12/24	26/40	64/95	38/60	76/110	90/100	102/140

TABLE 4: Summary of classification performance of neural network with selected features: second-level classification, two-step implementation.

	Scalar features	MFCCs	Spectrogram coefficients	Scalar features + MFCCs	Scalar features + spectrogram coefficients	MFCCs + spectrogram coefficients	Scalar features + MFCCs + spectrogram coefficients
Average correct classifications	94.33%	87.80%	80.69%	96.73%	94.27%	86.99%	94.27%
Standard deviation of correct classifications	14.98%	11.12%	27.56%	16.96%	19.64%	26.92%	19.94%
Number of features/hidden layer nodes	12/18/18	26/42/40	64/85/100	38/40/70	76/75/140	90/65/135	102/140/150

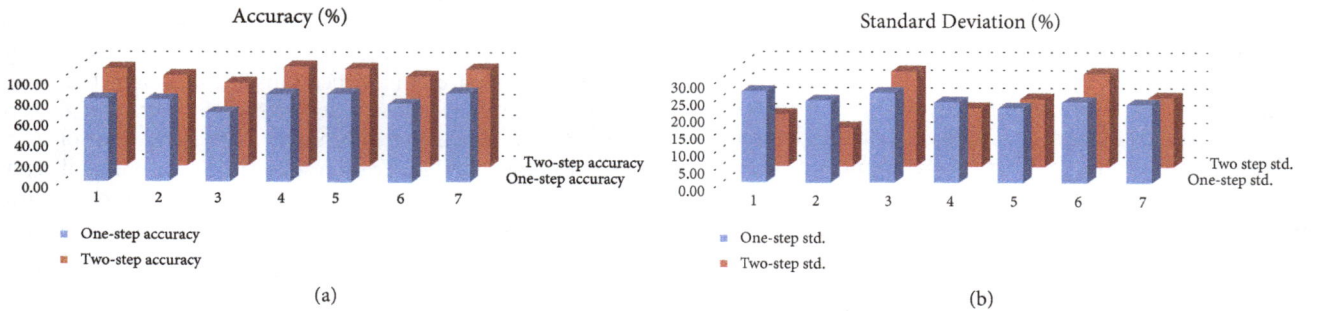

FIGURE 3: Comparison of second-level classification hierarchy results using one-step and two-step implementations.

of each secondary network corresponds to the percent of correct second-level classification *given that first-level classification is correct*. These results are tabulated in Table 4. It is interesting to point out that both Tables 3 and 4 confirm that a balanced choice of features for reasonably good classification accuracy and reduced complexity is either to use the scalar features only or to use the scalar features in combination with MFCCs. Another interesting result is that second-level classification in two steps exhibits much higher accuracy of classification compared to one-step implementation.

Figure 3 demonstrates the comparative results of classification accuracy to the second level of hierarchy using either one-step or two-step implementation. It is interesting to point out that the accuracy achieved in the latter case is much higher compared to the former, for all types of features that were used (scalar, MFCCs, spectrogram coefficients, and combinations among them). Furthermore, the standard deviation of results in the two-step implementation

case is much lower compared to the one-step implementation case; this implies that two-step implementation is more robust in terms of achieved accuracy compared to one-step implementation.

4.3. Confusion Matrices of First- and Second-Level Classification with Selected Features. Since scalar features are the most computationally effective yet yield satisfactorily accurate results, we consider the case of using only scalar features in order to combine both low computational cost and good enough accuracy. The computational effectiveness of using scalar features only is pointed out by the facts that (i) scalar features are easiest to calculate among all proposed features and (ii) the resulting neural network is fed with the minimum number of features and therefore exhibits the minimum number of hidden layer nodes. We also considered the case of scalar features combined with MFCCs (not presented herein) since the latter are the feature of choice most widely

TABLE 5: Second-level classification, one-step implementation, and scalar features only-confusion matrix.

	Airplane samples classified to:	Car samples classified to:	Pistol samples classified to:	Crow samples classified to:	Dog samples classified to:	Snake samples classified to:	Fire samples classified to:	Gale samples classified to:	Waterfall samples classified to:	Correctly classified samples
Airplane	1444	330	495	373	414	363	359	513	294	1444
Car	11	1125	0	0	0	0	1	2	15	1125
Pistol	0	0	990	0	0	0	0	13	0	990
Crow	0	0	0	1309	0	0	0	1	0	1309
Dog	5	15	0	3	1586	13	0	2	0	1586
Snake rattle	0	0	0	0	0	1123	0	0	0	1123
Fire	3	15	0	0	0	0	1125	0	0	1125
Gale	24	0	0	1	0	0	0	965	0	965
Waterfall	13	15	15	14	0	1	15	4	1191	1191
Total samples	1500	1500	1500	1700	2000	1500	1500	1500	1500	10,858/ 14,200 = 76.46%

TABLE 6: Second-level classification, two-step implementation, and scalar features only-confusion matrix results for second step classification of anthropogenic sounds to subclasses.

	Airplane samples classified to:	Car samples classified to:	Pistol samples classified to:	Correctly classified samples
Airplane	1485	30	30	1485
Car	4	1470	0	1470
Pistol	11	0	1470	1470
Total samples	1500	1500	1500	4425/4500 = 98.33%
Probability of correct classification of anthropogenic sounds during the first step (Table 2, row 3, column 3)				4408/4500 = 97.96%
Total accuracy of two-step classification for anthropogenic Sounds				96.32%

TABLE 7: Second-level classification, two-step implementation, and scalar features only-confusion matrix results for second-step classification of animal sounds to subclasses.

	Crow samples classified to:	Dog samples classified to:	Snake rattle samples classified to:	Correctly classified samples
Crow	1683	1	16	1683
Dog	17	1999	45	1999
Snake rattle	0	0	1439	1439
Total samples	1700	2000	1500	5121/5200 = 98.48%
Probability of correct classification of animal sounds during the first step (Table 2, row 2, column 2)				5062/5200 = 97.96%
Total accuracy of two-step classification for animal sounds				96.47%

used in the literature; however, numerical results demonstrated that using only scalar features is computationally much lighter and yields results that are comparable to those obtained by MFCCs or a combination of the two.

Confusion matrices are included in Tables 5–8, demonstrating the performance of the proposed platform using only scalar features. The network runs at these tables are different to the ones corresponding to the tables presented in Subsections 3.1 and 3.2. Table 5 illustrates the confusion matrix in the case of second-level classification in one step, that is, with one NN only with 9 outputs. The table clearly demonstrates

the error sources in the various subclasses. It is evident that the main source of errors in one-step classification is due to sounds mistakenly classified as airplanes.

Tables 6–8 illustrate the results of second-level classification in two-step implementation. For example, Table 6 displays the confusion matrix in the case of anthropogenic sounds. Assuming that an anthropogenic sound has already been correctly classified as such in the first step, Table 6 demonstrates the results of the second step only. This means that in order to calculate the overall accuracy of second-level classification with two-step implementation, we need to

TABLE 8: Second-level classification, two-step implementation, and scalar features only-confusion matrix results for second-step classification of geophysical sounds to subclasses.

	Fire samples classified to:	Gale samples classified to:	Waterfall samples classified to:	Correctly classified samples
Fire	1485	15	15	1485
Gale	0	1470	15	1470
Waterfall	15	15	1470	1470
Total samples	1500	1500	1500	4425/4500 = 98.33%
Probability of correct classification of geophysical during the first step (Table 2, row 1, column 1)				4446/4500 = 98.80%
Total accuracy of two-step classification for geophysical sounds				97.15%

multiply any given probability with the probability that an anthropogenic sound is correctly classified during the first step. The latter probability can be found by dividing the number of correctly classified anthropogenic sound samples to the total number of anthropogenic sound samples. These numbers are equal to 4408 and 4500 respective (row 3, column 3 of Table 2). Similar calculations are also included in Tables 7 and 8.

Table 6 demonstrates that the main source of errors in the case of anthropogenic sounds and two-step implementation is the classification of different sounds to the airplane class. This result agrees with the results tabulated in Table 5. Furthermore, Table 7 demonstrates that a similar trend is present in the case of animal sounds but for dog sounds. However, Table 8 illustrates that both fire and waterfall sounds are more prone to classification errors compared to gale sounds. Finally, the comparison of Tables 6–8 to Table 5 verifies that two-step implementation is much more accurate compared to one-step second-level classification.

5. Conclusions

STORM is an ongoing project aiming at developing a platform for safeguarding cultural heritage sites across Europe. Part of the project objectives is to deploy wireless acoustic sensor networks over different historical and archaeological sites across Europe (the Diocletian Baths in Rome, Italy; the Mellor Heritage site in Manchester, UK; and the Roman Ruins of Tróia in Portugal) that will be used to monitor the sites and alarm stakeholders in the case of potential hazardous events. In this context, the proposed sound classification platform is a first step towards the accomplishment of this goal. The literature review revealed a number of popular approaches for sound feature selection together with denoising techniques and classification methods. In this paper, we presented the development of an integrated classification platform and evaluated its performance while the proposed classifier is extended to include the capability of classifying sounds within a hierarchy of two levels. First-level classification, or classifying sounds into generic classes like anthropogenic, animal, and geophysical, is sometimes critical; the proposed platform has been shown to deliver highly accurate results in this case. Also, it has been shown that the proposed scalar features are simple and computationally light, yet very accurate. As long as second-level classification is concerned,

we showed that two-step classification may be more efficient compared to one-step implementation; in the presented numerical results, the achieved accuracy of the former was much higher compared to the latter. Furthermore, a confusion matrix analysis revealed that the main sources of errors are due to anthropogenic sounds mistakenly classified as geophysical sounds (first-level classification) or due to anthropogenic sounds mistakenly classified as airplanes (second-level classification). There is also a significant source of errors in second-level classification with other animal sounds mistakenly classified as dog sounds. In the future, we plan to apply the proposed classification approach in sound samples with varying signal-to-ratio values, as well as study the effect of noise on each sound feature separately and integrate our findings in the STORM platform at sensor and server level.

Conflicts of Interest

The authors declare that they have no conflicts of interest.

Acknowledgments

This work is partially funded by EU and national funds under the research project "STORM (Safeguarding Cultural Heritage through Technical and Organizational Resources Management)" Grant agreement no. 700191, cofunded by the Horizon 2020 Programme of the European Union.

References

[1] EC, "Towards an integrated approach to cultural heritage for Europe," European Commission, 2015, http://www.europarl.europa.eu/sides/getDoc.do?type=REPORT&reference=A8-2015-0207&language=EN.

[2] InHERIT, "Promoting cultural heritage as a generator of sustainable development, European Research Project, co-funded by the ERASMUS+ action of the European Union," 2015, project website: http://www.inherit.tuc.gr/en/home/.

[3] G. Mergos and N. Patsavos, "Cultural heritage as economic value: economic benefits, social opportunities and challenges of cultural heritage for sustainable development," Report of the InHERIT European Research Project, 2017, http://www.inherit.tuc.gr/fileadmin/users_data/inherit/_uploads/%CE%9F2_Book_of_Best_Practices-f.pdf.

[4] STORM, Safeguarding Cultural Heritage through Technical and Organizational Resources Management, co-funded by the

Horizon 2020 Programme of the European Union, 2016, project website: http://www.storm-project.eu/, Grant Agreement No. 700191.

[5] Carte del Rischio, "Territorial information system for the protection of cultural heritage," ISCR, Italy, 1992, project websites: http://www.cartadelrischio.it/eng/index.html; http://www.icr.beniculturali.it/pagina.cfm?usz=1&uid=16.

[6] HAR, "HAR – Heritage at Risk Programme, Historic England," 1998, project website: https://historicengland.org.uk/advice/heritage-at-risk/types/.

[7] EMERIC, "EMERIC – Expert System for the Monitoring and Management of the Natural Environment of Crete," 2006, project website: http://emeric.ims.forth.gr/.

[8] F. M. Mazzolani, "The PROHITECH research project," appears in Structural Analysis of Historic Construction, Taylor and Francis Ed., 2008, http://www.hms.civil.uminho.pt/sahc/2008/CH123.pdf.

[9] S. A. Mitilineos, S. M. Potirakis, N.-A. Tatlas, and M. Rangoussi, "High-level sound classification in the ESOUNDMAPS project," in *Proceedings of the 4th International Conference on Materials and Applications for Sensors and Transducers (ICMAST 2014)*, pp. 1–4, Bilbao, Spain, June 2014.

[10] S. A. Mitilineos, S. M. Potirakis, N.-A. Tatlas, and M. Rangoussi, "High-level sound classification in the ESOUNDMAPS project," *Key Engineering Materials*, vol. 644, pp. 83–86, 2015.

[11] M. Rangoussi, S. M. Potirakis, I. Paraskevas, and N.-A. Tatlas, "On the development and use of sound maps for environmental monitoring," in *128th Audio Engineering Society Convention*, London, UK, May 2010.

[12] R. Cai, Lie Lu, A. Hanjalic, Hong-Jiang Zhang, and Lian-Hong Cai, "A flexible framework for key audio effects detection and auditory context inference," *IEEE Transactions on Audio, Speech and Language Processing*, vol. 14, no. 3, pp. 1026–1039, 2006.

[13] S. Chu, S. Narayanan, and C.-C. J. Kuo, "Environmental sound recognition with time-frequency audio features," *IEEE Transactions on Audio, Speech and Language Processing*, vol. 17, no. 6, pp. 1142–1158, 2009.

[14] S. Despotopoulos, E. Kyriakis-Bitzaros, I. Liaperdos et al., "Pattern recognition for the development of Sound Maps for environmentally sensitive areas," in *International Scientific Conference eRA-7*, Piraeus, Greece, September 2012.

[15] A. J. Eronen, V. T. Peltonen, J. T. Tuomi et al., "Audio-based context recognition," *IEEE Transactions on Audio, Speech and Language Processing*, vol. 14, no. 1, pp. 321–329, 2006.

[16] I. Paraskevas, S. M. Potirakis, and M. Rangoussi, "Natural soundscapes and identification of environmental sounds: a pattern recognition approach," in *2009 16th International Conference on Digital Signal Processing*, pp. 473–478, Santorini-Hellas, Greece, July 2009.

[17] N.-A. Tatlas, S. M. Potirakis, S. A. Mitilineos, and M. Rangoussi, "On the effect of compression on the complexity characteristics of wireless acoustic sensor network signals," *Signal Processing*, vol. 107, pp. 153–163, 2015.

[18] B. Ghoraani and S. Krishnan, "Time-frequency matrix feature extraction and classification of environmental audio signals," *IEEE Transactions on Audio, Speech and Language Processing*, vol. 19, no. 7, pp. 2197–2209, 2011.

[19] J. M. Kates, "Classification of background noises for hearing-aid applications," *Journal of the Acoustical Society of America*, vol. 97, no. 1, pp. 461–470, 1995.

[20] N. Mesgarani, M. Slaney, and S. A. Shamma, "Discrimination of speech from nonspeech based on multiscale spectro-temporal modulations," *IEEE Transactions on Audio, Speech and Language Processing*, vol. 14, no. 3, pp. 920–930, 2006.

[21] E. Scheirer and M. Slaney, "Construction and evaluation of a robust multifeature speech/music discriminator," in *1997 IEEE International Conference on Acoustics, Speech, and Signal Processing*, pp. 1331–1334, Munich, Germany, April 1997.

[22] S. M. Potirakis, B. Nefzi, N.-A. Tatlas, G. Tuna, and M. Rangoussi, "A wireless network of acoustic sensors for environmental monitoring," *Key Engineering Materials*, vol. 605, pp. 43–46, 2014.

[23] I. Paraskevas, S. M. Potirakis, I. Liaperdos, and M. Rangoussi, "Development of automatically updated soundmaps for the preservation of natural environment," *Journal of Environmental Protection*, vol. 02, no. 10, pp. 1388–1391, 2011.

[24] J. J. Aucouturier, B. Defreville, and F. Pachet, "The bag-of-frames approach to audio pattern recognition: a sufficient model for urban soundscapes but not for polyphonic music," *The Journal of the Acoustical Society of America*, vol. 122, no. 2, pp. 881–891, 2007.

[25] S. Ntalampiras, I. Potamitis, and N. Fakotakis, "Acoustic detection of human activities in natural environments," *Journal of Audio Engineering Society*, vol. 60, no. 9, pp. 686–695, 2012.

[26] A. D. Mazaris, A. S. Kallimanis, G. Chatzigianidis, K. Papadimitriou, and J. D. Pantis, "Spatiotemporal analysis of an acoustic environment: interactions between landscape features and sounds," *Landscape Ecology*, vol. 24, no. 6, pp. 817–831, 2009.

[27] Y. G. Matsinos, A. D. Mazaris, K. D. Papadimitriou et al., "Spatio-temporal variability in human and natural sounds in a rural landscape," *Landscape Ecology*, vol. 23, pp. 945–959, 2008.

[28] K. D. Papadimitriou, A. D. Mazaris, A. S. Kallimanis, and J. D. Pantis, "Cartographic representation of the sonic environment," *The Cartographic Journal*, vol. 46, no. 2, pp. 126–135, 2009.

[29] C. E. Shannon, "A mathematical theory of communication," *Bell System Technical Journal*, vol. 27, pp. 379–423, 1948, 623–656.

[30] O. A. Rosso, S. Blanco, J. Yordanova et al., "Wavelet entropy: a new tool for analysis of short duration brain electrical signals," *Journal of Neuroscience Methods*, vol. 105, no. 1, pp. 65–75, 2001.

[31] S. M. Potirakis, G. Minadakis, and K. Eftaxias, "Analysis of electromagnetic pre-seismic emissions using Fisher information and Tsallis entropy," *Physica A: Statistical Mechanics and its Applications*, vol. 391, no. 1-2, pp. 300–306, 2012.

[32] C. Bandt and B. Pompe, "Permutation entropy: a natural complexity measure for time series," *Physical Review Letters*, vol. 88, no. 17, 2002.

[33] A. A. Bruzzo, B. Gesierich, M. Santi, C. A. Tassinari, N. Birbaumer, and G. Rubboli, "Permutation entropy to detect vigilance changes and preictal states from scalp EEG in epileptic patients - a preliminary study," *Neurological Sciences*, vol. 29, no. 1, pp. 3–9, 2008.

[34] X. Li, G. Ouyang, and D. A. Richards, "Predictability analysis of absence seizures with permutation entropy," *Epilepsy Research*, vol. 77, no. 1, pp. 70–74, 2007.

[35] S. Abe and Y. Okamoto, Eds., *Non-extensive Statistical Mechanics and its Applications*, Springer-Verlag, Heidelberg, Germany, 2001.

[36] C. Tsallis, "Non-additive entropy Sq and non-extensive statistical mechanics: applications in geophysics and elsewhere," *Acta Geophysica*, vol. 60, no. 3, pp. 502–525, 2012.

[37] A. Rényi, "On measures of entropy and information," in *Proceedings of the 4th Berkeley Symposium on Mathematics, Statistics and Probability*, pp. 547–561, University of California Press, Berkeley, LA, USA, 1961.

[38] S. G. Mallat, "A theory for multiresolution signal decomposition: the wavelet representation," *IEEE Transactions on Pattern Analysis and Machine Intelligence*, vol. 11, no. 7, pp. 674–693, 1989.

[39] D. L. Donoho, "De-noising by soft-thresholding," *IEEE Transactions on Information Theory*, vol. 41, no. 3, pp. 613–627, 1995.

[40] *505 Digital Sound Effects*, CD Box Set, Laserlight Digital, 2006.

7

Fog Computing-Based IoT for Health Monitoring System

Anand Paul ⓘ,[1] Hameed Pinjari,[1] Won-Hwa Hong ⓘ,[2] Hyun Cheol Seo ⓘ,[2] and Seungmin Rho ⓘ[3]

[1]*School of Computer Science and Engineering, Kyungpook National University, Daegu, Republic of Korea*
[2]*School of Architectural, Civil, Environmental and Energy Engineering, Kyungpook National University, Daegu, Republic of Korea*
[3]*Department of Media Software, Sungkyul University, Anyang, Republic of Korea*

Correspondence should be addressed to Anand Paul; paul.editor@gmail.com and Won-Hwa Hong; hongwonhwa@gmail.com

Academic Editor: Carmine Granata

Wireless sensor networks (WSNs) are widely used in the area of health informatics. Wireless and wearable sensors have become prevalent devices to monitor patients at risk for chronic diseases. This helps ascertain that patients comply by the treatment plans and also safeguard them during sudden attacks. The amount of data that are gathered from various sensors is numerous. In this paper, we propose to use fog computing to help monitor patients suffering from chronic diseases such that the data are collected and processed in an efficient manner. The main challenge would be to only sort out context-sensitive data that are relevant to the health of the patient. Just having a simple sensor-to-cloud architecture is not viable, and this is where having a fog computing layer makes a difference. This increases the efficiency of the entire system, as it not only reduces the amount of data that is transported back and forth between the cloud and the sensors but also eliminates the risk that a data center failure bears with it. We also analyze the security and deployment issues of this fog computing layer.

1. Introduction

With each passing day, the way the world interacts changes. The past few years have seen an increase in the usage of cloud computing for a huge number of applications. But now, the cloud does not cater to just simple applications and technological needs; the Internet of things (IoT) is proving to be the next technological trend when it comes to cloud computing. In healthcare applications, wireless sensor networks (WSN) have started playing a huge role in the way patients are being monitored. Wireless sensors in the form of wireless wearable accessories or devices are attached to a patient such that this information can be used for the monitoring process. The sensors can be of various forms and sizes [1, 2] as long as they are relevant to the need.

The wireless sensor networks generate a huge amount of data. These data that have been collected from all the devices connected to the network may be useful as well as redundant. All these unprecedented amounts of data can overwhelm the data storage systems and the data analysis applications. The

weeding out of irrelevant data has to be a context-sensitive process. Hence, the sensors would have to send the data collected to computing devices that are capable of performing tasks of analysis, aggregation, and storage. In many cases, each patient requires a high number of sensors, and hence, creating an infrastructure dedicated to an individual becomes inefficient. Hence, IoT provides an alternative approach in which sensor devices are used in a common infrastructure. These sensor devices can then forward the data to a cloud server.

For many healthcare applications, having a simple sensor-to-cloud architecture is not viable, especially due to the fact that most hospitals would not prefer patient data to be stored outside [3]. Also, there is always the bleak case of there being a network failure or a data center failure, which puts patients' health at risk. This is where fog computing aids healthcare applications.

Using only the cloud may cause delay during the transfer of data from the sensors to the cloud and the cloud to the hospitals or personal physicians. In healthcare, we have emergency response systems that require real-time operations in which efficiency and time play an important part; this may suffer due to delay caused by the cloud [4]. Hence, transfer of such immense amounts of data back and forth is not an efficient option not only due to latency issues but also due to security. The risks involved here are not only infringement of data but also risks to the health of patients. Hence, the classical centralized cloud computing architecture has to be extended to a distributed one. A distributed architecture refers to tasks being divided and then offloaded to more than one node. This is commonly referred to as edge computing.

In edge computing, the main computationally intensive operations are performed at the edge of the network instead of holding it in the cloud or on a centralized data warehouse. Edge computing uses computing resources near IoT devices for local storage and preliminary data processing. According to Cisco [5], by 2020, 50 billion devices will be connected to the Internet. Hence, edge computing will also require greater flexibility in order to manage this huge influx of devices. Edge devices cannot handle multiple IoT applications competing for their limited resources, which results in resource contention and increases processing latency. A distributed feature added to this will help in scalability and reduces the risk of exposure of data and hence increases security and eliminates most of the privacy concerns [6].

A fog computing layer integrates edge devices with cloud resources and hence extends the existing cloud infrastructure [7]. In this architecture, the application resides not only in the cloud but also in the devices closest to the patients and the infrastructure components between them. The term infrastructure refers to access points, gateways, and routers. The main objective of healthcare applications is to provide constant supervision on the health of a patient. Implementing this fog layer provides for the successful fulfillment of this requirement. Also, due to the fact that data are stored in data centers in the cloud, the problem of infrastructure, maintenance, upgrades, and costs is solved.

In this paper, we strive to improve current healthcare systems by implementing a context-sensitive fog computing environment. We discuss the various computational tasks involved in healthcare that will be performed in the fog layer, the cloud, or at the user devices and sensors. The security of the information passing the system will be improved as the exposure of data is limited due to the fact that it does not have to travel to and fro in the network. The services that the cloud performs can be distributed to other nodes in such a way that the overhead time taken for data to go back and forth from the cloud to devices is compensated or reduced. Distributing the services among various nodes overcomes some of the challenges that previous healthcare monitoring applications have faced.

The remainder of the paper is organized as follows. In Section 2, we discuss about fog computing, the expectations, and various challenges that have to be faced in the process of implementing fog computing in healthcare. The proposed research is discussed in Section 3. Section 4 analyzes the proposed scheme and the experimental results obtained. The conclusion to the work is given in Section 5.

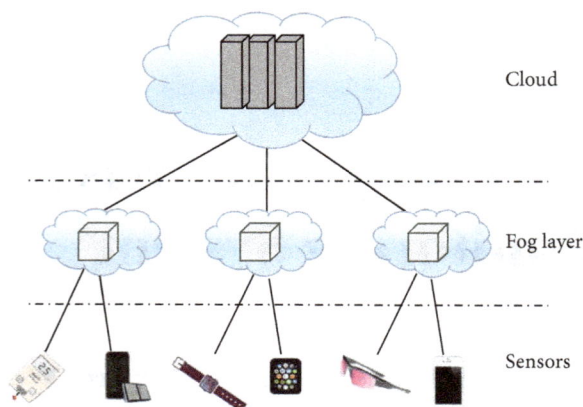

Figure 1: The layers in IoT.

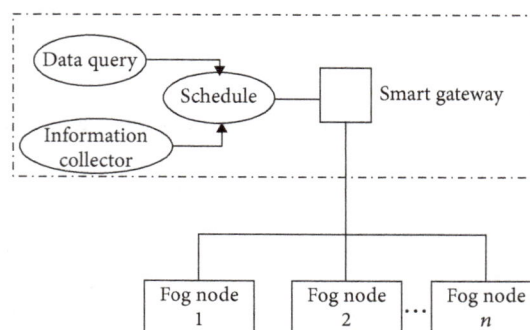

Figure 2: The architecture of a fog layer.

2. Related Work

In this section, we discuss the various technologies that will be used in the proposed research. Also, the research already done in this area will be reviewed in the upcoming subsections.

2.1. Fog Computing. Fog computing is highly virtualized and provides a medium for computing, storage, and networking between end devices and the cloud [7]. The main notion of fog computing is to migrate the tasks of data centers to fog nodes situated at the edge of the network. We refer to these fog nodes as the fog layer. As these devices that perform the tasks are at the edge of the network, it results in a higher data transfer rate and a reduced user response time.

Figures 1 and 2 show the architecture of a simple fog layer with respect to the cloud and the sensors.

2.2. Research in Fog Computing. The main applications where fog computing will prove very useful are time-sensitive applications where a huge amount of data has to be processed [8]. The following are the application areas that can benefit from fog computing.

(1) Healthcare applications: this is an area in which real-time processing plays an important role. Hence, data will have to be processed very fast, and the response time will have to be as less as possible [4, 9, 10].

(2) Augmented reality: for augmented reality (AR), overlaying useful information onto the physical world in real time is very important. Using fog computing will help achieve this aspect of augmented reality [11].

(3) Smart utility services: utility services like electricity, water, telephone, and so on can be managed by fog computing.

(4) Traffic management system: fog computing can increase the efficiency of the traffic signal system by reducing the latency. The interactions between vehicles, traffic signals, and access points can be enhanced by the fog [12].

(5) Caching and processing: fog computing can also be used for improving the performance of websites [13]. Certain websites have a lot of databases and data to be processed, for example, social networking sites and library or online shopping malls. These websites can use the fog layer for caching and preprocessing its data and hence reducing time and space complexity.

(6) Gaming: in the past few years, there has been a huge evolution in the gaming industry. Apart from games being computationally complex, they are mostly multiplayer these days and depend greatly on real-time processing.

(7) Decentralized smart building control: similar to smart utility services, even in the case of smart building, control fog computing will play a huge role in making it more efficient and secure.

Fog computing offers enormous advantages for delay-sensitive fog-based application. Chen et al. [14] implemented a prototype of a smart gateway for the use of WSN in healthcare systems at home. The system is able to transmit reports at real time in a low power embedded system. Hong et al. [15] presented Mobile Fog, which is a programming model for Internet applications that are geographically distributed and latency-sensitive.

2.3. Motivation. The use of smart devices has fortuitously been exercised in healthcare. These days, it is commonplace to find a range of healthcare gadgets that can be used by patients at home or even worn by them. The gadgets mostly encompass sensors. These sensors generally incorporate transducers and are capable of detecting electrical, thermal optical, chemical, and other signals [10].

The main motivation of this paper is to enhance health monitoring systems that are based on IoT devices such that the information collected from WSNs are processed efficiently, and the context-sensitive data that are relevant to the patient are considered. For this, we implement a fog layer that improves the latency of health monitoring systems and enables real-time health monitoring. By this, we strive to also ensure security for the information of the patients such that patient privacy is maintained, and also, tampering of data by third party is avoided.

2.4. Key Objectives of Healthcare Systems. Some of the key objectives of healthcare systems are [16]

(1) improved clinical decision making,

(2) reduced duplication of diagnostic testing, imaging, and history taking,

(3) better medication management,

(4) increased adoption of screening programs and preventive health measures.

2.5. Challenges in Implementing a Health Monitoring System. In order to successfully implement a health monitoring system, a number of challenges have to be addressed [16]:

(1) Safeguarding privacy and security

(2) Technical problems

(3) Organizational barriers

(4) Financial costs

(5) Different policies

(6) Training programs for practitioners and healthcare providers

3. Proposed Research

In this section, we present a tri-tier architecture for context- and latency-sensitive health monitoring using cloud and fog computing. The tri-tiers consist of cloud computing, fog computing, and sensors which work in conjunction with one another. Sensors consist of wearable or not wearable devices that are attached to the patients in the form of smart watches, fitness bands, smart phones, wearable glasses, and so on. The applications used for health monitoring will have components running in the edge devices situated in the fog layer, the wearable sensors, or the cloud. The edge devices may be controlled by the cloud and the fog layers. Information will flow across this tri-tiered infrastructure.

In context-sensitive health monitoring, personalized care can be given to each patient. Context can be classified into extrinsic and intrinsic context. In case of healthcare, extrinsic context is influenced by external factors like the environment that is surrounding the patient. Environmental sensors can be used to extract the extrinsic parameters of a user, and the intrinsic context can be extracted by biosensors. Both intrinsic and extrinsic sensors provide relevant information that may be used for monitoring patients' health. But depending on the disease of the patient, the type of data that is relevant differs. Hence, maintaining context-sensitive data processing has to also be done by the fog layer.

The schematic diagram of the proposed architecture is given in Figure 3. In this architecture of context- and latency-sensitive health monitoring systems, we ensure that all the key objectives of health monitoring and more are all accomplished.

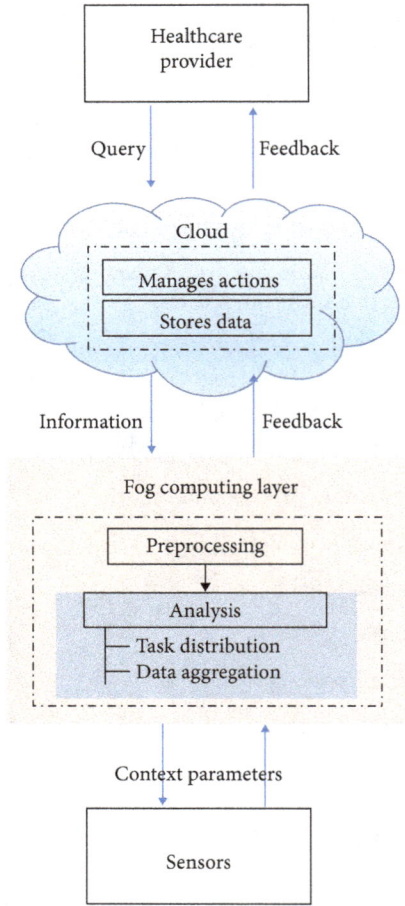

FIGURE 3: Schematic diagram of proposed architecture.

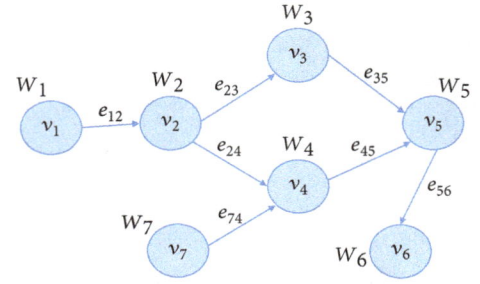

Task graph, $G = (V, E)$,
where $v = \{v_1, v_2, ..., v_7\}$ and $E = \{e_{12}, e_{23}, e_{24}, e_{35}, e_{45}, e_{56}, e_{74}\}$

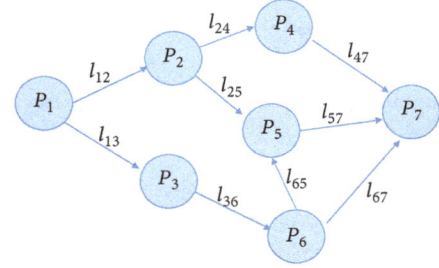

Processor graph, $H = (R, L)$,
where $R = \{P_1, P_2, ..., P_7\}$ and $L = \{l_{12}, l_{13}, l_{24}, l_{25}, l_{36}, l_{47}, l_{57}, l_{65}, l_{67}\}$

FIGURE 4: Graphical representation of the task and processor graphs.

mechanism for cloud-fog computing systems introduced by Pham and Huh [17]. Two graphs are first created: task graph and processor graph. Figure 4 gives a graphical representation of the task and processor graphs.

Let $G = (V, E)$ represent the task graph, where G is a directed acyclic graph (DAG); V is the set of vertices $\{v_1, v_2, ..., v_n\}$ that denote the parallel subtasks, and E is the set of edges where each edge $e_{ij} \in E$ implies that task v_i has a corresponding workload w_i that signifies the amount of work to be processed at a particular resource. Every edge e_{ij} has a corresponding weight c_{ij} which represents the amount of data that is transferred from v_i to $v + i$.

Let $H = (R, L)$ be a DAG that represents the processor graph, where R denotes the set of vertices $\{P_1, P_2, ..., P_n\}$ where each $P_i \in R$ is a processor at the cloud or fog. The edge $l_{ij} \in L$ denotes a link between processor P_i and P_j. Now, $R = N_{cloud} \cup N_{fog}$, where N_{cloud} and N_{fog} denote the set of cloud nodes and fog nodes, respectively.

The priority of computing a particular task is calculated as

$$\mathrm{pri}(v_i) = \begin{cases} \overline{w(v_i)} \\ \overline{w(v_i)} + \max_{v_j \in succ(v_i)} \left[\overline{c(e_{ij})} + \mathrm{pri}(v_j) \right], \end{cases} \quad (1)$$

$$\text{if } v_i \neq v_{exit}, \text{ if } v_i = v_{exit},$$

where $\overline{w(v_i)} = w(v_i)/\overline{w(v_i)}$, and $\overline{c(e_{ij})} = c_{ij}/\overline{BW}$; also, BW is the bandwidth.

3.1. Sensors Tier.
These are the devices that gather information from the patients. These sensors gather both extrinsic and intrinsic values. Extrinsic characteristics are the temperature, location, and so on. Intrinsic characteristics are the blood pressure, blood glucose level, heartbeat, and so on that are collected by the patient's wearable sensors. The patient can also enter data into his or her smart phone, and these data will then be made available for processing. The job of the sensors is to collect all this data and send them to the fog computing layer.

3.2. Fog Computing Tier.
This layer performs data analysis and aggregation of the data. The data and information collected by the edge devices are analyzed in this layer. This layer behaves as the server. Massive amounts of real-time data from sensors are sent to this tier. The fog layer then distributes the processing work to various edge devices connected to the fog layer, and hence, massive amounts of data are analyzed. The distribution of the processing work has to be done using an efficient task-scheduling algorithm.

(1) Work distribution: this task is performed via the smart gateway using a scheduler. In this paper, we distribute the tasks using the task-scheduling

TABLE 1: Comparison of percentages.

Physical topology	Average latency (ms)		Network usage (KBs)	
	Cloud only	*With fog layer*	*Cloud only*	*With fog layer*
Config 1	210.38	8.47	130	12
Config 2	210.78	8.47	351	22
Config 3	211.57	8.47	672	53
Config 4	1283.86	8.47	1061	98
Config 5	3225.91	8.47	1102	189

Once the priority for the various tasks has been calculated, these tasks are sent to nodes for execution. Now the choice of the nodes to do a particular task has to be figured out. For this, the time required by the processor at each node and the processing speed have to be taken into consideration. The earliest start time and earliest finish time will be used for making those calculations. For more details, please refer to the paper by Pham and Huh [17].

(2) Data aggregation: once tasks are distributed, the data have to be aggregated. Data aggregation consists of three main parts: schema mapping, duplicated detection, and data fusion. Schema mapping will ensure that the data are aggregated in such a way that it makes sense, and there is a flow to the data. Duplicated detection ensures that there will not be any redundancy in the data. Redundancy is a challenge because there are many nodes that perform the work, as a result, there may be overlap in data. Apart from duplicate detection, false data injection is also avoided to ensure security in the fog device. This is implemented by adding a local filter to the fog device. Data fusion is the final stage of data aggregation in which the final information is gathered and put together as one entity.

3.3. Cloud Computing Tier.
This is the layer that manages the various actions that are to be performed by the health monitoring system. A component of the monitoring app runs on the sensors which enables the sensors to collect data and send it to the fog layer. The decision of what task to perform is done with the help of the fog layer as described in the previous subsection. The cloud computing tier constantly supervises the health monitoring system.

3.4. Health Monitoring System.
Apart from the various tiers described above, the healthcare delivery system typically consists of four levels: the region, the institution, the clinical department or outpatient clinic, and the individual physician, nurse, or patient [18]. The flow of information between these four levels has to be efficiently managed. Hence, there are many important privacy and security challenges that have to be met.

4. Experimental Results and Analysis

In this section, we discuss the experiments performed so far and the analysis of the proposed system.

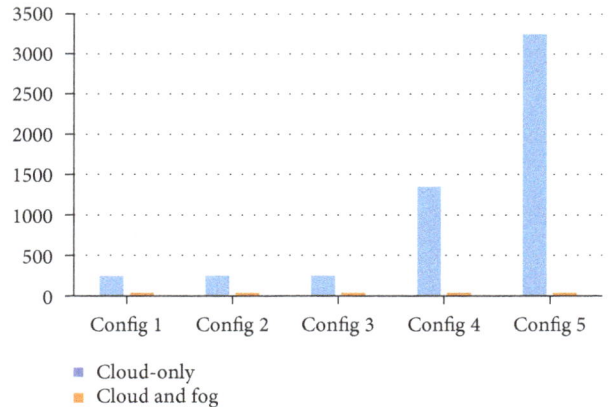

FIGURE 5: Average latency comparison.

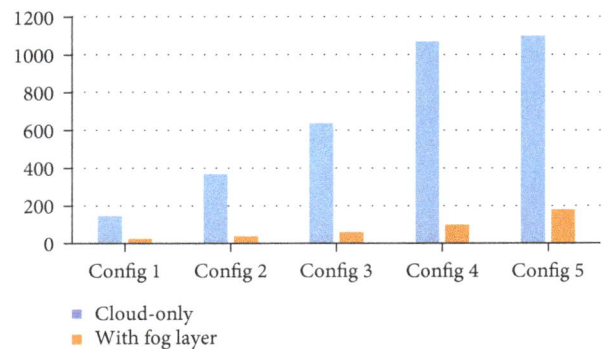

FIGURE 6: Network usage comparison.

4.1. Simulation of the Fog Environment.
For the purpose of comparing the use of fog computing versus just using the traditional cloud computing in healthcare, we used iFogSim toolkit to simulate the fog network [19].

The iFogSim simulates the specified configuration and gives the simulated outputs. This makes it convenient to observe end results when all required technology is not available. The simulator itself adds a bit overhead time. For this simulation, we conducted several test runs for 5 configurations of monitoring devices.

The average latency and network usage for the 5 configurations are given in Table 1. The table shows that the configurations of connected devices do not really affect the latency of our architecture that uses fog computing. The network usage of the architecture with fog computing is much lower than the architecture that uses only fog computing. The iFogSim toolkit is used to simulate the 5 different configurations. For each of the 5 configurations, the monitoring devices are varied. In Config 1, Config 2, Config 3, Config 4, and Config 5, they each have 4, 8, 16, 32, and 64 monitoring devices, respectively. So each configuration will give different results when simulated. The monitoring devices used in the configurations have a CPU length of 1000 million instructions, a network length of 20,000 bytes, and an average interarrival time of 5 ms. Figures 5 and 6 give the comparison of the latency and network usage for the various configurations. The simulation of configurations that use only the

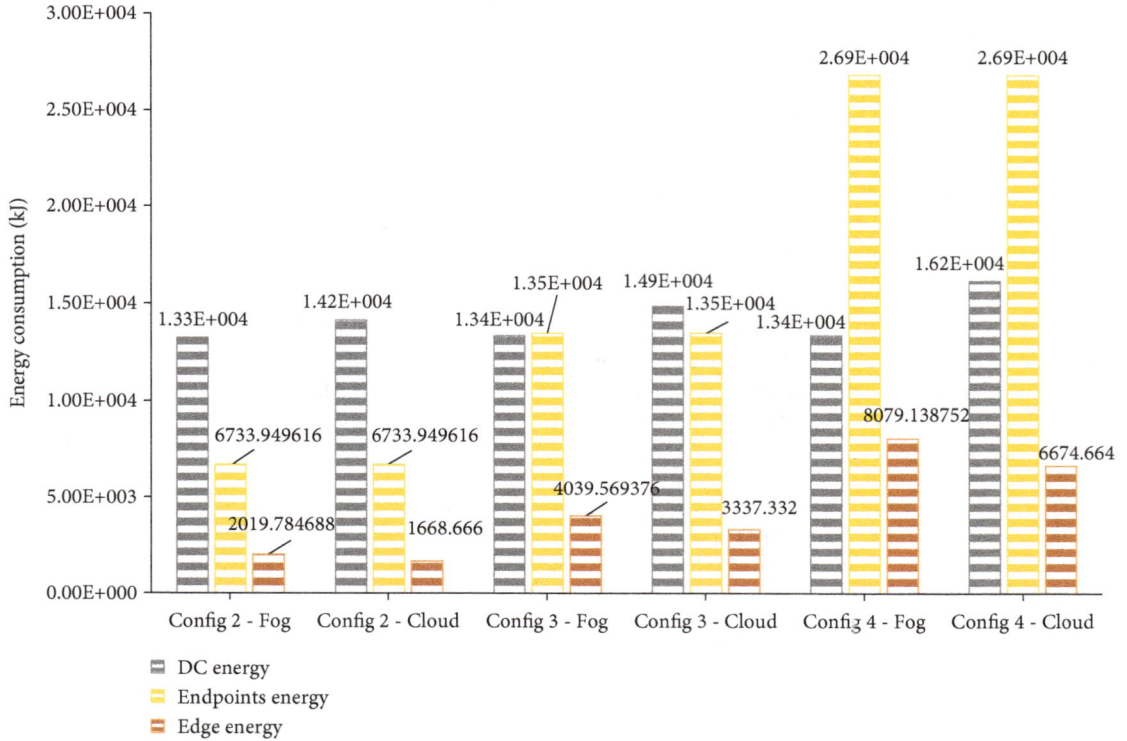

FIGURE 7: Comparison of energy comparison.

cloud is done using the CloudSim toolkit, and those that use the cloud in conjunction with the fog layer is done using the iFogSim toolkit. From both the charts, we can observe that the complexity of the fog layer does not affect the latency or the network usage. In fact, fog computing increases the efficiency of the entire system. From Table 1, we can observe that using fog layer indeed decreases the latency and network usage when compared to using only cloud computing. Figure 7 shows the energy consumption of fog computing versus only using the cloud. From the figure, it can be analyzed that the energy consumption for fog computing mainly takes places at the edge where most of the processing is done. On the other hand, in the case of cloud computing, the energy is mainly spent at the data centers or at the cloud.

4.2. Analysis

4.2.1. Latency. In our implementation of fog computing in health informatics, there will be transfer of data among the various tiers. The amount of data and the time taken will differ in different cases. Hence, the latency differs. Let us consider L_f as the latency when the evaluated data has to be returned to the IoT devices and L_e be the latency when the data is sent to the cloud. Then,

$$\begin{aligned} L_f &= t_s + t_r + e_f, \\ L_e &= t_s + e_f + e_e. \end{aligned} \tag{2}$$

In the above equations, t_s is the time taken from IoT sensors to the fog layer; t_r is the time taken for the data to return from the fog layer to the IoT devices; e_f is the evaluation time taken by the edge devices; and e_e is the evaluation time taken at the cloud.

These two equations will be used by us to analyze the latency times. This is very important because most applications of fog computing rely on the real-time processing capabilities of the network.

4.2.2. Computation. The computations in the fog layer should be real-time and latency-sensitive services. Many techniques for reducing the computation complexities have to be adopted. The data packets can be stored at the fog nodes for some time to avoid reloading of the same data. According to some renewal algorithms, these data packets may be replenished by new data packets. Intelligently, distributing data packets to the most efficient number of edge devices also plays a very important role.

4.2.3. Security Analysis. Introducing a fog layer into the cloud computing infrastructure reduces the security risk when it comes to data of the patients not getting lost due to failure in a data center. But at the same time, the data is stored in the cloud. This increases the threat to the privacy of patient information. In this scheme, we propose to secure the data of the patient by encrypting the data of the patient using a secret key.

5. Conclusion

With increase in the use of the IoT, there is a huge demand for data to be processed in real time and efficiently. The interest in implementing fog computing as a technology is growing rapidly as shown by the H2020 ICT Work Programme for 2016-17. In this paper, we have proposed a scheme for fog computing in health monitoring systems. The work shown in this paper is a work in progress. There are many factors that have hindered and slowed down our implementation process. These factors include not readily available software, complexity, and so on. In spite of all these obstacles, the research area is a very promising one and when implemented will prove to be a very useful technology.

The future work is to deploy this system to various edge devices and judge the behavior of the system.

Conflicts of Interest

The authors declare that they have no conflicts of interest.

Acknowledgments

This study was supported by the National Research Foundation Grant funded by the Korean Government (NRF-2017R1C1B5017464). This study was also supported by the National Research Foundation of Korea (NRF) funded by the Korean government (MSIP) (NRF-2016R1A2A1A05005459).

References

[1] D. Kang, Y. S. Kim, G. Ornelas, M. Sinha, K. Naidu, and T. Coleman, "Scalable microfabrication procedures for adhesive-integrated flexible and stretchable electronic sensors," *Sensors*, vol. 15, no. 9, pp. 23459–23476, 2015.

[2] N. M. Farandos, A. K. Yetisen, M. J. Monteiro, C. R. Lowe, and S. H. Yun, "Contact lens sensors in ocular diagnostics," *Advanced Healthcare Materials*, vol. 4, no. 6, pp. 792–810, 2015.

[3] F. A. Kraemer, A. E. Braten, N. Tamkittikhun, and D. Palma, "Fog computing in healthcare–a review and discussion," *IEEE Access*, vol. 5, pp. 9206–9222, 2017.

[4] Y. Cao, S. Chen, P. Hou, and D. Brown, "FAST: a fog computing assisted distributed analytics system to monitor fall for stroke mitigation," in *2015 IEEE International Conference on Networking, Architecture and Storage (NAS)*, pp. 2–11, Boston, MA, USA, August 2015.

[5] Cisco, "White paper: fog computing and the internet of things: extend the cloud to where the things are," Tech. Rep., Cisco, 2015.

[6] L. M. Vaquero and L. Rodero-Merino, "Finding your way in the fog: towards a comprehensive definition of fog computing," *ACM SIGCOMM Computer Communication Review*, vol. 44, no. 5, pp. 27–32, 2014.

[7] F. Bonomi, R. Milito, J. Zhu, and S. Addepalli, "Fog computing and its role in the internet of things," in *Proceedings of the First Edition of the MCC Workshop on Mobile Cloud Computing*, pp. 13–16, Helsinki, Finland, August 2012.

[8] S. Yi, C. Li, and Q. Li, "A survey of fog computing: concepts, applications and issues," in *Proceedings of the 2015 Workshop on Mobile Big Data*, pp. 37–42, Hangzhou, China, June 2015.

[9] K. Ha, Z. Chen, W. Hu, W. Richter, P. Pillai, and M. Satyanarayanan, "Towards wearable cognitive assistance," in *Proceedings of the 12th Annual International Conference on Mobile Systems, Applications, and Services*, pp. 68–81, Bretton Woods, NH, USA, June 2014.

[10] V. Stantchev, A. Barnawi, S. Ghulam, J. Schubert, and G. Tamm, "Smart items, fog and cloud computing as enablers of servitization in healthcare," *Sensors & Transducers*, vol. 185, no. 2, pp. 121–128, 2015.

[11] J. K. Zao, T. T. Gan, C. K. You et al., "Augmented brain computer interaction based on fog computing and linked data," in *2014 International Conference on Intelligent Environments*, pp. 374–377, Shanghai, China, 2014.

[12] F. Bonomi, R. Milito, P. Natarajan, and J. Zhu, "Fog computing: a platform for internet of things and analytics," in *Big Data and Internet of Things: A Roadmap for Smart Environments*, pp. 169–186, Springer, 2014.

[13] J. Zhu, D. S. Chan, M. S. Prabhu, P. Natarajan, H. Hu, and F. Bonomi, "Improving web sites performance using edge servers in fog computing architecture," in *2013 IEEE Seventh International Symposium on Service-Oriented System Engineering*, pp. 320–323, Redwood City, USA, March 2013.

[14] Y. Chen, W. Shen, H. Huo, and Y. Xu, "A smart gateway for health care system using wireless sensor network," in *2010 Fourth International Conference on Sensor Technologies and Applications*, pp. 545–550, Venice, Italy, July 2010.

[15] K. Hong, D. Lillethun, U. Ramachandran, B. Ottenwälder, and B. Koldehofe, "Mobile fog: a programming model for large-scale applications on the internet of things," in *Proceedings of the Second ACM SIGCOMM Workshop on Mobile Cloud Computing*, pp. 15–20, Hong Kong, China, August 2013.

[16] Institute of Medicine (US) Committee on Data Standards for Patient Safety, *Key Capabilities of an Electronic Health Record System: Letter Report*, National Academies Press, 2003.

[17] X.-Q. Pham and E.-N. Huh, "Towards task scheduling in a cloud-fog computing system," in *2016 18th Asia-Pacific Network Operations and Management Symposium (APNOMS)*, pp. 1–4, Kanazawa, Japan, October 2016.

[18] T. Benson, "Why general practitioners use computers and hospital doctors do not—Part 1: incentives," *BMJ*, vol. 325, no. 7372, pp. 1086–1089, 2002.

[19] H. Gupta, A. Dastjerdi, S. Ghosh, and A. Buyya, "iFogSim: A toolkit for modeling and simulation of resource management techniques in internet of things, edge and fog computing environments," 2016, http://arxiv.org/abs/1606.02007.

8

VANSec: Attack-Resistant VANET Security Algorithm in Terms of Trust Computation Error and Normalized Routing Overhead

Sheeraz Ahmed (ID),[1,2] **Mujeeb Ur Rehman,**[2] **Atif Ishtiaq,**[1] **Sarmadullah Khan** (ID),[3] **Armughan Ali,**[4] **and Shabana Begum**[5]

[1]*Iqra National University, Peshawar, Pakistan*
[2]*Career Dynamics Research Centre, Peshawar, Pakistan*
[3]*School of Computer Science and Informatics, De Montfort University, Leicester LE1 9BH, UK*
[4]*COMSATS Institute of Information Technology, Attock, Pakistan*
[5]*Islamia College University, Peshawar, Pakistan*

Correspondence should be addressed to Sheeraz Ahmed; sheerazahmed306@gmail.com

Academic Editor: Almudena Rivadeneyra

VANET is an application and subclass of MANETs, a quickly maturing, promising, and emerging technology these days. VANETs establish communication among vehicles (V2V) and roadside infrastructure (V2I). As vehicles move with high speed, hence environment and topology change with time. There is no optimum routing protocol which ensures full-pledge on-time delivery of data to destination nodes, and an absolutely optimum scheme design for flawless packet exchange is still a challenging task. In VANETs, accurate and on-time delivery of fundamental safety alert messages (FSAMs) is highly important to withstand against maliciously inserted security threats affectively. In this paper, we have presented a new security-aware routing technique called VANSec. The presented scheme is more immune and resistive against different kinds of attacks and thwarts malicious node penetration attempts to the entire network. It is basically based on trust management approach. The aim of the scheme is to identify malicious data and false nodes. The simulation results of VANSec are compared with already existing techniques called trust and LT in terms of trust computation error (TCE), end-to-end delay (EED), average link duration (ALD), and normalized routing overhead (NRO). In terms of TCE, VANSec is 11.6% and 7.3% efficient than LT and trust, respectively, while from EED comparison we found VANSec to be 57.6% more efficient than trust and 5.2% more efficient than LT. Similarly, in terms of ALD, VANSec provides 29.7% and 7.8% more stable link duration than trust and LT do, respectively, and in terms of NRO, VANSec protocol has 27.5% and 14% lesser load than that of trust and LT, respectively.

1. Introduction

Communication remains a main focus of interest in human beings. Hence, in results of continuous struggle, it became possible to replace one communication medium by other fastest communication means for sending and receiving information. Computer networks are a bunch of networked computing hardware devices interchanging data to the communicating networked devices through a data link. The link between nodes is fixed, that is, wired or with wireless media. The Internet is a prominent computer network. Wireless technology does not provide full security of information because the medium is open. To ensure security, encryption/

decryption techniques are used to identify the authorized users. Table 1 shows different types of wireless networks.

The wireless sensor network (WSN) is a self-organizing, infrastructureless network. WSN is an example of wireless networks using IEEE 802.15.4 protocol designed for low-rate WPANs and also for sensor networks. WSN consists of numerous small sensors with low cost, low battery power, and limited computational capabilities and low communication bandwidth. These sensor nodes are used to collect information as well as integrate and transmit data in a wireless fashion and handover it to the base station (BS) via a gateway node [1]. WSN is comprised of power components, radio transceiver, and computing and sensing devices. Sensors are

TABLE 1: Different types of wireless networks.

Type	Applications	Range	Standards
Personal area network (PAN)	Cable replacement for peripherals	Within reach of a person	Bluetooth, ZigBee, NFC IEEE 802.15
Local area network (LAN)	Wireless extension of wired network	Within a building or campus	IEEE 802.11 (Wi-Fi)
Metropolitan area network (MAN)	Wireless internetwork connectivity	Within a city	IEEE 802.16 (WiMAX)
Wide area network (WAN)	Wireless network access	Worldwide	Cellular (UMTS, LTE, etc.)

hundreds and thousands in number, communicating with each other through radio communication over an industrial, scientific, and medical (ISM) radio band.

To obtain information on location and positioning, local positioning algorithms and the global positioning system (GPS) can be employed [2]. The IEEE 802.11p standard known as wireless access in vehicular environments (WAVE) is a specially developed version to adapt vehicular ad hoc network (VANET) requirements and support intelligent transport systems (ITS). IEEE 802.11p is one of the fresh sanctioned amendments to the IEEE 802.11 standard to add wireless access in vehicular environments (WAVE). In this sense, IEEE 802.11p is denoted as WAVE.

Information and communication technology (ICT) plays a vital role in making the cities smarter in the future through intervehicle communication (IVC), using an infrastructure of Car4ICT using IEEE 802.11p based on dedicated short-range communication (DSRC) protocol [3]. Car4ICT infrastructure is a future technology which will facilitate users by easily accessing different applications like routing, uploading, and downloading data. It also provides data processing and storage facilities for the users. Such services are complex and require detailed knowledge to constitute it in big cities [3]. IEEE 802.11 is an accumulation of physical layer (PHY) specifications and media access control (MAC) for implementing WLAN in the 2.4, 3.6, 5, and 60 GHz frequency bands, maintained by the IEEE 802 LAN Standards Committee in 1997.

A mobile ad hoc network (MANET) is a network which has many autonomous mobile nodes which are free to move in any direction, also continuously modifying their locations in a self-configurable manner. It is an infrastructureless network; these nodes have the capacity to connect with Wi-Fi or any cellular infrastructure. VANET is an application of MANETs. VANET is a wireless ad hoc network, in which moving vehicles behave like mobile nodes and allow them to connect with each other via DSRC, and a protocol proposed for WAVEs is IEEE 802.11p for IVC. VANETs enable infrastructure-to-vehicle (I2V), or vehicle-to-infrastructure (V2I), and vehicle-to-vehicle (V2V) communication system [4, 5].

V2I communication is a wireless exchange of safety messages and access to the Internet between vehicles and roadway side units. A major concern of VANETs is to avoid vehicle collisions and get updates about road condition, weather information, traffic jam situation, and so on. In V2V infrastructure, when vehicles come in the communication range, it results in an automatic connection and establishes an ad hoc network. This enables sharing of position, speed, and direction data; again, DSRC connects with the global positioning system (GPS) resulting in a V2V communication system which provides a 360° view of vehicles within the communication range.

VANETs utilize movable vehicles and establish a wireless link among vehicles with features such as rapid changing topology, high computational ability, predictable mobility, and variable network density. VANET architecture consists of three parts: (i) an on-board unit (OBU) which is built in the vehicles or vehicles itself, (ii) an application unit (AU) person set in the car, that is, driver, and (iii) a roadside unit (RSU) installed on highways which constitutes the VANET system and provides a basis for an intelligent transportation system (ITS) [4, 5]. The researchers successfully advent a network with the collaboration of WSNs and VANETs named as vehicular sensor networks (VASNETs). Vehicles are mobile nodes in VASNET, and an important application for vehicular networks is cooperative collision warning (CCW) message disseminations, which uses V2V communication and hence achieves safety [6]. The basic VANET structure is shown in Figure 1.

VANET is an application of mobile ad hoc networks (MANETs) which differs from MANETs in a few ways like the following. (i) Power is a constraint in MANETs, but in the case of VANETs, power is not due to tremendous installed battery. (ii) Moving pattern: in VANETs, nodes move coherently, while in MANETs node moments are random. (iii) Mobility: the mobility ratio in VANETs is bigger than in MANETs [6]. VANETs have three main architectural categories, which are as follows. (i) Pure ad hoc mode: in this mode, only V2V communication exists and no other infrastructure takes part. (ii) Pure cellular or WLAN mode: in this mode, vehicles can easily access information from cellular towers and access points (APs). (iii) Hybrid mode: this mode can use and access data from cellular/WLANs as well as from pure ad hoc mode depending upon the information capacity and route congestions [4, 5].

VANETs have different characteristics, summarized as follows:

(i) *High mobility*: in VANETs, vehicles move at high velocity which causes the contraction of the mesh network. So, in such case, vehicle position identification is difficult and it also leads to poor security provision to node privacy.

(ii) *Rapidly changing network topology*: in VANETs, vehicles move randomly with high speed, so evidently

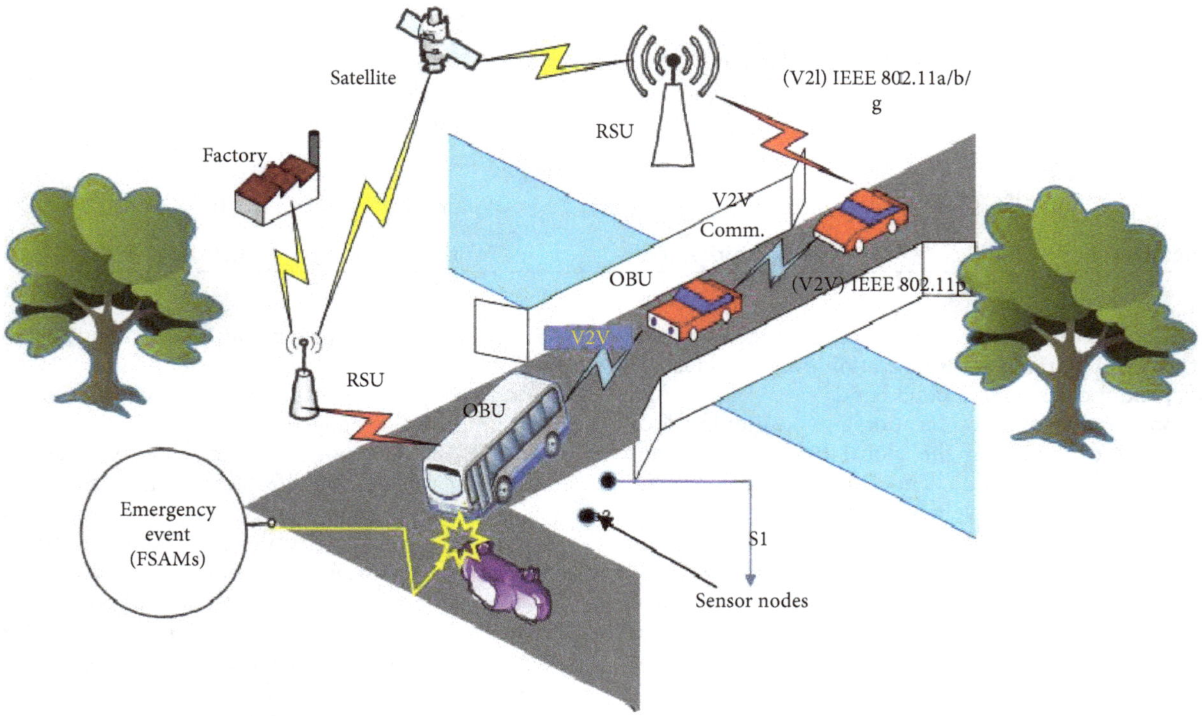

FIGURE 1: Basic VANET structure.

the position of vehicles will change most oftenly. The topology is dynamic and irregular. It encourages attacks in the network and makes it difficult to sort out misbehavior/attacks in the network [7].

(iii) *Availability of the transmission medium*: VANET size in the geographical point of view is boundless. VANET infrastructure can be designed for a city, cities, or as a whole for a country. The wireless medium is a universally available transmission medium, which is a big reward in IVC.

(iv) *Frequent exchange of information*: the VANET network is ad hoc in nature. In VANETS, nodes gather information from the neighbor vehicular nodes and also from RSUs. So, in this way, nodes exchange their information.

(v) *Attenuations*: DSRC is a digital transmission band facing problems in transmission frequencies; these are reflection, diffraction, and dispersion, various kinds of fading phenomena, and Doppler effect losses. Due to multipath padding propagation, delays occur [7].

(vi) *Time critical*: in VANETs, time period management is absolutely needed; it should be ensured that information reaches to the exact accurate node in the specified time, to enable the node for decision and execute action accordingly.

(vii) *Limited bandwidth*: VANETs use the DSRC band with a limited bandwidth of 27 MHz; the theoretical data rate is 27 Mbps.

(viii) *Energy storage and computing*: VANET is rich in energy, computing capability, and storage.

(ix) *Limited transmission power*: in a WAVE scenario, the transmission power is up to 1000 m and ensures data reachability to nodes. In congestion or accident situation, transmission power can be maximizes [7].

Security of VANETs is an important factor which protects information related to the driver and vehicle from unauthorized access and ensures privacy of the driver and vehicle. In VANET scenario, nodes are highly dynamic; in such networks, information security is a very tough job.

1.1. Security Requirements in VANETs. To ensure information security, different security goals should be fulfilled; the most common security requirements are confidentiality, data integrity, and availability. In addition, other security requirements are authentication, data check, and nonrepudiation [8]. So, collectively in VANETs, six security goals should be fulfilled. Keeping information hidden from unauthorized access is called *confidentiality*, protecting information from unauthorized changes is called *integrity*, and accessibility to the required information by an authorized user is called *availability* [7]. The process which belongs to the verification of information generated by the sanctioned user is called *authentication*. The transmitted message is confirmed and checked by the receiving node/vehicle; whether the received data is correct or having some false information is called *data check*. *Nonrepudiation* is the process in which the sender of a message cannot disown himself from the communication at the end of the communication session [8]. Data correlation can also be considered a security requirement which easily finds out bogus data, by correlation to finger out the similarity between the data received and the data transmitted. Making secure the position of vehicle and BS is also a concern with VANET security [7]. Entities that are involved in VANET security are drivers, OBUs, RSUs, and attackers. The driver is a key part of the VANET system taking decision in emergency situations providing safety to the vehicle and comfort to passengers.

Vehicle OBU may be a normal automated system or may be an attacker impersonating himself as a normal node, and similarly RSUs can be normal or may be a malicious node and can disrupt the normal network activities for the attacker's own benefits. Attackers can launch different kinds of techniques to interrupt normal network functions; attackers can be internal or external, and they have only one motive to benefit themselves [7]. Attackers can be of two types; they may be rational and irrational and can do active or passive attacks. Active attacks are detectable while passive attacks are not. The third party should be a trusted or semitrusted authority, or it may be a manufacturer of the vehicles which is also a key entity of the VANET system [8].

1.2. Possible Attacks that Are Vulnerable to VANET Security

(i) Attacks on availability: in such attacks, the attacker shuts down the entire network and the node has no access to the information.

(ii) Attacks on authentication: identification of vehicles is mandatory to rectify the genuine sender and receiver, confirm identity first to kick out intruders, and reduce the chance of information loss.

(iii) Attacks on confidentiality: the information should be confidential between the authorized users and kept hidden or encrypted from the intruders to avoid traffic analysis or snooping attacks.

(iv) Attacks on integrity: the intruder should change the data by deletion, insertion, and modification of data according to his requirements and benefits. Data integrity keeps away repudiation and replaying attacks.

(v) Attacks on nonrepudiation: the ability to confirm that the sender and receiver of the message are authentic users and at the end they cannot refuse to acknowledge [7, 9].

(vi) Another attack known as denial of service (DOS) or distributed denial of service (DDOS): it hijacks the network totally, slows down the entire process and interrupts the services of network. The intruders send many fake or bogus requests, reply to the network, and impersonate themselves as a normal vehicle OBU or RSU, and the network seems busy or out of reach, not responding to the genuine vehicles [10].

(vii) Identity revealing: disclosing details of the individual vehicle can put security at danger. Later character revealing must be avoided.

The various other types of attacks are like broadcast tampering, Sybil attacks, message suppression attacks, alteration attack, and wormhole attack [11]. Lots of research work are done on ITS, and nodes are equipped with communication technology. Messages are exchanged between nodes containing information regarding their current location and its surroundings. Different techniques are used in the VANET system to enhance its security. To reduce the accident ratios and ensure safe transportation, different approaches are used to identify the causes of traffic accidents in ITS.

National Databank Wegverkeer (NDW) is a database containing real-time data about the traffic network of the Netherlands. When a crash occurs, the factor can be easily found out from NDW. Another technique is event data recorder (EDR), a device built in the vehicle which collects violent information regarding the vehicle's speed, heading, and engine accelerator. The main aim of EDR is to get information about the event when the crash is faced by the vehicle system, that is, EDR provides postaccident information and causes of the accident can be easily investigated. EDR can also collect other kinds of data if appropriate sensor nodes are used [12].

IEEE 802.11p is a standard protocol for WAVE. In VANETs, vehicles are equipped with DSRC to broadcast messages to neighbor nodes. Neighbor nodes/vehicles are also equipped with DSRC or stationary stations located at the roadside. These messages contain information, like safety warnings and traffic information. IEEE 802.11p determines a set of two types of messages: cooperative awareness message (CAM) and decentralized environmental notification message (DENM) used in ITS [12]. CAM is broadcast and replicated again and again to all nodes in the neighborhood. CAM shows positioning and other basic status-related information of the communicating entities in the ITS system [12]. DENM is the second message presented by 802.11p. The message is also broadcasted to other ITS stations when a particular incident occurs to inform other vehicles. Wrong way driving, accident, and roadwork are the examples of such incidents.

On detection of hazardous events by the ITS station, it starts broadcasting without any delay a DENM message to other ITS stations in the region (a specific geographical area) which can be affected by the event. The message is continuously broadcasting repeatedly, till the event is over. When the specified event is over, a special DENM message is circulated to inform all nodes about the disappearance of the event [12].

An autonomous traffic management scheme which enables the vehicular network, to exchange data between vehicles, should be about the change of route in case of congestion, traffic jam condition, or any other emergency situation. The network is called VANET-based autonomous management (VAM) scheme.

In the presence of traffic light, VAM establishes coordination between vehicles and the light controller to overcome congestion [13].

To keep information security in VANETs, different approaches are used. Public key cryptography (PKC) is an asymmetric key algorithm, in which a key used for encryption of a message is not used for the decryption of that message. Encryption and decryption are done with two separate keys. In such algorithms, each node has a pair of cryptographic keys: one is public encryption key (PEK) and the other is private decryption key (PDK). The pair of cryptographic keys is generated by the real time application (RTA) technique periodically. Public keys are reached to each and every RSU in its operation area via a secure medium/channel.

Traditional wired networks are protected by several lines of defense such as firewalls and gateways. Security attacks on such networks may come from any direction and target all nodes. VANETs are susceptible to intruders ranging from passive eavesdropping to active spamming, tampering, and interfering due to the absence of basic infrastructure and centralized administration. The main challenge facing VANETs is user privacy. Whenever vehicular nodes attempt to access some services from roadside infrastructure nodes, they want to maintain the necessary privacy without being tracked down for whoever they are, wherever they are, and whatever they are doing. It is considered as one of the important security requirements that should be paid more attention for secure VANET schemes, especially in a privacy-vital environment [14].

2. Literature Review

ITS and VANETs have been under research for many years. But with the advancement in generation of communication technology, there is a need to refine the information exchange process and come up with better security and more fulfilling solutions against threats that meet the demands of the day. With the world moving steadily towards WAVE, there is a need to refigure the entire ITS system security and ensure that the VANET security process does not prove to be a bottleneck in the advancement of the ITS technology. We shall have a look at some of the earlier works done in the field of VANET security in order to eliminate or reduce the frequency of attacks in VANETs by malicious nodes.

In paper [15], researchers proposed a novel authentication mechanism for secure message transmission in a VANET scenario. The author has shown that an already existing technique of message authentication was based on a combined signature technique, in which the forwarding node used a combined signature algorithm via RSU and results in a huge transmission overhead message. Due to a combined signature scheme, RSU sometimes transmits fake authenticated messages toward nodes. To avoid such issues, the authors proposed an aggregate message authentication code technique which verifies the integrity and authenticity of messages and thinned communication overhead.

Pseudo-RSUs were installed in the neighborhood of RSU to stop false information dissemination, to ensure exchange of rectified authenticated messages. The authors proposed a technique based on results obtained from simulations and security parameters which reduced considerably the communication overhead and enhanced the validity of disseminated information, which validated the authors' suggested technique. In [16], the authors fabricated a technique which has studied security aspects of V2V communication utilizing a radiofrequency (RF) transceiver. The main part of the VANET environment is position-based information of the vehicular node. The use of an RF transceiver improved the trust on received data about the vehicle's location. The suggested model of authors followed the rule of "Trust on what you observed, confirm what you listen." The basic motive of the scheme was to find a vehicular communication system best suited in minimum cost and more effective in data distribution, as well as to ensure passengers' safety, security, and comfortability.

The RF transceivers verify reported data in the network and approve the position of the neighbor's vehicles and that of the malicious vehicle too and hence ensured the security of the network. The authors suggested a scheme which enhanced VANET security through precluding malicious entities from penetration into the network, hence reducing the chances of putting invalid data about the position information of vehicles.

In paper [17], researchers designed a novel technique of detection, named greedy detection for VANETs (GDVANs); it was for the purpose of reducing greedy behavior frequency in VANETs. VANETs' basic motive was to assure road passengers' safety and enhance transportation quality. Multiple attacks were launched in VANETs; among them, one was denial of service (DOS) attack, which interrupted authorized clients from available information.

The authors proposed a technique incorporating two phases: suspicious phase and decision phase. The suspicious phase followed the concept of linear mathematical regression, where a fuzzy logic decision scheme was followed by a decision phase. The advantage of the designed scheme was that the network nodes had the capacity of execution and no change was required in the standard IEEE 802.11p protocols at any stage. Moreover, the technique had the ability of greedy behavior-type threat detection and found a potentially compromised node list, utilizing three newly defined metrics. The authors justified and validated their proposed scheme from results obtained from experiments or simulation.

In [18], the authors demonstrated that VANETs basically had the opportunity of safe wireless communications with threat avoidance capability, but still, security threats in VANETs are a disputing task, like access control, integrity of data, confidentiality, nonrepudiation, availability, and data privacy. The paper suggested a model which was about VANETs protecting against threats, labeled as an attack-resistant trust (ART) management algorithm. ART had not only detection capability of malicious data and node but also the ability to deal with malicious attacks. In VANETs, ART judged the trustworthiness of both data and mobile nodes. Especially, assessment of data trust was done on the basis of sensed and collected data from various vehicles; judgment of node trust was done in two ways, that is, functional trust and recommendation trust, which reveal how probably a node could accomplish its functionality and how trustworthy the recommendations from a node for other nodes would be, respectively. The authors validated their proposed scheme ART, via experimental data they analyzed. Moreover, the scheme ART had broad applications in VANET background, to enhance traffic experimentation in terms of secure mobility, with reinforced reliance. Agarwal et al. [19] developed a theory to assure security inside educational institutions, medical institution/health care centers, residential places, and so on, through conversion of stodgy vehicles into connected vehicles to prevent careless driving. In the designed model, entry and exit points (gates) were defined. Authors suggested wireless hardware-type "GPS" arrangement to supervise moving vehicles, velocity, and region of entry. At the entryway, orthodox vehicles obtain a device from guards on duty and return the device back upon exiting to authorized guards on duty.

When the devices were activated, a communication mean/path is set up among security depots and drivers inside the specified region to avoid rule violation. For vehicles inside that particular region which have a speed threshold, on crossing the threshold value warning messages were disseminated between the vehicle operator and the system. In the depot, receiving unit holds previous record of each individual drive separately; in terms of any misconduct penalizing action taken versus the handed driver. The scheme proposed by authors was judged on trial bases and, over race, was cut down up to sixty percent securing residential human areas; these characteristics validated the scheme efficacy.

In [20], researchers had considered VANET a complicated network, in which all vehicular node moments were in random manner. In VANETs, the node position changes, so data dissemination was a problem; also, creation of new links took place each time for data packet transmission. So, in such scenario, an attack could wind up all communications running among vehicular nodes. According to authors' conclusion, the Sybil attack was one among other different attacks in VANETs, due to which packet loss occurred. In this paper, the authors discuss impacts of Sybil attempts on VANET communication protocols. Further, researchers examined and scrutinized the verity of VANET routing hierarchies and found the AODV routing scheme to be more efficient in terms of attacks launched in VANET fencing. In the existence of attack in VANETs, the AODV algorithm used simulator QualNet 5, whose output results were satisfactory, but more advancement in routing hierarchy was still required.

Researchers in [21] exhibited that VANET security was the most research-adopted area due to its quality of providing better protection to drivers, vehicles, and so on. Vehicles in VANETs move with maximum acceleration, and also network topology dynamically changes which makes it hard to wipe out false invalid nodes totally and ensure dispersion of data among nodes safely. Hence, in the authors' view, information privacy and security in VANETs were the most vital research-inquired tasks.

In the paper, the authors exemplified different security threats to VANETs and pointed out possible remedy algorithms to mitigate those attacks. The authors had categorized those defensive mechanisms and analyzed them on a dissimilar performance point of view. Eventually, research workers found different research subjects based upon VANET security threats and incited scientists to work on these topics and discover an efficient method to resolve threats and attacks in VANETs.

Research work in [22] presented a detection problem of DOS attacks happening in VANETs. The authors' primary contribution was to conceptualize a new security model based on a game pattern for DOS attacks in VANETs. Secondly, researchers expressed two conditions about game theory, strategic-type game, and extensive-type game. Thirdly, authors had studied DOS attacks on the basis of practical suppositions, utilizing the actual mobility models based on an actual map. Finally, authors analyzed their designed model and validated it through a simulation process. Moreover, authors stated about their contribution in research that no such type of game-related model was designed earlier. Researchers concluded their research work analysis that they will solve DOS attack problems arising in VANETs.

In [23], researchers showed that with the growth of security techniques in VANETs, threats also grow relatively. Authors proposed a trust-based management algorithm called threshold adaptive control technique; the technique was mainly used to detect malicious and selfish nodes, and they fixed themselves inside the network intelligently. Authors showed that previous detection techniques failed up to some extent in detecting these intelligent malicious nodes. Authors have designed an adaptive detection threshold technique, which motivates the attackers to act well, and finally, the designed technique catches the malicious behavior and hence was able to detect the malicious nodes immediately.

From their simulation results, authors concluded that their designed technique had best detection ratio more than 80% even in high ratio of attackers present inside the network. Also it handovers high data packets among nodes even when VANETs are dense.

In [24], the authors proposed a trust-based framework for communication in VANETs that is capable of accommodating traffic from different applications. Their scheme assigned a trust value to each road segment and one to each neighborhood, instead of each car. It scaled up easily and was completely distributed. Experimental results demonstrated that their framework outperformed other well-known routing protocols since it routed the messages via trusted vehicles.

In [25], authors showed that only authentication of nodes was not enough for secure data transmission in the VANET network, because sometimes even authentic nodes disseminated fake information and on/off attacks lead network application to threats of various attacks. To avoid such threats and attacks, authors proposed a technique called logistic trust mechanism, which has the ability to detect and identify malicious false messages and nodes. According to authors, to detect an attack, the first correct event should be identified as data depends on the events. The proposed scheme identifies the correct event first through information collected from trusted sources and also from the receiver observation itself. On the basis of this information, in logistic trust algorithm, the behavior of the nodes was identified through the receiver's own observation which was complemented by the opinions of other nodes. Authors proposed a scheme which had 99% accuracy in detection of malicious nodes and messages, which shows the efficacy of their proposed technique.

VANET security-oriented schemes are summarized in Table 2 given below with various parameters addressed in schemes, area of applications, techniques utilized, and deficiencies or research gaps present in these schemes.

3. Objectives of Research

In our research work, we proved our proposals with the assistance of a mathematical model and a flow chart. Our mathematical model and designed flow charts evidently validated our research. In our research work, we evaluate the performance of our design scheme VANSEC to trust [24] and logistic trust (LT) [25] schemes with respect to vehicle density using a MATLAB simulator to model all the driving environment and networking details of VANETs.

In the last phase of our research, we conducted a relative comparison. We compared our suggested VANSEC protocol with existing VANET algorithms, and comparative investigations are made and presented. The parameters we choose for our research work are TCE, EED, ALD, and NRO. In our presented scheme, the latency and TCE are dragged to minimal values and show enhanced efficacy with respect to other algorithms in terms of compared parameters.

The main objectives of our research work are as follows:

(i) To propose a protocol that can work efficiently, ensuring improvement in VANET security, and which should be scalable for the network in the future

(ii) Ensure data confidentiality, data integrity, and data availability for the clients in a VANET scenario

(iii) Propose an efficient technique to make the intruders' attempts thwart against data modification through data an insertion or deletion process

(iv) Adopting/applying different security mechanisms/ protocols through which the VANET system becomes much secure as well as provides better performance in terms of delay, higher PDR, small packet loss ratio, and efficient utilization of energy resources

4. Research Methodology

The process we implemented includes three vehicles and RSUs communicating with each other via IEEE 802.11p and IEEE 802.11 a/b/g. The scenario we put in our design is a hidden node for some other nodes moving towards each other. V3 and V1 are unaware of each other, because vehicle V3 is out of range to vehicle V1, that is, V1 and V3 are not in range of each other. Both vehicles are hidden from each other. Vehicle V2 is in range of V3 as well as of V1 via DSRC. However, there is also an RSU in access of all the vehicles.

In a narrow road scenario, V3 broadcasts an alert about its speed and position to inform nearby vehicles through DSRC and sends an alert towards the RSU. Vehicle V2 received the alert and propagated the alert to its nearby vehicles as shown in Figure 2.

On reception of alert by V1 from V2 and also from RSU, V1 goes for registration or authentication verification process, to make sure that the message was issued from an authentic source or from a malicious node. From Figure 2, there is communication among vehicles which is called ad hoc mode, while with the addition of an infrastructure it is switched into infrastructure mode. The VANET security model is confined and explained with the help of a flow chart shown in Figure 3.

In the initialization process, vehicles and RSU register themselves to a registration server. The registration server verifies its authentication from a verification server to avoid penetration of malicious node and make the system secure at the primary level. There are three vehicles (V1, V2, and V3) and an RSU participating in the session; V1 receives a FSAM from V3 through V2. V1 inquired the same alert message FSAMs from RSU to confirm whether the received FSAM from V3 is correct. The decision-making block will check the similarity index. If the alert FSAMs received from both entities are the same, then it will inform the driver about the validity of node V3 also informing ConVai (confirm validity) message exchange about the validity of node V3 correctness, where ConVai exchange confirms the confidentiality of FSAMs to avoid snooping and traffic analysis. Integrity of FSAMs is checked to handle modification, masquerading, repudiation, and replay of attempts of false nodes. Also, it ensures on-time availability of FSAMs for requesting vehicles. After meeting the minimum acceptable threshold value of ConVai exchange, node 3 and other nodes meet the same criteria and are declared valid. These valid nodes are enlisted in the list of correct true nodes and allowed for communication or broadcasting FSAMs in the network.

If received FSAMs from RSU and V3 do not match and decision is blocked, then it is switched into another block for further verification about V3. This helps to look over node V3 position availability or unidentified position. If the position is identified, then FSAM is forwarded to the next block, to check FSAM confidentiality and for further investigation about FSAMs which is confirmed by ConVai exchange. After position validity and FSAM correctness, node V3 validity is endorsed and allowed for broadcasting FSAMs in the network. If node V3 position is invalid, then FSAM is discarded; again if the position of the node is valid but FSAM does not hold confidentiality check properly, FSAM is pumped into the discard bin.

Sensor nodes are also dispersed on highways which also gather data about events; Cluster Head (CH) forwards FSAMs towards ConVai exchange which are filtered here. If the received FSAMs from CH and RSUs are same notifying V3, then V3 and other nodes are assumed valid and are allowed for broadcasting. In case of any dissimilarity among the received FSAMs from different entities, they are pushed towards the discard bin which is shown in the flowchart diagram. All of these FSMs received from different sources alerts

TABLE 2: Summery of related work on VANET security.

Scheme	Technique	Area of applications	Parameters addressed	Deficiencies
Secure message delivery and authentication [15]	Aggregate message authentication code (MAC)	VANET security	Reduced communication overhead, improved authenticity	Packet loss End-to-end delay
Believe what you see, verify what you hear [16]	Detection and correction of error	VANET security	False position information detection, quick and fast data dissemination	Safety warnings, electronic toll collection, blind curve problem, etc.
Greedy detection for VANETs (GDVANs) [17]	Linear regression, mathematical concept, and fuzzy logic	VANET security	Prevention of DOS attack, no modification in IEEE 802.11p; greedy behavior detection	Duration between two successive transmission, transmission time, connection attempts made by node
Attack-resistant trust (ART) management [18]	Two separate metrics: data trust and node trust	VANET security	Ensure trustworthiness of data and node, cost effective in terms of comm. overhead	Misbehavior detection and trust management
GPS-based wireless hardware system [19]	Conversion of conventional vehicles into connected vehicles	VANET security	Limited speed threshold, effective in terms of safety, avoids accident	Cloud computing, advanced features needed to make the system smart and more realistic
AODV [20]	Routing protocol	VANET security	Sybil attack in VANETs	AODV with features of anti-Sybil attack
Various security threats and possible defensive mechanisms [21]	Threat investigation	VANET security	False info. dissemination, black hole attack, impersonation, man in the middle attacks, etc.	Security check when changing RSU by vehicles, secure private data like e-mail, IP address changed to pseudonym, etc.
Game theory model [22]	Reaction game mechanism	VANET security	DOS attack reaction problem	Costly and complex
Detection of intelligent malicious behavior [23]	Adaptive detection threshold	VANET security	High detection ratio, high packet delivery ratio, etc.	Investigates other adversaries, mobile certification authority
Trust [24]	Trust value assignment	VANET security	Routed the messages via trusted vehicles	Misbehavior detection and trust management
Logistic trust (LT) [25]	Authenticated node, correct event detection	VANET security	90% accurate with 2% error possibility in information	No specified attacks, that is, ballot stuffing and bad mouth attack.

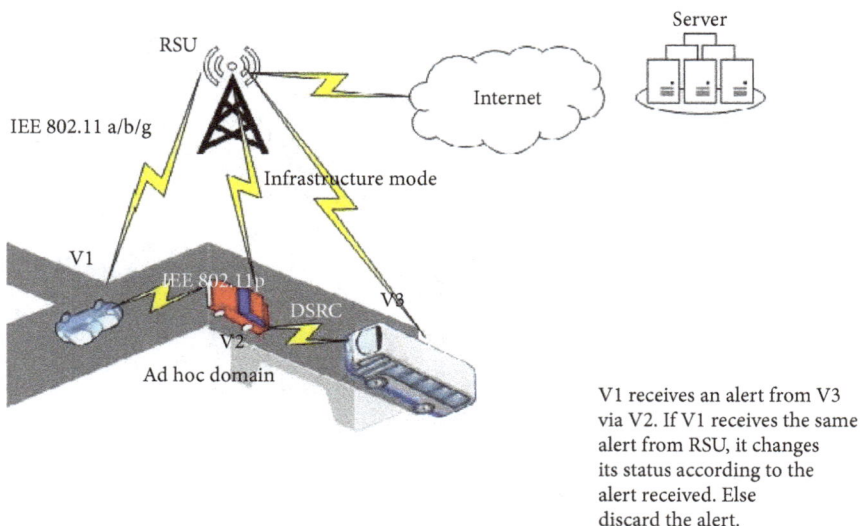

V1 receives an alert from V3
via V2. If V1 receives the same
alert from RSU, it changes
its status according to the
alert received. Else
discard the alert.

FIGURE 2: Vehicle-to-vehicle communication.

are forwarded to *ConVai* exchange for judgment, to check whether these alerts satisfy the *ConVai* exchange minimum accepted threshold value.

If it holds, it enhances V3's trustiness. If the received FSAMs satisfies the *ConVai* exchange minimum acceptable threshold value, then the exchange notifies and informs the driver, neighbor RSU, and CH if anything about FSAM's validity ensures V3 trustiness and accuracy. However, if the *ConVai* exchange threshold value does not meet the required criteria (certain mathematical value), then it alerts all RSU, CH, and vehicular nodes participating in the current communication session that the given FSAMs broadcast by V3 are invalid.

The alert is also forwarded to drivers to make them sure about the malicious node penetration. All these node CH and vehicles held for next FSAMs alert the message, and fake formulated FSAMs are moved toward the discard block. This reduces the level of V3 trustiness and enables other nodes to be aware about the falsehood of received FSAMs from V3 and ensures to remember the bad experience for a long time. Moreover, vehicle V3 is forbidden to pump any alert in the network because the system declares it invalid and a fake node. However, if node V3 is declared a true one, the experience of validity is also remembered for long time and enhances V3 trustiness in the entire network. It is a brief description of our flowchart shown in Figure 3 which made our efforts of the VANSEC model for VANET security useful. In the future much, work is also possible in the area of VANET security in routing protocols and thwarting different attack launches by attackers for their own benefits.

4.1. Algorithm of VANSEC Communication Model. The algorithm below exhibits that input nodes will broadcast or issue an alert message received by output nodes and act according to the received alert messages if found authentically verified through *ConVai* exchange.

5. Mathematical Modelling of VANSec Protocol

As mentioned earlier, the VANET system is a threat from various attacks. Here, we will study them mathematically and understand how they work. In the VANSec security model on reception of any consequences from the source node, the destination nodes have different ways of confirmation about the validity of received FSAMs. Two verification techniques are listed below.

In the first technique, the receiver node checks the status of the sender/source node and verifies the status of the received FSAM's validity. Secondly, the receiver goes through a comparison phase where the receiver relates and compares the results collected from the source node and neighbor nodes of the source; if both have the same opinion about the received FSAMs, then the sender is considered a valid/ true node. Our designed VANSec model comprises multiple events, so the occurrence of incorrect events should also be possible. The result reported from a source needs to be confirmed before exchanging information in the network. About the event accuracy, the VANSec model collects enough evidence to list the event valid/invalid and correct false information to avoid nodes from misguidance.

Our work provides a basis for all kinds of trust models, and we also used this idea in our proposed model. The accuracy of any occurred phenomenon is recorded and based on the observation of participating nodes (from event occurrence to the reported event). So a valid node forwards a valid event towards the receiving node, and with the passage of time, more nodes are also aware about the event to occur. However, the trustiness of the discussed technique may face failure when a valid node in the VANSec model furbishes invalid/fake information. To avoid fake information dissemination, a mass metric procedure (MMP) is used to confirm actual true or valid report and contradicting report. Mass is used to measure the weight of an object. For example, you are measuring the mass of your body when you step on to a

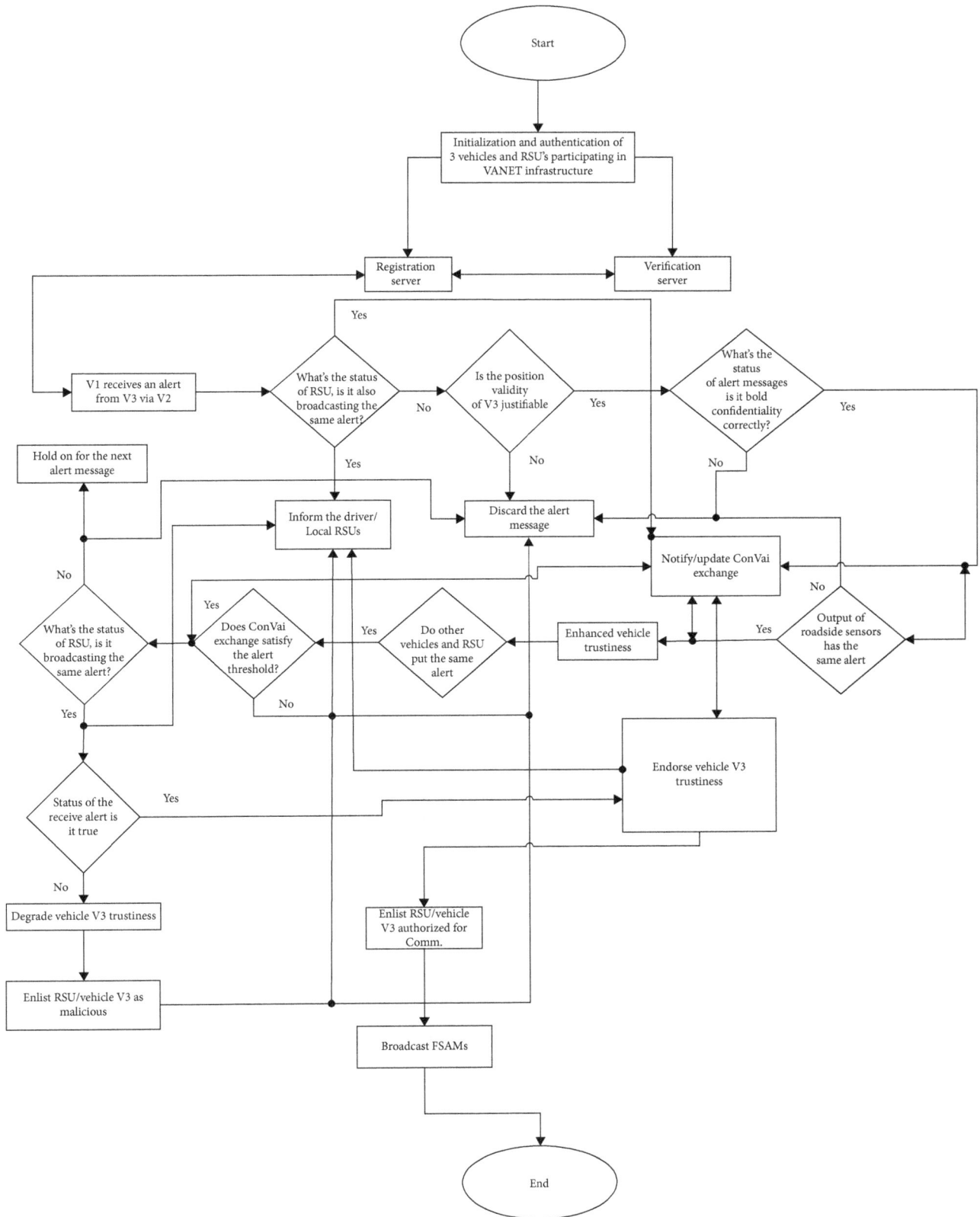

FIGURE 3: Flow chart of VANSEC security model.

```
Input: V1, V2, V3, RSUs.
Output: Only authorized node (vehicle/RSU) Broadcast information.
    1: Nodes participating in communication session are {V1, V2, V3, and RSU}
    2: V3 ← broadcasts an alert ← {V1 not in range while V2 and RSU receive the broadcast}
    3:    V1 in range of V2 ← Receives broadcast from V2
    4:    V1 ← receives an alert from RSU and Local sensors
    5:        V1 compares ← Alert {V2, RSU and Sensors}
    6:        V1← Verifies authenticity {V2, V3 and RSU}
    7:    If vehicle/RSU not registered
    8:    then
    9:    Mismatch
   10: Notify ← ConVai exchanged and verification Server {Discard the alert message}
   11: Also Notify ← V3 Trustiness degraded {V3 are black listed}
   12:    else
   13:       Match ← Registered and Authentic {V3, V2, RSU}
   14: Update ← ConVai exchange {Satisfied basic security goals}
   15: Enlist ← Endorse V3/RSU trustiness
   16: Allow ← V3 V2 and RSU
   17: Broadcast ← Alert if any
   18:    end if
```

ALGORITHM 1

scale. As we want to assign weights in affirming reports and contradicting reports, hence we have utilized this parameter. These weights are used to make a decision on relaying the messages if the equation is true. MMP is used in the decision-making process to allow the sender for communication or stop it. In (1), M_v is the valid mass metric whereas $M\neg_v$ represents contradiction.

$$\frac{M_v}{M_v + (M\neg_v)} < 1 - \xi. \tag{1}$$

The system took the source and event reporter's own confidence in the report received and then followed the received report for further action. Further, neighbor validity is also updated after looking over the results of the taken decision. If (1) becomes true, then the node is allowed for communication. However, this approach is unsafe in live safety applications, where dissemination of invalid activity may be disastrous. So, it is important to understand the nature of the reporter node well before making any decision. In trust-based approaches, the node-computed trust is a function of their own observations and opinion of neighbor nodes.

Our scheme VANSec is more immune and resistive against different kinds of attacks and thwarts malicious node penetration attempts to the entire network. It is basically based on trust management approach. The aim of the scheme is to identify malicious data and false nodes. In the designed scheme, at the beginning the node has information about the network behavior and nature. It investigates event accuracy from information received or from its own analysis. In the VANSec model, when a node undergoes unusual changes, it forwards these changes to surrounding nodes through a broadcast message and alerts nodes to switch into a safe

mode. It is also possible that malicious nodes misguide other nodes through falsified FSAMs and drive the network for its benefits.

In the VANSec model, any node that receives FSAMs goes into a verification phase to understand the nature of information received before taking any decision. Therefore, a process is required to judge the correctness of received information, while the destination node holds a series of consequences achieved from received information and sender to verify the message's validity. Before any judgment about the accuracy of received information from the sender, a trust/confidence value for that sender's authenticity is established. The confidence value for the Sth sender at time interval n can be written as $C_S(n)$ where, for the message correctness about a consequence verification, the mass metric is used shown in (1), where $C_S(n)$ comes true if it follows (2).

$$0 \leq Cs(n) \leq 1. \tag{2}$$

The node has two containers: information containing a consequence is marked as P-container and is also represented by a binary digit 1, and a bin with no consequence is marked with NP-container in which also a binary digit 0 is lap to the NP-container. Average confidence values are computed from these containers utilizing sender confidence. Suppose P-container has S sender and NP-container has Q sender, the average confidence of each container at time interval n is given as

$$C_1(n) = \sum_{i=1}^{S} \frac{C_i}{S},$$

$$C_o(n) = \sum_{j=1}^{Q} \frac{C_j}{Q}, \tag{3}$$

where $C_1(n)$ is the average confidence of an event and $C_0(n)$ is the average confidence of no consequence. The normalized confidence of the node from each pot is called the mass metric of the given container which is shown in (4), where $m_i(n)$ is the mass metric of the ith node for bin 1 and $m_j(n)$ for the jth node for pot 0.

$$m_i(n) = \frac{c_i(n)}{C_1(n)},$$
$$m_j(n) = \frac{c_j(n)}{C_o(n)}. \tag{4}$$

When a node confirms a consequence in its previous report and later it cannot deny from its previously submitted report; similarly if the node denies a consequence, then one cannot confirm the same event; hence, masquerading is not allowed then. The mass metric confidence for each pot is computed to judge whether the consequence is true based on information received. The average mass metric confidence for these pots is given below:

$$C^1_{\text{avg}}(n) = \sum \frac{m_i(n) * c_i(n)}{S},$$
$$C^0_{\text{avg}}(n) = \sum \frac{m_j(n) * c_j(n)}{Q}. \tag{5}$$

In the decision-making process in VANSEC, the node utilizes the average mass metric confidence value to determine whether the consequence occurred or not. Hence, authentic source notifies an accurate consequence which does not threaten the decision. From (5), it is clear that $C^1_{\text{avg}}(n) - C^0_{\text{avg}}(n) > 0$ $Q > 0$, while $C^1_{\text{avg}}(n) > (n) > \min$ accepted M_t when $Q = 0$ where $0 < C_{\text{avg}}(.) < 1$ and $0 < \min$ accepted $M_t < 1$ which are the decision-making rules.

Any observation violating from the true consequence is considered malicious or eccentric; otherwise, it will be a genuine analysis. Our VANSEC model collects evaluator/judge and neighbor responses and also enlists misbehavior activity for a long time.

5.1. Evaluator/Judge Response. Evaluator/judge response is the response of a specific evaluator with a given sending source. Evaluator response is expressed in terms of eccentric ratio (ER), which evaluates whether the sender is malicious or honest. ER is defined as follows and is represented by $\Omega_n(s)$ where s is the sender/generator of packets in time interval n. $f_n(s)$ is the incorrect packet and $w_n(s)$ is the total number of packets generated by the source.

$$\text{ER} = \frac{\text{Modified or incorrect packets}}{\text{Total packets generated by source}} \Rightarrow \Omega_n(s) = \frac{f_n(s)}{W_n(s)}. \tag{6}$$

If the ER is analyzed and it crosses a particular threshold, say "Ψ," then a flag raises up indicating the source as a malicious entity and activity related to that node which is marked as untrue or false. If the ER value remains below the threshold "Ψ," then it fails in detection of malicious nodes. ER is

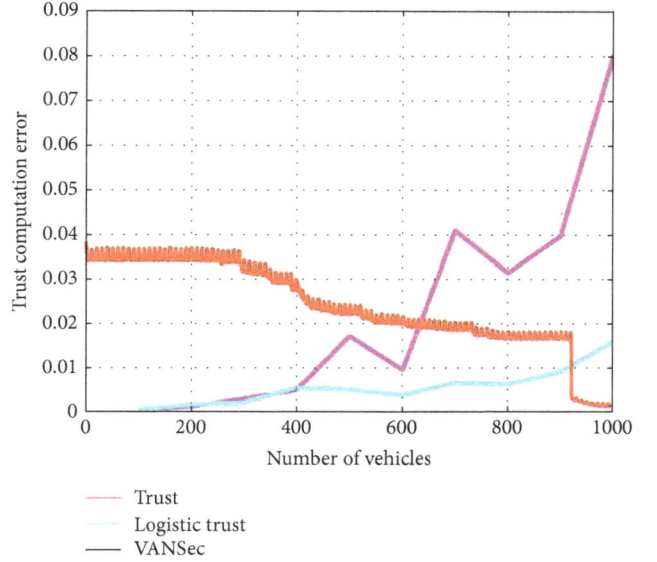

FIGURE 4: Trust computation error versus number of vehicles.

also a decision-making ratio. To identify the node's true nature, an average ER is used.

$$\overline{\Omega n(s)} = \frac{1}{n} \sum_{i=1}^{n} \frac{f_i(s)}{W_i(s)}. \tag{7}$$

Any node willing to establish a communication session with the source computes the ER first, so in the future the node easily broadcasts a list of trusted honest nodes and malicious nodes to its neighbor nodes. On reception of the list, the destination node updates its list of honest and malicious entries. Let us suppose that x is the node recognition entity, then λ_x is the number of malicious values received about x where H_x is the honest count. The receiving node establishes two parameters τ_x and π_x expressed with a relation given in (8). To estimate value for node x, these parameters are used.

$$\tau_x = H_x + 1, \pi_x = \lambda_x + 1. \tag{8}$$

The reputation tally for node x is

$$\text{Best tally}_X = \frac{\tau_x}{\tau_{x+\pi_x}}, \tag{9a}$$

$$\text{Worst tally}_X = \frac{\pi_x}{\tau_{x+\pi_x}}. \tag{9b}$$

From (9a) and (9b) Best Tally for x can be computed as

$$\text{Best tally}_X = 1 - \text{worst tally}_X. \tag{10}$$

5.2. Neighbor Response. Neighbor response is a response faced by neighbor nodes of a particular sender. It is an expectation obtained from neighbor nodes of a given sender in terms of binary values (0, 1). When the response of the neighbor node is binary digit 0, it means that the specific sender is malicious, while if it is one, then it points an honest node, where $Z_n(s)$ is the total number of received zeros and $O_n(s)$

TABLE 3: Trust computation error per 200 vehicles.

Protocol	200	400	600	800	1000	Average	% improvement
VANSec	0.03453	0.0280	0.0210	0.0166	0.00194	0.02041	4.463
Trust	0.001757	0.00719	0.0270	0.0718	0.0894	0.03942	7.30
L. trust	0.001757	0.00445	0.00546	0.00890	0.0119	0.0054	1.00

is the total number of received ones. Then, the neighbor response $N_n(s)$ is calculated as

$$N_n(s) = \frac{Z_n(s) + 1}{Z_n(s) + O_n(s) + 2}. \tag{11}$$

If there is no advice received about the sender and neighbor nodes, then the neighbor response $N_n(s)$ will assume a value of 0.5. Our proposed VANSec model uses characteristic confidence (CC) to filter out incorrect advices. CC uses the idea of resemblance and coherency or uniformity of advices for specific neighbors, where resemblance $R_n(L)$ is calculated between evaluator (J) and sender (L) of the data. Resemblance is calculated using the Jaccard similarity (JS) tally or score [26].

$$R_n(L) = \frac{1}{s} \sum_{g=1}^{s} \frac{A_L \cap A_{yg}}{A_L \cup A_{yg}} \tag{12}$$

Let A_L be the L sender's advice where evaluator J has its own analysis. Other advices from geographically closed nodes $y_1 \dots y_d$ are $A_{y1} \dots A_{yd}$ used to compare and calculate JS tally. In order to analyze the behavior of L, a time average of resemblance tally (score) is computed which is

$$\text{Res}(n)[J, L] = \epsilon * R_n(L) + (1 - \epsilon) * R_{n-1}(L). \tag{13}$$

The CC in the VANSEC model also takes uniformity of advices for the current source. Suppose $I_n(L)$ is the advice value for sender L at time slot n, then $I_{n-1}(L)$ will be adviced at time $n - 1$ for that sender. So the total value of advice or recommendation from L at time n will be $\|AL\|$.

Hence, uniformity will be expressed through (14):

$$\beta_n(L) = \sum_{i}^{A_L} \frac{I_n(L) \oplus I_{n-1}(L)^{\circ}}{\|A_L\|}. \tag{14}$$

To establish characteristic confidence for the VANSEC model, the time average of uniformity/coherency is used which is calculated below:

$$\overline{\beta}_n(J, L) = \varphi * \beta_n(L) + (1 - \varphi)\beta_{n-1}(L). \tag{15}$$

Equation (15) shows average uniformity calculated among the evaluator (J) and source. Now combining both resemblance and uniformity to establish a CC for the VANSec model,

$$CC_n(J, L) = \theta_1 * R_{es(n)}[J, L] - \theta_2 * \overline{\beta}_n(J, L) +_3 * R_{es(n-1)}[J, L]. \tag{16}$$

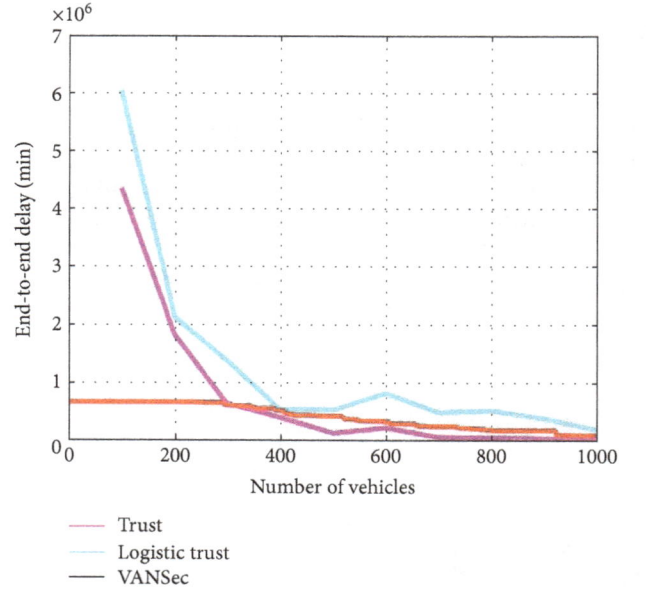

FIGURE 5: End-to-end delay (minutes) versus number of vehicles.

So from (16), we can easily calculate CC for a specific node and judge the nature of advice or recommendation reliability. In the VANSec model, if the value of CC falls below a particular threshold, say Υ, then the advice value for that specific node is filtered to be utilized in complete confidence trust estimation. After filtering out false untrue advices, the neighbor L and evaluator J responses along with the fanion (flag) $P_n(s)$ are collected in the VANSec model

$$t_n(J, L) = \frac{1}{1 + e^{L.\vec{c}.C_o}}, \tag{17}$$

where $L = [\Omega_n(s), C_S(n), P_n(s), t_{n-1}(J, S)]$, $\vec{c}$ is the mass metric associated with each of the abovementioned parameters, and C_o is the bias and is chosen on the basis of initial confidence trust of the nodes. If initial trust assigned to a node is 0.3, then C_o will be approximately -0.85. Once the value of $\vec{c}$ is found, then a new confidence value should be computed and updated using (17). If the new calculated confidence trust value falls below threshold Δ, then the node is considered malicious, and fanion $P_n(s)$ is raised; similarly, if the confidence/trust is above threshold Δ_L, then the node is marked as a true one and $P_n(s)$ goes down.

TABLE 4: End-to-end delay (minutes) per 200 vehicles.

Protocol	200 v	400 v	600 v	800 v	1000 v	Average	% improvement
VANSEC	0.67	0.55	0.39	0.29	0.17	0.414	1.00
Trust	2.8	0.31	0.09	0.031	0.031	0.6524	1.576
L. trust	2.34	0.75	0.52	0.473	0.274	0.8714	2.104

The performance rate will be identified in terms of valid optimistic rate (VOR), invalid optimistic rate (IOR), and consequence detection probability (CDP). VOR is the probability of identifying an invalid/false node as invalid or untrue, while IOR is the probability of pointing an honest node as a malicious node. CDP is the probability of identifying the true result.

VOR is mathematically shown in the following equation:

$$VOR = \frac{P(I/I)}{P(I/I) + P(H/I)}. \tag{18}$$

where $P(I/I)$ is the probability of identifying a node as an intruder such that the given node is also intruding or malicious, while $P(H/I)$ identifies an intruder or false malicious node as a true or valid node. Similarly, IOR can be calculated via using the following relation.

$$IOR = \frac{P(I/H)}{P(H/H) + P(I/H)}. \tag{19}$$

6. Simulation, Results, and Discussion

We compared our scheme to the present and tested schemes in terms of different performance metrics like TCE, EED, ALD, and NRO. To verify our VANSEC scheme to be efficient than the existing techniques, a comparison is done using simulation.

6.1. Trust Computation Error. Trust computation error (TCE) is the mean square error between the predicted/calculated and known/observed or actual trust value assessment of the vehicles. TCE can also be found through tracking the root mean square (RMS) of the calculated trust computed for all nodes. Figure 4 shows the execution of VANSec technique which is most favorable and has optimal performance than the trust scheme with logistic trust (LT). Keeping in view Table 3, VANSEC has consistency among the values of TCE with an increase of 200 vehicles in each step, while in case of trust and logistic trust techniques there is no such consistency among the values of TCE with 200 vehicles per step increase recorded.

Moreover, the TCE contributes to an interpretation that trust estimation in VANSec is more active, precise, and authentic, while in case of trust and logistic trust techniques, TCE values are not so active to properly handle altered data by misbehavior node data size, which may be a possibly malicious vehicle forwarding fake information to the destination vehicle. From Table 3, our proposed methodology of VANSec shows that our scheme is 11.6% and 7.3% more efficient in terms of TCE than the LT and trust schemes are,

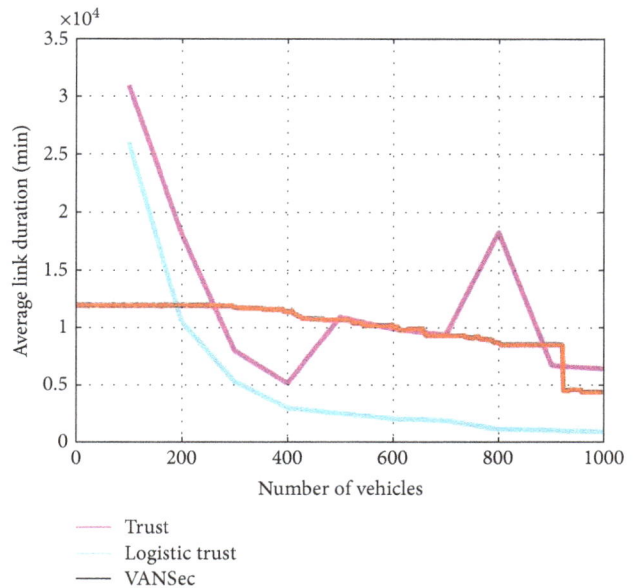

FIGURE 6: Average link duration (minutes) versus number of vehicles.

respectively, while the trust scheme is 4.3% more efficient in terms of TCE than LT. The enhancement in the performance of VANSec is due to the fact that our model calculates trust for all nodes randomly and identifies malicious node from their negative feedback. VANSec performed well in the presence of a huge number of false or malicious node concentrations. The reason for the good performance in a malicious environment is the feature of feedback metric credibility in the VANSec algorithm.

6.2. End-to-End Delay. The time taken by FSAMs to travel in a VANETs/VANSec model from the source vehicle to the destination vehicle is called end-to-end delay. Due to high mobility scenarios in VANSec, on-time delivery of FSAMs may be delayed. To prevent latency in packet delivery, delay-tolerant networks (DTNs) are favored to be used, in order to minimize end-to-end delay in VANETs. Figure 5 depicts that the performance of the VANSec algorithm is better than trust and logistic trust techniques. Figure 5 and Table 4 show close consistency along with an increase of 200 vehicles in each step. The table values for VANSec with the increase in number of vehicles also depict a consistent reduction in packet end-to-end latency.

Such coherent gradual reduction in end-to-end delay declares VANSec more logical than trust and logistic trust approaches. It is clearly depicted from the table that there is

TABLE 5: Average link duration (minutes) per 200 vehicles.

Protocol	200 v	400 v	600 v	800 v	1000 v	Average	% improvement
VANSEC	1.20	1.139	1.008	0.8775	0.4438	0.9337	2.594
Trust	1.06	0.3062	0.2139	0.1217	0.0978	0.3599	1.00
L. trust	1.824	0.5221	0.9957	1.833	0.6466	1.1642	3.234

no such consistency in the values of EED which are recorded with the increase in number of vehicles. From Table 4, it is concluded that average EED delay in the case of VANSec technique is approximately 0%. The trust and LT schemes face 57.6% and 5.2% longer delay, respectively, than the VANSec algorithm does, whereas the trust scheme has 52.4% more EED than the LT scheme does.

So, from these simulation results, the VANSec algorithm has enormous performance rather than trust and LT techniques. VANSec's outperformance than the rest of the two algorithms is due to the fact that our scheme considerably needed less information about the network behavior and route discovery process, which remarkably reduced the network overhead and suggested best for dynamic and ascendable networks.

6.3. Average Link Duration. Average link duration is the communication link lifespan estimation established among source and destination vehicles to exchange FSAMs. In VANETs, a path choice is an important parameter for good performance and better data rate. But in VANETs, link duration depends on various parameters like transmission range of the vehicle, intervehicle distance, vehicle density, and vehicle velocity which made link duration stability a challenging job. We used average link duration because link duration depends on the verity of parameters.

Figure 6 depicts our scheme VANSec to be more stable and reliable. Also, Table 5 reveals that our proposed scheme has stable link duration. For each step, there is a uniform increase in number of vehicles of 200 vehicles per step. From Table 5, we concluded that our designed VANSec technique provides 29.7% and 7.8% more reliable and stable ALD than trust and LT techniques, respectively. However, LT ALD is 21.9% more than the trust algorithm. So, an increase in the number of vehicle VANSec preserves link stability and very little gradual change noticed in the average link values. It means that ALD in the VANSec scheme is more reliable and stable.

However, the remaining schemes trust and LT undergo sudden change in ALD with increase in vehicle density and small consistency which are observed in ALD values. So comparison results show that our proposed VANSec scheme has better efficiency in terms of average like duration, and very little packets are lost. Such ambiguity in our VANSec protocol's better efficiency is that our algorithm chooses and prefers more stable and reliable routes/links among nodes for data transmission which has high link stability timing interval.

6.4. Normalizing Routing Overhead. Normalized routing overhead is a ratio of transmitted routing packets divided

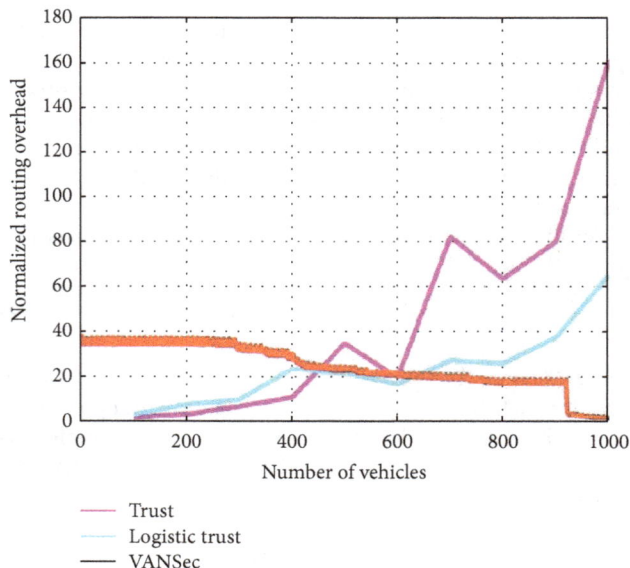

FIGURE 7: Normalized routing overhead versus vehicle density.

by the number of data packets delivered at the destination node. Figure 7 depicts an overhead returned by VANSec, trust, and logistic trust. Effects of overhead of these schemes are shown with the increase in vehicle density, respectively, depicted in Figure 7 and Table 6.

In Figure 7, the vehicle's density is adjusted at 1000 vehicles. We notice that our algorithm VANSEC has significant reduction in load with an increase in number of vehicles. In the VANSec scheme, overhead/load gradually reduces with the increase in vehicle density, while in other two algorithms, overhead enormously increases with increase in number of vehicles.

From Figure 7 and Table 6, it is concluded that overhead recorded in VANSec is nearly 0%, while trust and LT schemes in comparison with VANSec face 27.5% and 14% more NRO overhead, respectively; also, trust has 13.5% more load or NRO than LT does. So, in conclusion, VANSec is 27.5% and 14% more efficient than trust and LT protocols, respectively.

The particular improvement in our scheme is due to the fact that our designed scheme considerably reduces route request (RReq) query to conceive routes and choose the most stable and reliable route for transmission of data packets. This results in minimal route failure and considerably small number of control messages; that is, overhead is required to detect a route for information exchange. Table 6 shows a gradual reduction in NRO values with 200 increase in vehicle

TABLE 6: Normalized routing overhead per 200 vehicles.

Protocol	200 v	400 v	600 v	800 v	1000 v	Average	% improvement
VANSec	34.53	28.01	21.07	16.63	1.94	20.5	1.00
Trust	1.781	14.39	54.07	143.6	178.8	78.53	3.83
L. trust	7.028	17.8	21.85	35.61	47.7	26.1	1.28

density per step, while trust and LT procedures favor sudden change in NRO with 200 increase in vehicle density per step. From this analysis, our scheme outperforms the rest of the two schemes.

7. Conclusion

VANET is a subclass and an application of MANETs. Early VANET networks were a car-to-car (C2C) communication network basically designed for data exchange among vehicles. Later on, the feature of vehicles to roadside infrastructure were also added to the VANET network to make the system more efficient for data exchange to ensure safety of vehicles and humans and avoid unpleasant situations. VANET is a building key block of the ITS framework also known as intelligent transportation networks (ITNs). VANET is basically a design for the disseminations of cooperative awareness messages (CAMs) in the network for long distances among the vehicles and RSUs in range. For V2V communication, the IEEE 1609 WAVE protocol stack was designed on IEEE 802.11p WLAN standard utilizing a frequency band of 5.9 GHz for DSRC. Researchers proposed the verity of routing schemes aiming at enhancing the performance of vehicle information interchange among source and destination vehicles in the VANET system by taking into account various performance parameters. From comparison of different routing algorithms, we demonstrated that if a scheme is better in one response, it faces certain challenges in another response.

To avoid hazardous circumstances, FSAMs or any other emergency messages required priority based on time dissemination among vehicular nodes and roadside infrastructure and assurance of its flawless delivery at a receiving node is a most critical task. In case of such critical situation link failure occurs, the packets of FSAMs may face delay and once can face the worst tragic situation in sense of loss of precious lives and property.

In our research work, we have studied a variety of routing techniques including but one an analysis of designed technique VANSec, with already existing techniques trust and logistic trust in terms of different performance metrics like TCE, ALD, EED, and NRO with respect to an increase in vehicle density. VANSec is compared with trust and LT schemes because the modelling done in our scheme and the parameters considered closely match with the environment catered in those schemes along with the same parameters taken. In terms of performance metric TCE, VANSec is 11.6% and 7.3% efficient than LT and Trust are, respectively, while the trust scheme is 4.3% efficient than LT. From the EED comparison, we found VANSec to be 57.6% more efficient than trust and 5.2% than LT; also, trust schemes faced

52.4% more delay than LT did. Similarly, in terms of ALD, VANSec provides 29.7% and 7.8% more stable link duration than trust and LT did; however, LT is 21.9% more efficient ALD than trust. In terms of NRO, our proposed VANSec protocol has 27.5% and 14% lesser load than trust and LT, while trust has approximately 13% more NRO than LT. From these observations, we concluded that performance of our designed scheme in terms of these parameters is more valuable and authentic than the trust and LT algorithms. Our research shows that the VANSec scheme has better stability period, less latency, and improved data rate over trust and LT schemes.

Disclosure

This work is an extension of our already published paper in conference "ARV2V: Attack Resistant Vehicle to Vehicle Algorithm for Trust Computation Error in VANETs."

Conflicts of Interest

The authors declare that they have no conflicts of interest.

References

[1] D. Puccinelli and M. Haenggi, "Wireless sensor networks: applications and challenges of ubiquitous sensing," *IEEE Circuits and Systems Magazine*, vol. 5, no. 3, pp. 19–31, 2005.

[2] M. A. Matin and M. M. Islam, "Overview of wireless sensor network," in *Wireless Sensor Networks - Technology and Protocols*, IntechOpen, 2012.

[3] O. Altintas, F. Dressler, F. Hagenauer, M. Matsumoto, M. Sepulcre, and C. Sommery, "Making cars a main ICT resource in smart cities," in *2015 IEEE Conference on Computer Communications Workshops (INFOCOM WKSHPS)*, pp. 582–587, Hong Kong, 2015, IEEE.

[4] D. Patel, M. Faisal, P. Batavia, S. Makhija, and M. Mani, "Overview of routing protocols in VANET," *International Journal of Computer Applications*, vol. 136, no. 9, pp. 4–7, 2016.

[5] V. Duduku, V. Ali Chekima, F. Wong, and J. A. Dargham, "A survey on routing protocols in vehicular ad hoc networks," *International Journal of Innovative Research in Computer and Communication Engineering*, vol. 3, no. 12, 2015.

[6] M. J. Piran, G. Rama Murthy, G. Praveen Babu, and E. Ahvar, "Total GPS-free localization protocol for vehicular ad hoc and sensor networks (VASNET)," in *2011 Third International*

Conference on Computational Intelligence, Modelling & Simulation, pp. 388–393, Langkawi, Malaysia, 2011, IEEE.

[7] M. N. Rajkumar, M. Nithya, and P. HemaLatha, "Overview of VANETs with its features and security attacks," *International Research Journal of Engineering and Technology*, vol. 3, no. 1, 2016.

[8] A. Luckshetty, S. Dontal, S. Tangade, and S. S. Manvi, "A survey: comparative study of applications, attacks, security and privacy in VANETs," in *2016 International Conference on Communication and Signal Processing (ICCSP)*, pp. 1594–1598, Melmaruvathur, India, 2016, IEEE.

[9] R. Barskar and M. Chawla, "Vehicular ad hoc networks and its applications in diversified fields," *International Journal of Computer Applications*, vol. 123, no. 10, pp. 7–11, 2015.

[10] A. Jain and D. Sharma, "Approaches to reduce the impact of DOS and DDOS attacks in VANET," *International Journal of Computer Science*, vol. 4, no. 4, 2016.

[11] A. Suman and C. Kumar, "A behavioral study of Sybil attack on vehicular network," in *2016 3rd International Conference on Recent Advances in Information Technology (RAIT)*, pp. 56–60, Dhanbad, India, 2016, IEEE.

[12] R. Boon, *Post-Accident Analysis of Digital Sources for Traffic Accidents*, University of Twente, Enschede, Netherlands, 2014.

[13] S. Gupte and M. Younis, "Vehicular networking for intelligent and autonomous traffic management," in *2012 IEEE International Conference on Communications (ICC)*, pp. 5306–5310, Ottawa, ON, Canada, 2012, IEEE.

[14] C.-T. Li, M.-S. Hwang, and Y.-P. Chu, "A secure and efficient communication scheme with authenticated key establishment and privacy preserving for vehicular ad hoc networks," *Computer Communications*, vol. 31, no. 12, pp. 2803–2814, 2008.

[15] H. Liu, Y. Chen, H. Tian, T. Wang, and Y. Cai, "A novel secure message delivery and authentication method for vehicular ad hoc networks," in *2016 First IEEE International Conference on Computer Communication and the Internet (ICCCI)*, pp. 135–139, Wuhan, China, 2016, IEEE.

[16] P. Wararkar and S. S. Dorle, "Transportation security through inter vehicular ad-hoc networks (VANETs) handovers using RF trans receiver," in *2016 IEEE Students' Conference on Electrical, Electronics and Computer Science (SCEECS)*, pp. 1–6, Bhopal, India, 2016, IEEE.

[17] M. N. Mejri and J. Ben-Othman, "GDVAN: a new greedy behavior attack detection algorithm for VANETs," *IEEE Transactions on Mobile Computing*, vol. 16, no. 3, pp. 759–771, 2017.

[18] W. Li and H. Song, "ART: an attack-resistant trust management scheme for securing vehicular ad hoc networks," *IEEE Transactions on Intelligent Transportation Systems*, vol. 17, no. 4, pp. 960–969, 2016.

[19] Y. Agarwal, K. Jain, and O. Karabasoglu, "Turning conventional vehicles in secured areas into connected vehicles for safety applications," in *2016 IEEE Information Technology, Networking, Electronic and Automation Control Conference*, pp. 538–542, Chongqing, China, 2016, IEEE.

[20] Mujeeb Ur Rehman, Sheeraz Ahmed, Sarmad Ullah Khan, Shabana Begum, and Atif Ishtiaq, "ARV2V: Attack resistant vehicle to vehicle algorithm, performance in term of end-to-end delay and trust computation error in VANETs," in *2018 International Conference on Computing, Mathematics and Engineering Technologies (iCoMET)*, pp. 1–6, Sukkur, Pakistan, 2018, IEEE.

[21] A. S. Al Hasan, M. Shohrab Hossain, and M. Atiquzzaman, "Security threats in vehicular ad hoc networks," in *2016 International Conference on Advances in Computing, Communications and Informatics (ICACCI)*, pp. 404–411, Jaipur, India, 2016, IEEE.

[22] M. N. Mejri, N. Achir, and M. Hamdi, "A new security games based reaction algorithm against DOS attacks in VANETs," in *2016 13th IEEE Annual Consumer Communications & Networking Conference (CCNC)*, pp. 837–840, Las Vegas, NV, USA, 2016, IEEE.

[23] C. A. Kerrache, A. Lakas, and N. Lagraa, "Detection of intelligent malicious and selfish nodes in VANET using threshold adaptive control," in *2016 5th International Conference on Electronic Devices, Systems and Applications (ICEDSA)*, pp. 1–4, Ras Al Khaimah, UAE, 2016, IEEE.

[24] K. Rostamzadeh, H. Nicanfar, S. Gopalakrishnan, and V. C. M. Leung, "A context-aware trust-based communication framework for VNets," in *2014 IEEE Wireless Communications and Networking Conference (WCNC)*, pp. 3296–3301, Istanbul, Turkey, 2014.

[25] S. Ahmed and K. Tepe, "Misbehaviour detection in vehicular networks using logistic trust," in *2016 IEEE Wireless Communications and Networking Conference*, pp. 1–6, Doha, Qatar, 2016, IEEE.

[26] S. Ahmed and K. Tepe, "Evaluating trust models for improved event learning in VANETs," in *2017 IEEE 30th Canadian Conference on Electrical and Computer Engineering (CCECE)*, pp. 1–4, Windsor, ON, Canada, 2017, IEEE.

9

A Novel Non-Line-of-Sight Indoor Localization Method for Wireless Sensor Networks

Yan Wang⑩, Xuehan Wu, and Long Cheng⑩

Department of Computer and Communication Engineering, Northeastern University, Qinhuangdao 066004, China

Correspondence should be addressed to Yan Wang; ywang8510@gmail.com and Long Cheng; chenglong8501@gmail.com

Academic Editor: Grigore Stamatescu

The localization technology is the essential requirement of constructing a smart building and smart city. It is one of the most important technologies for wireless sensor networks (WSNs). However, when WSNs are deployed in harsh indoor environments, obstacles can result in non-line-of-sight (NLOS) propagation. In addition, NLOS propagation can seriously reduce localization accuracy. In this paper, we propose a NLOS localization method based on residual analysis to reduce the influence of NLOS error. The time of arrival (TOA) measurement model is used to estimate the distance. Then, the NLOS measurement is identified through the residual analysis method. Finally, this paper uses the LOS measurements to establish the localization objective function and proposes the particle swarm optimization with a constriction factor (PSO-C) method to compute the position of an unknown node. Simulation results show that the proposed method not only effectively identifies the LOS/NLOS propagation condition but also reduces the influence of NLOS error.

1. Introduction

The rapid development of microelectromechanical system (MEMS) technology, sensor technology, wireless communication, and low-power embedded technology promotes the progress and development of wireless sensor networks (WSNs). WSNs consist of a large number of inexpensive microsensor nodes deployed in a monitored region. The sensor nodes are connected to each other by self-organization and multihop communications [1]. Sensor nodes consist of sensors, digital processing units, a wireless communication module, and a power module. They can collaboratively sense, gather, and process the information of the perceived objects in a monitored region and then send the information to the sink node. WSNs are widely used in traffic management, environmental monitoring, medical care networks, logistics management, and other fields and profoundly influence the social life of people [2].

One of the most important issues for WSNs is localization technology [3]. The localization technology is the essential requirement of constructing a smart building and smart city. WSN-based localization methods can be categorized as range-based localization methods and range-free localization methods. In range-based localization methods, different measurement techniques for localization can be classified as time of arrival (TOA), time difference on arrival (TDOA), received signal strength (RSS), and angle of arrival (AOA). The range-free localization methods do not need to measure the distance or angle between the nodes [4]. These methods can estimate position based on the network connectivity and the distribution of the history measurements. The range-free localization methods can be divided into multihop estimation-based localization and pattern matching-based localization.

For the TOA-based localization method, the signal velocity is known in advance. It measures the travel time of the signal from the beacon node to the unknown node, and the distance between two nodes is equal to the product of the signal velocity and the travel time. However, this method requires high-precision time synchronization between two nodes. As light synchronization error can significantly affect the ranging error. Therefore, the TOA method requires additional hardware to ensure the time synchronization. The TDOA method requires two different transceivers on a node so that the node can transmit two signals with different velocities at the same time. It estimates the distance by measuring the two signals' arrival time difference between the beacon node and the unknown node. The requirement of time accuracy of the TDOA method is lower than the TOA method, but it still has high requirements for hardware. The RSS method is one of the least expensive ways to locate an unknown node because it does not need additional hardware. The RSS method measures the signal power loss value from a beacon node to an unknown node, and it converts the power loss value to the distance through a signal propagation model. The AOA method measures and calculates the angles between beacon nodes and an unknown node and then estimates the position of the unknown node based on the angle between two nodes.

In this paper, we investigate the TOA-based localization method in an indoor environment. Obstacles can result in NLOS propagation in harsh indoor environments, and the accuracy of localization will drop sharply. We first propose an NLOS identification method based on residual analysis. The propagation condition can be identified by it. Then, the localization objective function is established using the LOS measurements. In addition, the particle swarm optimization with a constriction factor method is proposed to find the optimal solution of the localization function. The optimal solution is the estimated position of the unknown node. The main contributions of this paper are given as follows:

(1) The NLOS identification method does not need prior knowledge of the NLOS error. In addition, it can identify the NLOS measurements when the number of LOS measurements is larger than the number of NLOS measurements.

(2) The proposed NLOS correction method can mitigate the effect of the NLOS error.

(3) The proposed method not only uses TOA measurements but also uses other signal features such as TDOA and RSS easily. Therefore, it is not constrained by different physical measurement techniques.

The rest of the paper is organized as follows. Section 2 analyzes the NLOS localization technology for WSNs. Section 3 introduces the proposed NLOS identification methods based on residual analysis and a localization method based on an intelligent optimization algorithm. In Section 4, the simulation results of the proposed algorithm are presented, and the performance of the proposed algorithm is analyzed. The conclusions are presented in Section 5.

2. Related Work

Compared with traditional positioning systems, WSN-based localization systems can be quickly deployed and can adapt to various harsh environmental conditions. They have the characteristics of low power consumption, low cost, and strong expansibility. In addition, the Global Positioning System (GPS) technique, which is widely used at present, has the characteristics of high energy consumption, high cost, and large volume compared with WSNs [5]. Thus, WSN-based localization systems have broad application prospects, and they can be used in environmental monitoring, medical care networks, military applications, target tracking, intelligent traffic management, and other fields. The development of WSN-based localization technology has promoted an industrial revolution that influences the social life of people.

Because the WSN localization technology has remarkable superiority, both researchers and designers are paying more attention to it and devoting more effort to improving the positioning accuracy. In [6], a residual test method is proposed to determine the number of LOS and identify the propagation condition synchronously. This method can identify the NLOS with high accuracy. In [7], the authors proposed a routing algorithm that is widely used in centralized range-based localization schemes. Experimental results show that the algorithm provides distance estimates with low estimation error. However, the algorithm requires a large amount of calculation. A novel localization algorithm based on an approximate convex decomposition (ACDL) is proposed [8]. It relies only on network connectivity information. The hop count distance between nodes can provide a good approximation of the Euclidean distance. In [9], the authors design a localization method with outlier detection, and the ranges with large errors can be eliminated explicitly before computing the location. However, the method must define verifiable graphs in which all edges should be verifiable. To obtain a low complexity, the authors proposed a modification of the gradient descent method and an accurate multilateration localization algorithm for wireless sensor networks [10]. Only when using the RSSI to estimate the distances between nodes, the proposed algorithm can obtain better convergence properties and a lower computational load in the presence of significant range error.

NLOS propagation is ubiquitous in practical indoor environments. NLOS propagation will contribute a positive additional excessive delay to the measured value. NLOS error is the main source of the localization error. To improve the positioning accuracy in practical conditions, NLOS identification and mitigation methods are widely investigated. A residual weighting algorithm (Rwgh) is proposed in [11]. The sum of squared residuals of a least squares estimation is used as the indicator to show the accuracy of the calculated node coordinates. Least squares multipoint location is applied on all possible combinations of the distance measurements. Then, the authors compute the estimated location and used it as a weighted combination of these intermediate estimates. The RANSAC algorithm is an iterative method to estimate the position from a set of measurements that contains NLOS error [12]. A reasonable result is produced only with a certain probability, so RANSAC is a nondeterministic algorithm in this sense. The probability can be increased as more iterations are allowed. In [13], the authors proposed a distributed multiple-model estimator for simultaneous localization and tracking (SLAT) with NLOS mitigation. The difficulties of exponentially growing terms for centralized multiple-model estimation can be overcome if the fusion is carried out in a distributed manner. An NLOS mitigation technique based on convex SDP optimization is proposed in [14]. Especially in severe NLOS environments, the proposed SDP estimator outperforms the other algorithms substantially. In [15], a novel algorithm is presented by the authors to solve NLOS propagation. The algorithm depends only on the features extracted from the received waveform. In addition, there is no need to formulate an explicit statistical model for the features.

3. System and Range Measurement Model Description

In this section, we first analyze the TOA measurement model in LOS and NLOS propagation conditions, respectively. Then, we propose an NLOS identification method based on residual analysis, according to the characteristics of the NLOS error. Finally, we improve the existing NLOS localization method by using particle swarm optimization with a constriction factor.

3.1. TOA Measurement Model. The TOA method measures the travel time of a signal from the beacon node to the unknown node. The true distance of TOA is modeled as follows:

$$d = c \cdot t, \tag{1}$$

where c is the speed of the signal, d is the distance between the two nodes, and t is the travel time of the signal between the two nodes.

Because the travel time t cannot be completely synchronous in LOS propagation conditions, it consists of measurement error. The time estimation of TOA is as follows [16]:

$$\widehat{t} = t + n_{it}, \tag{2}$$

where t is the true travel time of the signal between the two nodes; n_{it} is the measurement error modeled as a zero-mean white Gaussian process with variance σ_{it}^2. The distance between the ith beacon node and the unknown node in LOS propagation conditions is as follows [17]:

$$\widehat{d}_i = c \cdot (t + n_i) = c \cdot t_i + n_i = d_i + n_i, \tag{3}$$

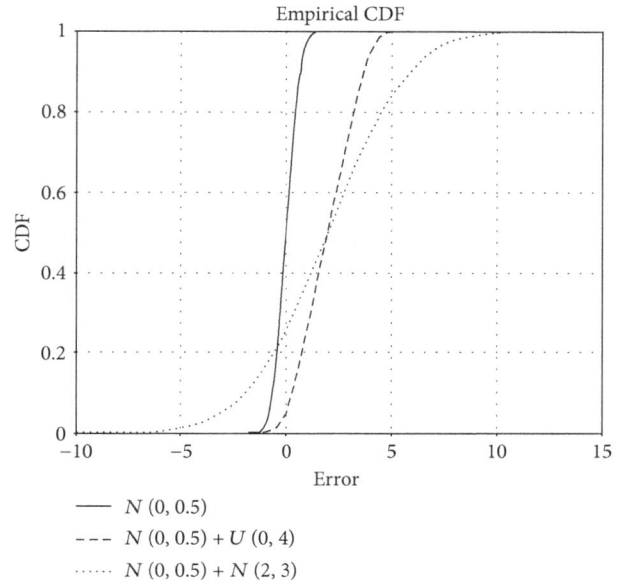

FIGURE 1: The CDF for measurement noise and NLOS error.

where d_i is the true distance between the two nodes; n_i is the measurement error modeled as a zero-mean white Gaussian process with variance σ_i^2.

In practical conditions, the existence of obstacles will result in NLOS conditions. Such obstacles will admit a positive error component to the estimated distance. Considering the NLOS error, the distance between the ith beacon node and the unknown node in NLOS propagation conditions is modeled as follows [18, 19]:

$$\widehat{d}_i = d_i + n_i + n_{\text{NLOS}}, \tag{4}$$

where n_{NLOS} is the NLOS error and it is the positive bias error, and n_{NLOS} is uniformly distributed ($n_{\text{NLOS}} \sim U(0, B_{\max})$). Because the causes of NLOS error and measuring error are different, NLOS error is assumed to be independent of the measuring error [20].

Figure 1 shows the cumulative distribution function (CDF) of measurement noise and NLOS error. The measurement noise n_i obeys a Gaussian distribution, that is, $n_i \sim N(0, 0.5)$. The NLOS error is a uniform distribution or a Gaussian distribution, that is, $n_{\text{NLOS}} \sim U(0, 4)$ and $n_{\text{NLOS}} \sim N(2, 3)$.

3.2. An NLOS Identification Method Based on Residual Analysis. NLOS propagation is ubiquitous in practical conditions and has a large influence on measurements. To obtain more accurate measurements, approaches to reduce the influence that NLOS error admits to localization accuracy must be considered. NLOS error has distinct characteristics compared with the measuring error: (1) NLOS error is always positive. (2) The standard deviation of the distance measurement in NLOS propagation conditions is larger than that in LOS propagation conditions. (3) NLOS error exhibits high randomness.

Considering the characteristics of the NLOS error, NLOS identification methods based on residual analysis can be used to determine and eliminate the NLOS measurements. The basic approach of the residual analysis method can be expressed as follows:

Step 1. There are N different beacon nodes in the field. Combine the measurements provided by these beacon nodes. M different combinations of distance measurements can be obtained.

$$M = \sum_{i=3}^{N} C_N^i. \tag{5}$$

Step 2. Use the maximum likelihood method to compute the estimated location of each combination. The estimated position of the kth combination is $\widehat{X}_k$. The details of the maximum likelihood method are shown in Appendix. Calculate the residual of each measurement as follows:

$$\mathrm{Re}\, s_i(k) = \left| \widehat{d}_i - \left\| \widehat{X}_k - X_i \right\| \right|, \tag{6}$$

where $\widehat{d}_i$ is the distance from the unknown node to the ith beacon node, and X_i is the coordinate of the ith beacon node.

Step 3. Accumulate the residuals of each measurement as follows:

$$\mathrm{CRes}_i = \sum_{k=1}^{M} \mathrm{Res}_i(k). \tag{7}$$

We can obtain N cumulative residuals CRes_i, $i = 1, \ldots, N$.

Step 4. Sort the cumulative residuals from large to small; the measurements with the largest residual can be regarded as NLOS measurements.

By using the above steps, we can determine the NLOS measurements, and the rest of the measurements can be regarded as measurements in LOS propagation conditions.

3.3. PSO with a Constriction Factor-Based Localization Method. The probability density function of the measurement in an LOS condition can be expressed as follows [21]:

$$f_{\mathrm{LOS}}(d_i) = \frac{1}{\sqrt{2\pi\sigma_i^2}} \exp\left(-\frac{\left(\widehat{d}_i - d_i\right)^2}{2\sigma_i^2}\right), \tag{8}$$

where $\widehat{d}_i$ is the measured distance from a beacon node to an unknown node, $d_i = \sqrt{(x_i - x)^2 + (y_i - y)^2}$ is the true distance between the ith beacon node and an unknown node, (x_i, y_i) is the coordinate of the ith beacon node, and (x, y) is the true location of the unknown node. We use the LOS measurements to establish the objective function as follows:

$$(\widehat{x}, \widehat{y}) = \arg\max\left\{\prod_{i=1}^{L} f_{\mathrm{LOS}}(d_i)\right\}, \tag{9}$$

where L is the number of LOS measurements.

To solve the position function directly, not only is a large amount of calculation required but the difficulty of finding an analytical solution is also encountered. Therefore, we use the particle swarm optimization with a constriction factor (PSO-C) method to determine the optimal solution. PSO is based on simulating a simplified model of social interaction. PSO is easy to implement and does not require gradient information, so it is widely used in scientific research and engineering practice.

The basic principle of the algorithm is as follows: assume that a swarm includes M particles. The search space is a D-dimensional vector. The location of the ith particle in the swarm is $X_i = (x_{i1}, x_{i2}, \ldots, x_{iD})$. The velocity is $v_i = (v_{i1}, v_{i2}, \ldots, v_{iD})$. The experienced best location of a particle is $p_i = (p_{i1}, p_{i2}, \ldots, p_{iD})$, where $1 \le i \le m$. The experienced best location of all particles in the swarm is $p_g = (p_{g1}, p_{g2}, \ldots, p_{gD})$. The location and velocity of the particles change according to the equation as follows:

$$v_{iD}^{k+1} = K\left[v_{iD}^k + c_1\xi\left(p_{iD}^k - x_{iD}^k\right) + c_2\eta\left(p_{gD}^k - x_{iD}^k\right)\right], \tag{10}$$

$$x_{iD}^{k+1} = x_{iD}^k + v_{iD}^{k+1}. \tag{11}$$

K is a constriction factor and is a function of c_1 and c_2.

$$K = \frac{2}{\left|2 - \varphi - \sqrt{\varphi^2 - 4\varphi}\right|}, \qquad \varphi = c_1 + c_2 > 4, \tag{12}$$

where c_1 and c_2 are learning factors, and where $c_1 = c_2 = 2.05$. ξ and η are two uniform random numbers in $[0, 1]$, that is, $\xi, \eta \in U(0, 1)$. The velocity of the particles is limited to a maximum range $V_{\max}$. $V_{\max}$ determines the search ability of particles in the search space.

The pseudocode of the PSO-C strategy is shown as follows:

```
1. Initialize the basic parameters of PSO-C
2. Generate an initial population X = {X1, … , XM}and its velocities V = {V1, … , VM}randomly
3. Calculate the fitness values of the population F = {f1, … , fM}
4. Set S to be the pbest = {p1, … , pM} for each particle
5. Set the particle with best fitness to be pg
6. For t = 1 to tmax do
7. For i = 1 to M do
8. Update the velocity of particle Xi using equation (10)
9. Update the location of particle Xi using equation (11)
10. Compute the fitness values of the new particle Xi
11. If the fitness value of Xi is better than the fitness values of pbi
12. Then, set Xi to be pi
13. End if
14. If the fitness value of Xi is better than the fitness values of pg
15. Then, set Xi to be pg
16. End if
17. End for
18. End for
```

PSEUDOCODE 1

The p_g is the estimated position of an unknown node.

4. Simulation and Experiments Results

In this section, we evaluate the performance of our proposed NLOS localization algorithms. We compare the proposed method with RANSAC [12], ML [22], and Rwgh [11] methods. The N beacon nodes and one unknown node are randomly deployed in a $30\,\mathrm{m} \times 30\,\mathrm{m}$ square space. One obstacle is randomly deployed in the field. The communication range of sensor node is 50 m. The measurement error n_i is modeled as a zero-mean white Gaussian process with variance σ_i^2. The NLOS error n_{NLOS} obeys the uniform distribution $(n_{\mathrm{NLOS}} \sim U(0, B_{\max}))$. The simulation results are obtained through 1000 Monte Carlo runs. The default parameter values in the simulation are shown in Table 1. We consider the average localization error (ALE) as the performance metric.

$$\mathrm{ALE} = \frac{1}{M} \sum_{i=1}^{M} \sqrt{(\hat{x}(i) - x(i))^2 + (\hat{y}(i) - y(i))^2}, \quad (13)$$

where $M = 1000$, $[x(i), y(i)]$ is the true location of the mobile node, and $[\hat{x}(i), \hat{y}(i)]$ is the estimated location for the ith Monte Carlo run.

First, the identification success rate of the proposed method is evaluated. Figure 2 shows the identification success rate versus the number of beacon nodes. In this simulation, the standard variance of the measurement noise in the LOS condition σ_i is varied from 0.1 to 0.5, and the number of beacon nodes is varied from 5 to 10. The results show that as the number of beacon nodes increases, the success rate of the proposed method increases. In addition, as the value of σ_i increases, the success rate of the proposed method decreases because, as the value of σ_i increases, the measurements will be disturbed by measurement noise more seriously.

TABLE 1: The default parameter values.

Parameters	Symbol	Default values
Number of beacon nodes	N	8
The standard deviation of the measurement noise	σ_i	1
The NLOS errors	$N(\mu_{\mathrm{NLOS}}, \sigma_{\mathrm{NLOS}}^2)$	$\mu_{\mathrm{NLOS}} = 2, \sigma_{\mathrm{NLOS}} = 7$
The NLOS errors	$U(0, B_{\max})$	$B_{\max} = 7$

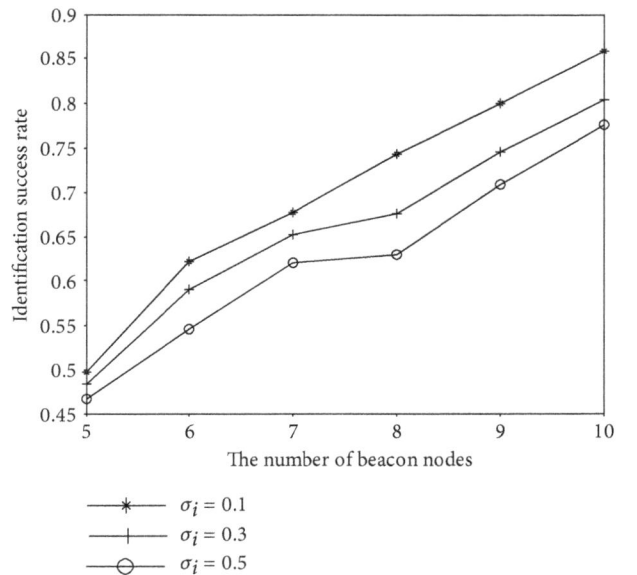

FIGURE 2: The identification success rate versus the number of beacon nodes.

When the NLOS error obeys the uniform distribution $(n_{\mathrm{NLOS}} \sim U(0, B_{\max}))$, the identification success rate versus the maximum bias of NLOS error $B_{\max}$ is determined as shown in Figure 3. In this simulation, the standard variance

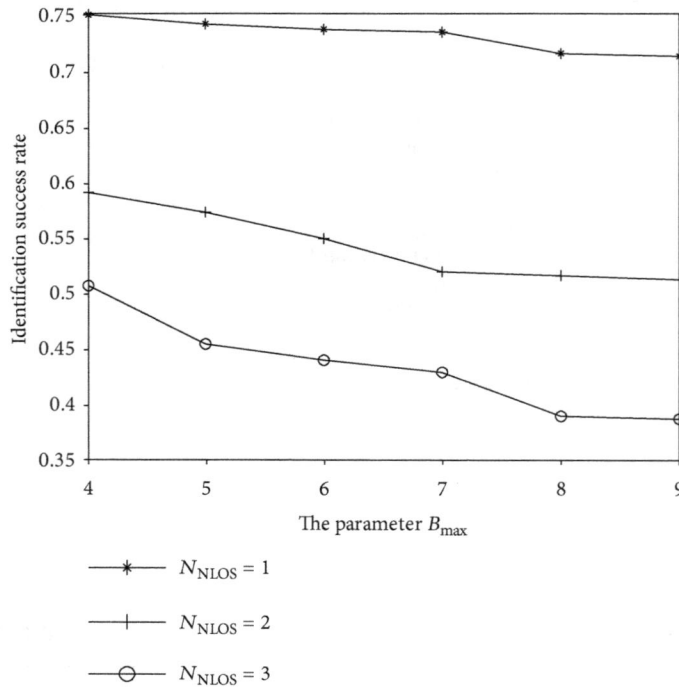

FIGURE 3: The identification success rate versus B_{max}.

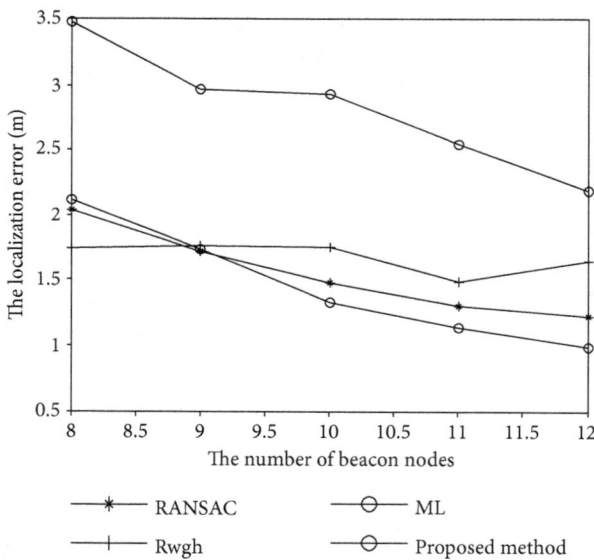

FIGURE 4: The localization error versus the number of beacon nodes.

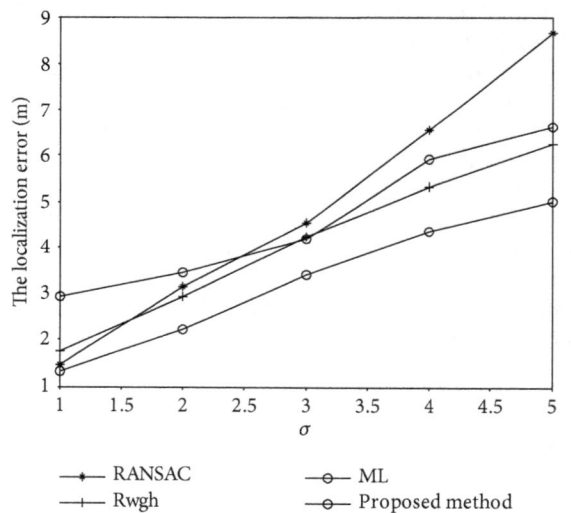

FIGURE 5: The localization error versus σ_i.

of the measurement noise in the LOS condition σ_i is 1. The results show that when the number of measurements in the NLOS condition is equal, the success rate is less affected by B_{max}. However, as the number of measurements in the NLOS condition increases, the success rate of the proposed method decreases.

In Figure 4 we evaluate the impact of the number of beacon nodes on the localization error. The results show that the localization error decreases as the number of beacon nodes increases. In addition, the ML method has the largest

localization error. When the number of beacon nodes is relatively fewer, such as 8, the Rwgh method works best. When the number of beacon nodes is 9, the localization errors of the Rwgh method, the RANSAC method, and the proposed method are approximately the same. However, the localization error of the proposed method declines faster, and this method has the highest localization accuracy when the number of beacon nodes increases.

Figure 5 shows the relation between the localization error and the standard variance of the measurement noise σ_i. σ_i is varied from 1 to 5. The results show that as the value of σ_i

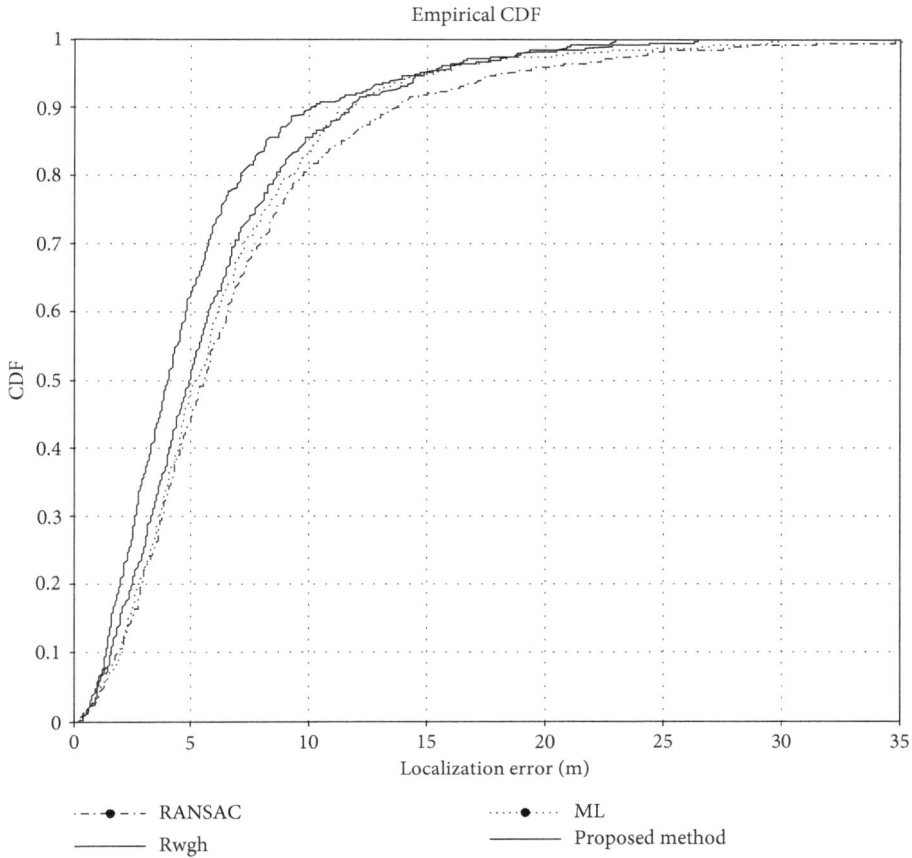

FIGURE 6: The CDF of localization error.

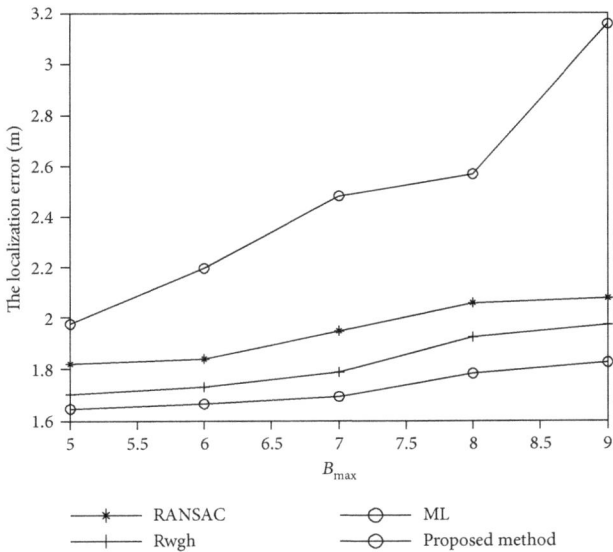

FIGURE 7: The localization error versus $B_{\max}$.

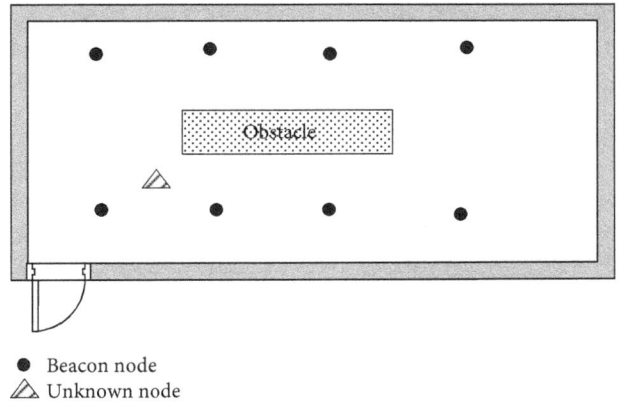

FIGURE 8: The floor plan for the test bed.

increases, the localization error increases. In addition, the proposed method has the highest localization accuracy compared with the other methods. In comparison with ML, RANSAC, and Rwgh methods, the localization accuracy of the proposed method increases to 29.36%, 33.05%, and 20.37%, respectively.

Figure 6 shows the CDF of the localization error when the NLOS error obeys the uniform distribution $n_{\mathrm{NLOS}} \sim U(0, 6)$. We can see that the 80% localization error of the proposed method is less than 7.122 m, and the CDF trends toward one with a localization error of less than 22.7 m. In comparison with the 80% localization error of the Rwgh, ML, and RANSAC methods, 7.2 m, 8.1 m, and 6.1 m are achieved, respectively.

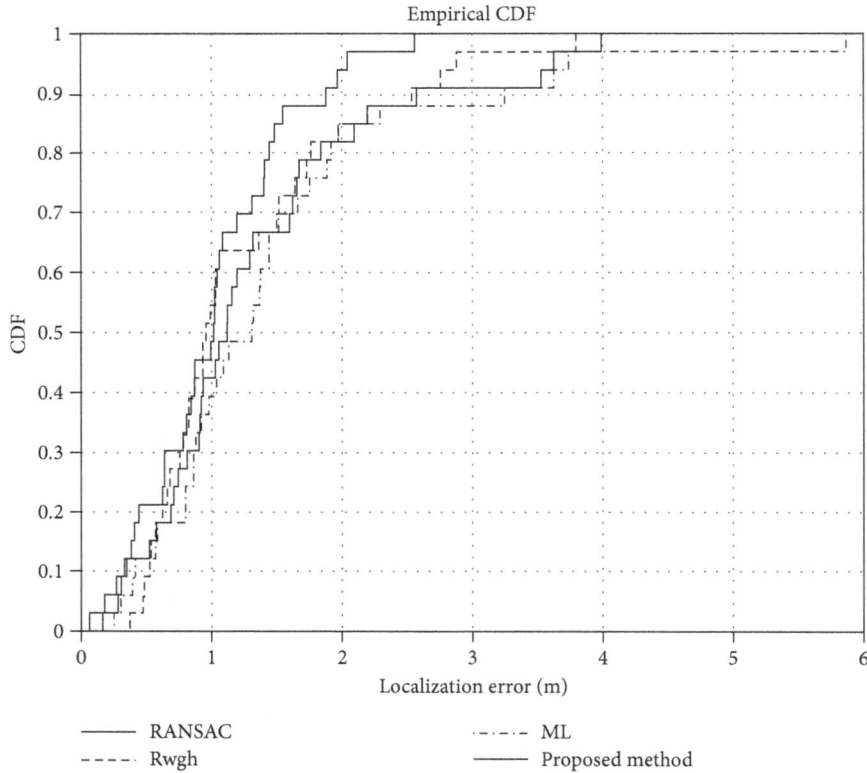

FIGURE 9: The CDF of localization error in realistic experiment.

Figure 7 shows the impact of the parameter $B_{\max}$ on the localization error. As the value of $B_{\max}$ increases, the localization errors of all algorithms are increasing. However, the ML method is seriously affected by $B_{\max}$. The proposed method achieves the lowest localization error. In comparison with ML, RANSAC, and Rwgh methods, the localization accuracy of the proposed method increases to 30.46%, 11.68%, and 5.64%, respectively.

In order to verify the effective of the proposed localization method, we perform the realistic experiment in the indoor environment. As shown in Figure 8, eight beacon nodes and one unknown node are deployed in the room. The beacon node and unknown node are installed up to 1.7 m above the ground. The experimental equipment is chirp spread spectrum (CSS) localization system.

Figure 9 shows the CDF of the localization error in realistic experiment. It can be seen that 80% localization error of the proposed method is less than 1.879 m. The CDF trends toward one with a localization error of less than 2.559 m. The average localization error of the proposed method is 1.0205 m. The average localization errors of RANSAC, Rwgh, and ML are 1.3491 m, 1.2603 m, and 1.5036 m, respectively.

5. Conclusion

The NLOS problem is one of the most challenging problems for wireless sensor networks. It can seriously reduce localization accuracy. In this paper, the TOA measurement

model is first introduced. We then proposed an NLOS identification method based on residual analysis to solve the problem caused by the NLOS error. In addition, the particle swarm optimization with a constriction factor algorithm is proposed to find the optimal solution of the location estimate of an unknown node. Simulation results show that this method can reduce the influence of NLOS error and improve the positioning accuracy, especially when the number of beacon nodes is relatively large. In future work, the proposed method could be extended to the distributed localization method. At the same time, we will modify the residual analysis method and apply it to the mobile localization to improve the effectiveness of particle filter.

Appendix

In this section, we introduce the maximum likelihood localization method. We assume that the position of the beacon node is denoted as $[(x_1, y_1), \ldots, (x_N, y_N)]$. The position of an unknown node is $\theta = [x_u, y_u]^{\mathrm{T}}$. $\widehat{d}_i$ is the measurement distance for the ith beacon node.

$$(x_1 - x_u)^2 + (y_1 - y_u)^2 = \left(\widehat{d}_1\right)^2,$$
$$\vdots \qquad\qquad (A.1)$$
$$(x_N - x_u)^2 + (y_N - y_u)^2 = \left(\widehat{d}_N\right)^2.$$

The above equation can be represented by a linear equation $\mathbf{A} \cdot \boldsymbol{\theta} = \mathbf{B}$, where $\mathbf{A}$ and $\mathbf{B}$ are given by

$$\mathbf{A} = 2 \begin{bmatrix} (x_1 - x_2) & (y_1 - y_2) \\ (x_1 - x_3) & (y_1 - y_3) \\ \vdots & \vdots \\ (x_1 - x_N) & (y_1 - y_N) \end{bmatrix},$$

$$\mathbf{B} = \begin{bmatrix} \left(\widehat{d}_2\right)^2 - \left(\widehat{d}_1\right)^2 - \left(x_2^2 + y_2^2\right) + \left(x_1^2 + y_1^2\right) \\ \left(\widehat{d}_3\right)^2 - \left(\widehat{d}_1\right)^2 - \left(x_3^2 + y_3^2\right) + \left(x_1^2 + y_1^2\right) \\ \vdots \\ \left(\widehat{d}_N\right)^2 - \left(\widehat{d}_1\right)^2 - \left(x_N^2 + y_N^2\right) + \left(x_1^2 + y_1^2\right) \end{bmatrix}. \tag{A.2}$$

We can obtain the estimated position of the unknown node as follows:

$$\widehat{\boldsymbol{\theta}} = \left(\mathbf{A}^{\mathrm{T}}\mathbf{A}\right)^{-1}\mathbf{A}^{\mathrm{T}}\mathbf{B}. \tag{A.3}$$

Conflicts of Interest

The authors declare that there is no conflict of interest regarding the publication of this paper

Acknowledgments

This work was supported by the National Natural Science Foundation of China under Grant no. 61803077, the Natural Science Foundation of Hebei Province under Grant no. F2016501080, and the Fundamental Research Funds for the Central Universities of China under Grant nos. N172304024 and N152302001.

References

[1] P. H. Tsai, R. G. Tsai, and S. S. Wang, "Hybrid localization approach for underwater sensor networks," *Journal of Sensors*, vol. 2017, Article ID 5768651, 13 pages, 2017.

[2] S. Xie and Y. Wang, "Construction of tree network with limited delivery latency in homogeneous wireless sensor networks," *Wireless Personal Communications*, vol. 78, no. 1, pp. 231–246, 2014.

[3] S. Subedi and J. Y. Pyun, "Practical fingerprinting localization for indoor positioning system by using beacons," *Journal of Sensors*, vol. 2017, Article ID 9742170, 16 pages, 2017.

[4] Y. Ahmadi, N. Neda, and R. Ghazizadeh, "Range free localization in wireless sensor networks for homogeneous and non-homogeneous environment," *IEEE Sensors Journal*, vol. 16, no. 22, pp. 8018–8026, 2016.

[5] K. Subbu, C. Zhang, J. Luo, and A. Vasilakos, "Analysis and status quo of smartphone-based indoor localization systems," *IEEE Wireless Communications*, vol. 21, no. 4, pp. 106–112, 2014.

[6] Y. T. Chan, W. Y. Tsui, H. C. So, and P. C. Ching, "Time-of-arrival based localization under NLOS conditions," *IEEE Transactions on Vehicular Technology*, vol. 55, no. 1, pp. 17–24, 2006.

[7] J. Cota-Ruiz, P. Rivas-Perea, E. Sifuentes, and R. Gonzalez-Landaeta, "A recursive shortest path routing algorithm with application for wireless sensor network localization," *IEEE Sensors Journal*, vol. 16, no. 11, pp. 4631–4637, 2016.

[8] W. Liu, D. Wang, H. Jiang, W. Liu, and C. Wang, "An approximate convex decomposition protocol for wireless sensor network localization in arbitrary-shaped fields," *IEEE Transactions on Parallel and Distributed Systems*, vol. 26, no. 12, pp. 3264–3274, 2015.

[9] Z. Yang, C. Wu, T. Chen, Y. Zhao, W. Gong, and Y. Liu, "Detecting outlier measurements based on graph rigidity for wireless sensor network localization," *IEEE Transactions on Vehicular Technology*, vol. 62, no. 1, pp. 374–383, 2013.

[10] C. Müller, D. I. Alves, B. F. Uchôa-Filho, R. Machado, J. B. S. Martins, and L. L. de Oliveira, "Improved solution for node location multilateration algorithms in wireless sensor networks," *Electronics Letters*, vol. 52, no. 13, pp. 1179–1181, 2016.

[11] P. C. Chen, "A non-line-of-sight error mitigation algorithm in location estimation," in *Proceedings of the wireless communication and networking conference*, pp. 316–320, New Orleans, LA, USA, 1999.

[12] I. Rasool and A. H. Kemp, "Statistical analysis of wireless sensor network Gaussian range estimation errors," *IET Wireless Sensor Systems*, vol. 3, no. 1, pp. 57–68, 2013.

[13] W. Li, Y. Jia, J. Du, and J. Zhang, "Distributed multiple-model estimation for simultaneous localization and tracking with NLOS mitigation," *IEEE Transactions on Vehicular Technology*, vol. 62, no. 6, pp. 2824–2830, 2013.

[14] R. Vaghefi, J. Schloemann, and R. Buehrer, "NLOS mitigation in TOA-based localization using semidefinite programming," in *2013 10th Workshop on Positioning, Navigation and Communication (WPNC)*, pp. 1–6, Dresden, Germany, March 2013.

[15] U. Hammes and A. M. Zoubir, "Robust mobile terminal tracking in NLOS environments based on data association," *IEEE Transactions on Signal Processing*, vol. 58, no. 11, pp. 5872–5882, 2010.

[16] L. Cheng, C.-D. Wu, Y.-Z. Zhang, and Y. Wang, "An indoor localization strategy for a mini-UAV in the presence of obstacles," *International Journal of Advanced Robotic Systems*, vol. 9, no. 4, p. 153, 2017.

[17] X. Yu, C. Wu, and L. Cheng, "Indoor localization algorithm for TDOA measurement in NLOS environments," *IEICE Transactions on Fundamentals of Electronics, Communications and Computer Sciences*, vol. E97.A, no. 5, pp. 1149–1152, 2014.

[18] X. Wang, L. Ding, and S. Wang, "Trust evaluation sensing for wireless sensor networks," *IEEE Transactions on Instrumentation and Measurement*, vol. 60, no. 6, pp. 2088–2095, 2011.

[19] C. Yang, B. Chen, and F. Liao, "Mobile location estimation using fuzzy-based IMM and data fusion," *IEEE Transactions on Mobile Computing*, vol. 9, no. 10, pp. 1424–1436, 2010.

[20] L. Cheng, Y. Wang, H. Wu, N. Hu, and C. Wu, "Non-parametric location estimation in rough wireless environments for

wireless sensor network," *Sensors and Actuators A: Physical*, vol. 224, pp. 57–64, 2015.

[21] Y. Wang, L. Cheng, and N. Hu, "Bayes sequential test based NLOS localization method for wireless sensor network," in *The 27th Chinese Control and Decision Conference (2015 CCDC)*, pp. 5230–5234, Qingdao, China, May 2015.

[22] R. M. Vaghefi and R. M. Buehrer, "Cooperative sensor localization with NLOS mitigation using semidefinite programming," in *2012 9th Workshop on Positioning, Navigation and Communication*, pp. 13–18, Dresden, Germany, March 2012.

10

An Energy Efficient Routing Protocol based on Layers and Unequal Clusters in Underwater Wireless Sensor Networks

Fang Zhu [1] and Junfang Wei[2,3]

[1]School of computer and communication engineering, Northeastern University at Qinhuangdao, Northeastern University, Qinhuangdao 066004, China
[2]School of Resources and Materials, Northeastern University at Qinhuangdao, Northeastern University, Qinhuangdao 066004, China
[3]Key Laboratory of Dielectric and Electrolyte Functional Material Hebei Province, School of Resource and Materials, Northeastern University at Qinhuangdao, Qinhuangdao 066004, China

Correspondence should be addressed to Fang Zhu; sky050607@sina.com

Academic Editor: Abdellah Touhafi

Underwater Wireless Sensor Networks (UWSNs) have drawn tremendous attentions from all fields because of their wide application. Underwater wireless sensor networks are similar to terrestrial Wireless Sensor Networks (WSNs), however, due to different working environment and communication medium, UWSNs have many unique characteristics such as high bit error rate, long end-to-end delay and low bandwidth. These characteristics of UWSNs lead to many problems such as retransmission, high energy consumption and low reliability. To solve these problems, many routing protocols for UWSNs are proposed. In this paper, a localization-free routing protocol, named energy efficient routing protocol based on layers and unequal clusters (EERBLC) is proposed. EERBLC protocol consists of three phases: layer and unequal cluster formation, transmission routing, maintenance and update of clusters. In the first phase, the monitoring area under the water is divided into layers, the nodes in the same layer are clustered. For balancing energy of the whole network and avoiding the "hotspot" problem, a novel unequal clustering method based on layers for UWSNs is proposed, in which a new calculation method of unequal cluster size is presented. Meanwhile, a new cluster head selection mechanism based on energy balance and degree is given. In the transmission phase, EERBLC protocol proposes a novel next forwarder selection method based on the forwarding ratio and the residual energy. In the third phase, Intra and inter cluster updating method is presented. The simulation results show that the EERBLC can effectively balance the energy consumption, prolong the network lifetime, and increase the amount of data transmission compared with DBR and EEDBR protocols.

1. Introduction

Nowadays, Underwater Wireless Sensor Networks (UWSNs) are attracting more attentions from academia and industry because of their broad application fields such as environmental monitoring, disasters prevention, auxiliary navigation, resource exploration and so on [1–7]. UWSN is composed of the base station, sink nodes and ordinary sensor nodes. The sensor nodes are randomly deployed from surface to bottom of water. They collect and transmit the information to the sink nodes. Sink nodes receive the information form sensor nodes, and then send the information to the base station. Sink nodes are usually deployed on the surface of water. The base station processes the data and supports the final decision through data analysis. Figure 1 shows the architecture of UWSNs. UWSNs are similar to terrestrial Wireless Sensor Networks (WSNs). However, considering the physical layer technology and propagation medium, the challenges are very different from terrestrial WSNs. In the terrestrial WSNs, data are transmitted through radio signals. But it is not suitable for underwater environment, due to rapid decay and high energy consumption in the water. Usually in the water, the acoustic signal is adopted as communication medium. The speed of acoustic signal is significantly

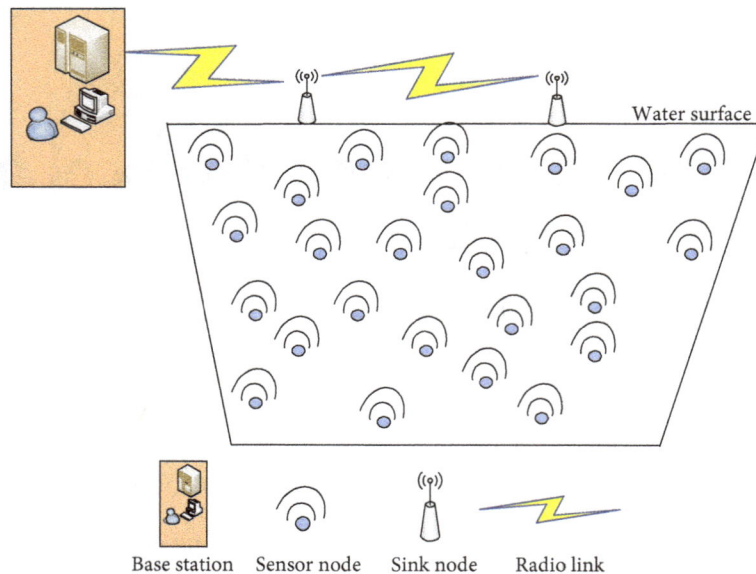

FIGURE 1: Architecture of UWSN.

slower than radio signal, which causes the high propagation delay. Moreover, multipath fading channel causes high bit error ratio and the low bandwidth. These characteristics lead to many problems such as retransmission, high energy consumption and low reliability. Therefore, design of routing protocols suitable for UWSNs becomes a challenge task.

Many routing protocols have been proposed for UWSNs in last few years [8–29]. These protocols can be classified into two categories: localization-based and localization-free routing protocols. In the section 2, we introduce two types of protocols in detail. In the localization-based protocols, each node requires to know its location. However, location-known is a hard issue for UWSNs because GPS device is not suitable for working in the water and the position of node is changing at any time with water current. Localization-free routing protocols are more suitable for UWSNs than localization-based routing protocols, because the nodes do not need to know their location in localization-free routing protocols.

In this paper, we propose a localization-free routing protocol, named energy efficient routing protocol based on layers and unequal clusters (EERBLC). This routing protocol aims to solve the problems of high energy consumption, long propagation delay and high error rate. In this protocol, the monitoring area is layered according to the distance to the surface, and then nodes are clustered in each layer. In this process, unequal clusters are formed to balance energy. The size of cluster closer to the surface is smaller than far away from the surface. EERBLC protocol presents layer algorithm, cluster head election algorithm and cluster size calculation algorithm. Cluster member nodes collect and send the data to cluster head node in each cluster. Cluster head nodes perform data aggregation and forward the data to sink nodes by multi-hop path. Furthermore, a novel approach of selecting the next-hop based on link quality and residual energy is proposed during the data forwarding. Finally, the method of maintenance and update of cluster is presented in EERBLC protocol.

The rest of this paper is organized as follows. Some well-known routing protocols proposed for UWSNs are summarized in Section 2. Section 3 describes the proposed routing protocol EERBLC in detail. Section 4 presents the performance evaluation of EERBLC. Finally, Section 5 illustrates the conclusion of this study.

2. Related Works

UWSNs have been under research over last ten years. Many routing protocols for UWSNs have been proposed by some researchers. As mentioned above, these protocols can be classified into two categories: localization-based and localization-free.

2.1. Localization-Based Routing Protocols. A vector-based forwarding (VBF) routing protocol is proposed in literature [14]. In this protocol, it supposes that each sensor node knows its own location. Sensor nodes forward data packets with the location information of source, destination and a range field. Only those nodes in a pipe with a given radius can be selected as forwarders. The pipe is halved by a vector from the source node to the sink node. VBF reduces the number of forwarding nodes, so the overhead of whole network is reduced. However, it is sensitive to the radius of pipe. Based on the VBF protocol, many improved protocols are proposed [15–17]. Some of these protocols [15, 16] improve the delivery ratio, and some [17] improve the energy consumption. Focused Bream Routing (FBR) is presented to reduce unnecessary flooding in literature [18]. During the selection of forwarder, FBR uses different transmission power levels. In the FBR protocol, the mechanism of RTS/CTS leads to long end-to-end delay and excessive energy consumption. A novel multipath grid-based geographical routing protocol (EMGGR) for UWSNs is proposed in literature [19]. EMGGR routing protocol assumes that the geographic area of the network is partitioned into 3D logical grids and sensor nodes are deployed in some cells. It selects gateways based on their locations and remaining energy level. Some protocol similar to EMGGR are proposed such as NGF [11], GGFGD and GFGD [20]. Literature [21] proposes the depth-controlled routing (DCR). In DCR, each sensor node with network topology controller can adjust the its depth to organize the network topology when the greedy geographic routing fails. DCR is the first geographic routing protocol for UWSNs that considers the sensor node vertical movement ability to move it for topology control purpose.

2.2. Localization-Free Routing Protocols. In the localization-based routing protocols, full-dimensional location information is needed. However, in the underwater environment, the location information is hard to acquire. So, many localization-free routing protocols were proposed.

A classical localization-free routing protocol called depth-based routing (DBR) is presented in literature [22]. This protocol assumes that each node equipped with a depth sensor. DBR selects next forwarder based on depth information. DBR is a practical routing protocol. However, it has some serious problems. First, if the nodes have same depth, they will forward the data packet at same time. Even if the depths are not exactly the same, because of the long propagation delay in UWSNs, the same packet can be forwarded by different nodes. Particularly, in density areas, large number of redundant data packets will be generated which leads to high energy consumption. Second, in the case of sparse areas, it is possible that no sensor node can be selected as forwarder due to its greedy mode. Third, because each node needs to keep data packet for a certain time after receiving data packet, it causes long end-to-end delay. These problems affect the performance of DBR. To improve DBR's energy consumption problem, Energy-efficient depth-based routing protocol (EEDBR) for UWSNs is proposed in literature [23]. EEDBR improvs the performance in terms of the network lifetime, energy consumption and end-to-end delay. Hop-by-hop dynamic addressing-based routing protocol (H2-DAB) is proposed in literature [24]. In H2-DAB, each node needs to be assigned a HopID. The HopID of node represents hop count from current node to sink nodes. The nodes with smallest HopID are selected as forwarders by source node. That causes the problem of void region. Moreover, the inquiry request and reply mechanism results in long end-to-end delay and extra energy consumption. To improve the delivery ratio, some localization-free protocols are proposed such as CARP and E-CARP [12, 25]. In literature [26], a cluster-based routing protocol (CBKU) for UWSNs is proposed. CBKU uses the improved K-means algorithm for clustering to avoid energy unbalanced. To prolong the lifetime of UWSNs, some location-free protocols are proposed based on cluster [27–29].

3. Design of Proposed Protocol (EERBLC)

3.1. Motivation of EERBLC. Reducing energy consumption is the main objective of a routing protocol for UWSNs. In underwater environment, acoustic signal is adopted as communication medium, which leads to more energy

consumption. The energy of sensor nodes is limited and hard to be supplied. Hence, energy efficient and energy balance are primary design objectives in a routing protocol. Cluster-based routing protocols have been proposed in the terrestrial wireless sensor networks [30–33]. After clustering, member sensor nodes collect and send data to the cluster head, cluster head sends the data to the sink node after data fusion. It has been proved that cluster-based protocols are very effective on saving energy. In UWSNs, transmission requires more energy than receiving. Therefore, reducing the number of transmissions is useful in reducing the energy consumption. Cluster head aggregates and fuses data can effectively reduce the number of transmission. Because long distance communication leads to more energy consumption, multi-hop path routing method is adopted to save the energy in our protocol. Cluster heads forward the data to sink nodes by other cluster heads. Therefore, our work aims to design a routing protocol more suitable for UWSNs based on cluster.

Some protocols for UWSNs have introduced clustering techniques [34–38], but none of them takes into account the "hotspot" problem. "Hotspot" problem is that cluster heads especially the nodes near the surface forward data more frequently than others, which results in premature death of these nodes. The "hotspot" problem affects whole network lifetime. Therefore, EERBLC protocol aims to improve the clustering technique to solve the hotspot problem. The idea is that the nodes near the surface are not clustered, and each node of these can be selected as forwarder; At the same time, unequal cluster technology is applied to avoid the "hotspot" problem. The layer is closer to the surface, the number of cluster is lager and size of cluster is smaller. Furthermore, during the selection of the forwarder, energy balance should be considered.

Due to the harsh environment, the bit error rate is very high and delivery ratio is very low in UWSNs. High quality links can improve bit error rate, delivery ratio, and energy consumption. Therefore, the selection of reliable routing path with good link quality is very important issue. During the selection of next forwarder, the link quality and residual energy are both taken into account in EERBLC protocol.

3.2. Network Structure and Assumption. The network structure is shown in Figure 1. EERBLC adopts multi-sinks mode to increase the reliability of the network and the data delivery ratio. This mode also can reduce the energy consumption of the nodes around the sinks. Sink nodes are distributed evenly on the water surface. Each sink node is equipped with a radio modem and acoustic modem. Radio modem is used to communicate between sink nodes and the data center. Acoustic modem is used to communicate between sensor nodes under the water.

To make the clustering more effective, Underwater sensor nodes are deployed in the form of layer from bottom to surface in a 3D space. The sensor nodes are deployed at different layers by a bouncy control mechanism. The numbers of layers depended on the depth of the water and the communication range between the layers. If the average depth of the ocean is about 2.5–3 km and the communication rang is

500 m, about 5-6 layers are needed. Each sensor node is equipped with an acoustic modem for communicating with each other in the underwater. Underwater sensor nodes drift with water current in the horizontal direction, the vertical movement can be negligible.

All underwater sensor nodes can sensor and collect data. In each layer, the nodes are clustered except in the highest layer where the nodes can directly communicate with sink nodes. In each cluster, cluster members send the data to the cluster head. Cluster head aggerates and forwards the data to other cluster head. The data are sent to the sink by multi-hop path. Considering the energy balance, clustering is not needed in the highest layer, where all nodes can be selected as forwarder.

In this paper, we assume the energy of sink nodes is unlimited because the battery of sink nodes can be replaced. Meanwhile, we assume the underwater sensor nodes have equal initial energy and same communication range. And the energy of underwater sensor nodes is one-time. Each underwater sensor node equipped with the depth sensor. Once any one of sink nodes receives the data packets, delivery is considered to be successful.

3.3. Energy Consuming Model. Due to the characteristics of underwater acoustic channel, the energy consuming model of UWSNs is quite different from the energy consuming model of terrestrial WSNs. The formula (1) shows the calculation method.

$$E(d, f) = P_0 d^k a^d \qquad (1)$$

Where P_0 is the power threshold that the data can be received by the nodes, d is the transmitting distance, k is the coefficient of energy expansion, $E(d, f)$ is the lowest energy consumption to send the data to the destination. a is defined as formula (2).

$$a = 10^{a(f)/10} \qquad (2)$$

Where a is concerned with frequency, which is defined as formula (3).

$$a(f) = 0.11 \frac{f^2}{1 + f^2} + 44 \frac{f^2}{4100 + f^2} + 2.75 \times 10^{-4} f^2 + 0.003 \qquad (3)$$

Where f is the frequency of the carrier acoustic signal in Hz and $a(f)$ is in dB/m.

$E_{DA}(l)$ means the energy consumed by fusing l bits of data, which is shown as formula (4).

$$E_{DA}(l) = E_{EA} * l \qquad (4)$$

Where E_{EA} is the energy consumed by fusing one bit of data, generally it can be taken as 5nJ/bit.

3.4. EERBLC Protocol. As mentioned above, in the EERBLC protocol, the underwater sensor nodes are deployed in form of layer. For example, if the depth of monitoring area is 1000 m and the communication rang of nodes is 250 m, the whole network needs 4 layers of nodes. The distance is less than or equal to 250 m between two layers. In each layer, the sensor nodes are randomly deployed. The nodes compete to be cluster head at the same layer.

There are three phases in the EERBLC protocol, including cluster formation, transmission routing and maintenance and update of clusters.

3.4.1. Cluster Formation. The multi-hop transmission causes an unbalanced energy consumption. Cluster heads closer to the sink nodes take on more forwarding tasks than other cluster heads. To solve this problem, the unequal cluster technology is adopted in EERBLC protocol. Clusters will have equal cluster sizes at the same layer. The size of cluster with higher depth is larger. In the first layer, cluster is not formed and each node can forward the data to sink nodes directly. The structure of the network in EERBLC protocol is shown in Figure 2. In order to balance energy consumption between nodes in each cluster, the cluster heads selection mechanism is based on three parameters: residual energy, the degree and the layer number.

After sensor nodes are deployed, each node calculates its own layer number. The formula (5) shows the calculation method of layer number.

$$Ln = \left(\frac{n_{i\_depth}}{R} \right) + 1 \qquad (5)$$

Where Ln is layer number of node n_i, n_{i_depth} is the depth of n_i, R is the maximum communication rang of node.

There are three steps in the cluster setup phase. The first step of this phase is information collection phase, whose duration is set as T1. At the beginning of this phase, each node broadcasts a N_HELLO message which includes node id, layer number, residual energy. Once receiving the N_HELLO message, each node records the message of neighbors at the same layer. The message from nodes at other layers is dropped. The format of the N_HELLO packet is shown in Table 1.

The second step is the cluster head competition phase, which starts when T1 expires. In this phase, the sensor node whose waiting time Tc expires becomes cluster head, and broadcasts a competition message N_COM. The competition message N_COM maintains node id, layer number, residual energy and degree. Degree is the number of neighbors. The format of the N_COM packet is shown in Table 2.

If the node S_i received N_COM message from other nodes in the same layer before Tc expires, node S_i drops competition and becomes an ordinary node. Tc is different for each node, which is calculated as following formula (6).

$$Tc = \left(1 - \frac{\text{residual energy}}{\text{initial energy}} \right) * T2 + P \qquad (6)$$

Where T2 is the duration of the competition, P is the random value in [0.5, 1], which is used to avoid the

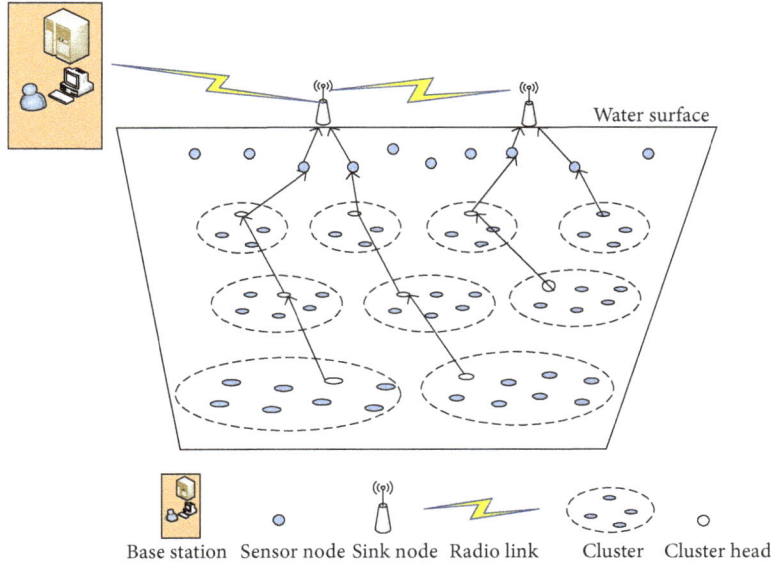

FIGURE 2: Structure of the network in EERBLC.

TABLE 1: Format of N_HELLO packet.

Type of packet	Node id	Layer number	Residual energy

TABLE 2: Format of N_COM packet.

Type of packet	Node id	Layer number	Residual energy	Degree

communication conflict when residual energy of nodes is the same. It can be seen from formula (6) that the nodes with higher residual energy have shorter waiting time and have a greater chance to be cluster head. Once the waiting time Tc of a node expires, it is selected as candidate cluster head and broadcasts N_COM message within competition communication rang.

In order to balance energy of whole network, unequal clusters are formed in different layers. The nodes closer to the surface have heavier forwarding load. Hence, the cluster size is smaller in the layer which is closer to surface. In order to generate unequal clusters, these nodes need to calculate their own competition radius Rc. The formula (7) shows the computing method of Rc.

$$Rc = \beta \frac{energy_c}{energy_{init}} * \left(\alpha \frac{Ln}{Ltn} * R \right) \qquad (7)$$

Where Ln is the layer number of node, Ltn is total layer number of whole network, $energy_c$ is current residual energy, α and β is weighted value. It is shown from formula (7) that nodes with bigger layer number have lager competition radius. And the nodes with more energy have lager competition radius at the same layer.

If a node S_j receives a N_COM message from node S_i before its waiting time Tc expires, it adds the S_i to its candidate cluster head list and becomes a non-cluster head node. If the node receives many N_COM messages from several

TABLE 3: Format of N_JOIN packet.

Type of packet	Head node id	Member id	Residual energy

cluster heads, it requires to select a cluster head from its candidate cluster head list. To optimize the selection method of cluster head, we design a multi-objective optimization technique to calculate the cost value for each candidate. The cost value is calculated as formula (8).

$$\cos t(i) = \gamma \frac{energy_{res\_i}}{energy_{init}} + \theta \frac{\deg ree_{\_i}}{N}, \gamma + \theta = 1 \qquad (8)$$

Where $energy_{res\_i}$ is the residual energy of candidate S_i, $\deg ree_{\_i}$ is the number of neighbors of candidate S_i, N is the total number of all nodes. γ and θ are the parameters which are used to adjust the weight of energy and degree. The node with maximum cost value will be selected as cluster head by node S_j. If there are several cluster head nodes with the maximum value, the first node with maximum value in the list is selected as its cluster head. After selection of cluster head, node S_j replies a N_JOIN message to the selected head node. N_JOIN maintains id of head node, id of S_j, residual energy. The head node adds S_j to its member list. The format of the N_JOIN packet is shown in Table 3.

The procedure of cluster formation is shown as algorithm 1.

3.4.2. Transmission Routing Algorithm. After cluster formation phase, each cluster head assigns a TDMA schedule to each member of its cluster. Each member node transmits its data at its allocated time slot. The cluster head aggregates and forwards the data to the sink nodes. If the node resides at the highest layer which number is 1, the data are transmitted to the sink nodes directly. In other case, the data are forwarded by cluster heads through multi-hop path.

```
1: Procedure cluster formation
2: for each node Ni
3: calculate the layer number LN
4: Ni < - LN
5: end for
6: for each node Ni
7: broadcast N_HELLO
8: end for
9: for each node Ni received N_HELLO packet
10: if node's LN in received packet = Ni's LN
11: record message of neighbors
12: else drop the packet
13: end if
14: end for
15: for each node Ni
16: if LN ≠ 1
17: Ni calculate Tc
18: end if
19: end for
20: for each node Ni
21: while Tc does not expire
22: end while
23: if did not receive N_COM packet
24: calculate competition radius Rc
25: broadcast N_COM packet within Rc
26: else CH(Sj) < -N_COM packet
26: end if
27: end for
28: for each node received N_COM packet
29: calculate cost (i) value
30: send the N_JOIN packet
31: end for
32: for each head node
33: record the member nodes
34: end for
35: end procedure
```

ALGORITHM 1: Cluster formation algorithm.

Each cluster head records the information of head id, residual energy and forwarding ratio when it hears data packets from other heads at its upper layer which layer number is one less than its layer number. Cluster head only selects the nodes at upper layer as next forwarder. Considering the energy efficient and delivery ratio, the selection of next hop is based on residual energy and forwarding ratio. The cost value of selection is computed as following formula (9).

$$H\_\cos t(i) = \mu \frac{energy_{res\_Hi}}{energy_{init}} + \varepsilon fr, \mu + \varepsilon = 1 \quad (9)$$

Where $energy_{res\_Hi}$ is the residual energy of head node S_i, fr is the forwarding ratio of S_i, μ and ε are the weight coefficient.

Cluster head calculates cost value of each cluster head according to the information recorded. For improving deliver ratio, two head nodes with the biggest value are selected as next forwarder. Cluster heads forward data packet including data number, source id, head id, next hop id1, next

hop id2, data, residual energy and forwarding ratio. The format of data packet is shown as Table 4.

Each data packet from the same node has a unique number. The source node id and data number represent a unique data packet in whole network. The head id represents the head node of forwarding this data packet. The next hop id1 and next hop id2 represent the next forwarder nodes which cost value is the two largest among the all neighbor head nodes at upper layer. The cost value of next hop id1 is larger than next hop id2. Residual energy is the current energy of head node of forwarding data. Forwarding ratio is the delivery ratio of forwarding head node.

If a cluster head receives no messages from other head nodes at upper layer, it will send the data to neighbor heads at same layer. The head nodes received data packet will compare its id to next hop id1 and next hop id2. If node's id is next hop id1 and the data packet is not recorded, it will forward the data packet immediately. Otherwise, if node's id is next hop id2 and data packet is not recorded, it will forward the data packet after waiting for a certain time. During the waiting time, the forwarder will drop the data packet if it receives the same data packet. The forwarding node records 10 data packet forwarded recently. If node's id is neither next hop id1 nor next hop id2, the data packet will be dropped. The procedure of next forwarder selection is shown as algorithm 2.

3.4.3. Maintenance and Update Method of Clusters. The structure of underwater wireless sensor network is dynamic. The nodes move randomly with water current. Therefore, the structure of clusters may change at any time. A recovery mechanism is proposed in this paper.

For member nodes, when the member node fails to send the data to its cluster head, that means it has moved out of its cluster. In this case, it will monitor data packets from cluster heads. Once hearing a data packet of cluster head, it will join this cluster according to the information in data packet.

Each cluster head calculates the average energy of the its cluster. When its energy is lower than average energy, cluster is reformed in this layer and routing information is updated.

For the whole network, at each round of data collection, sink nodes expect the delivery ratio of next round. Once the delivery ratio is less than 70% of expected, the sink nodes will broadcast reformation message to the nodes. The underwater sensor nodes will rebuild the clusters.

4. Simulations and Discussions

In this section, the performance of EERBLC is evaluated and compared with DBR and EEDBR. The simulations are implemented using MATLAB 7.0. The same number of nodes are used in all simulations for fair comparison of EERBLC, DBR and EEDBR. 400 sensor nodes are deployed in a 500 m x 500 m x 500 m 3-D area. Initial energy of each sensor node is 5 joules. Power consumption of node in sending and receiving of data is 2 W and 0.1 W, respectively. Transmission range of each sensor node is 100 m. In the simulations of EERBLC, depth of each layer is defined at 80 m, and 80 nodes are randomly deployed in each layer. All nodes

TABLE 4: Format of data packet.

Type of packet	Data number	Source node id	Head id	Next hop id1	Next hop id2	Data	Residual energy	Forwarding ratio

```
1: Procedure forwarder selection
2: for each head node Hi
3:    for each head Hj ∈ neihgbors ∩ Hi's LN-1 = Hj's LN
4:       calculate H_cost(j)
5:    end for
6:    select two nodes as forwarders with two largest H_cost
7:    send N_DATA packet to forwarder
8: end for
9: for each node Ni received data packet
10:   if next hop id1 in received packet = Ni's id
11:      if the data packet ∉ records
12:         record data packet
13:         forward data packet
14:      else drop the data packet
15:      end if
16:   else if next hop id2 in received packet = Ni's id
17:      if the data packet ∉ records
18:         record data packet
19:         waiting for a certain time
20:         if not receive the same packet
21:         forward data packet
22:         else drop the data packet
23:         end if
24:      else drop the data packet
25:   end if
26:   else drop the data packet
27:   end if
28: end for
```

ALGORITHM 2: Next forwarder selection algorithm.

TABLE 5: Parameters of simulations.

Parameters	Values
Monitoring area	500 m x 500 m x 500 m
Number of nodes	400
Initial energy of nodes	5 joule
Data packet size	200
Communication rang	100 m
Number of sink nodes	2
$\alpha, \beta, \gamma, \theta, \mu, \varepsilon$	1, 1, 0.7, 0.3, 0.6, 0.4

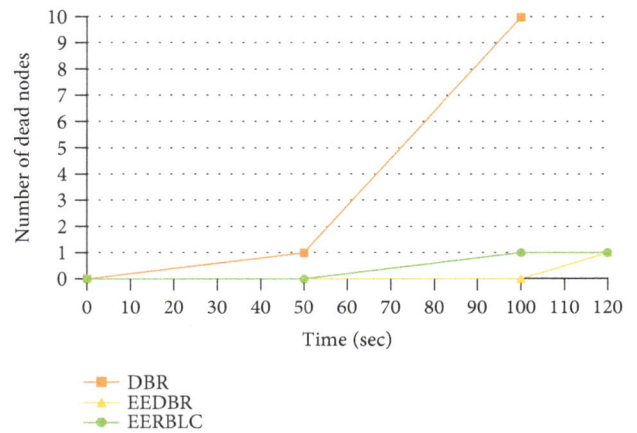

FIGURE 3: Comparison of stability period.

move with water currents from 2-3 m/sec in the horizontal directions. Two buoys are deployed at the surface and used as sinks in order to collect the data packets from the sensor nodes. Every sensor node can hold 10 data packets in its buffer. Size of a data packet is 200 bytes, size of N_HELLO message is 5 bytes, size of N_COM message is 6 bytes, size of N_JOIN message is 5 bytes. Frequency of the carrier acoustic signal is 10 kHz. Simulation parameters are given in Table 5. The final simulation results are taken as an average of 5 different results.

In this work, the performance of protocols is evaluated in terms of stability period, network lifetime, throughput, delivery ratio, energy consumption and end-to-end delay. Stability period is defined as the duration till the first node dies in the whole network. For evaluating the performance of stability, the comparison simulations of DBR, EEDBR and EERBLC are implemented. All parameters of simulations are shown as Table 5. The final simulation results are shown in Figure 3, which are taken as an average of 5 different results. In the DBR protocol, the first nodes died at 50s approximately. In the EEDBR and EERBLC protocol, the first nodes both died at 100 s approximately. In the DBR, a large number of redundant data packets are generated, and residual energy of node is not considered. That leads to premature death of nodes. In the EEDBR and EERBLC, energy is a major

consideration factor during the forwarding data. Especially, EERBLC uses clustering technology to reduce the number of forwarding data. Hence, the time of first dead node is later than DBR.

Network lifetime is defined as the duration till the all nodes die in the whole network. Figure 4 demonstrates the network lifetime of three protocols in random topologies. The simulation results show that lifetime of DBR, EEDBR and EERBLC are about 1600s, 1800s and 2000s, respectively. Due to the redundant data and retransmitting, the lifetime of DBR is shortest in three protocols. In EEDBR, problems of energy and redundant data are improved to some extent, so its lifetime is longer than DBR. Clustering technology used in EERBLC saves the energy. And energy balance is taken into account during the routing. So, EERBLC has the longest lifetime among the three protocols.

Throughput is defined as the number of data packets successful received at sink nodes. Figure 5 shows the throughput simulation results of three protocols. The results show that DBR receives about 95,000 packets, EEDBR receives about 86,000 packets, and EERBLC receives about 70,000 packets during 2000s. DBR receives most packets due to large number of redundant data. EEDBR reduces the number of redundant data to some extent, so throughput of EEDBR is less than DBR. In EERBLC, because cluster heads aggregate data,

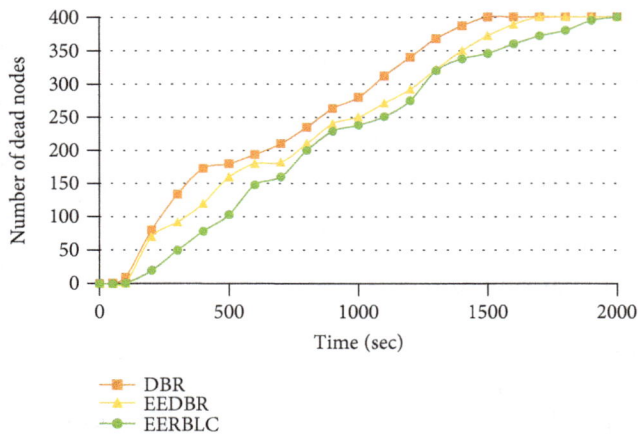

FIGURE 4: Comparison of network lifetime.

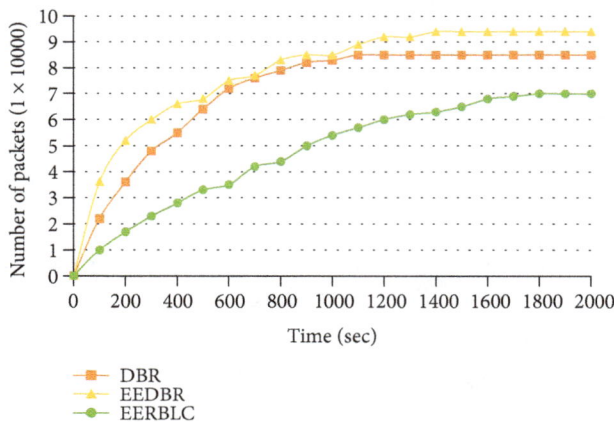

FIGURE 5: Comparison of throughput.

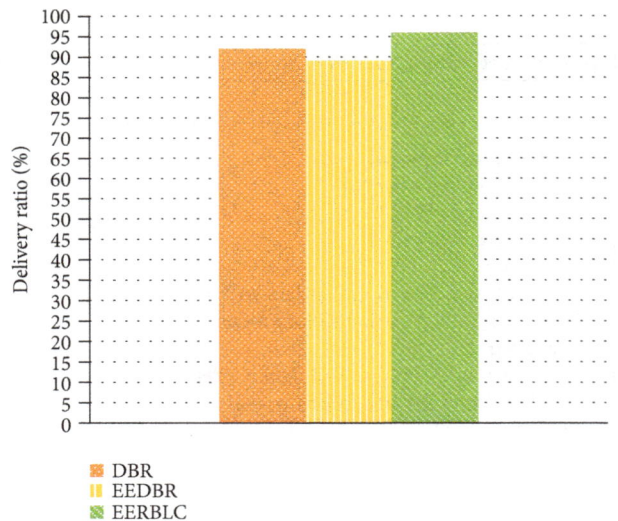

FIGURE 6: Comparison of delivery ratio.

the number of data packets is reduced significantly. Hence the throughput of EBECRP is less than DBR and EEDBR.

Delivery ratio is defined as the ratio of the number of packets successfully received at the sink nodes to the number of packets transmitted from the source nodes. The comparison simulations are implemented to evaluate the performance of delivery ratio in DBR, EEDBR and EERBLC. The results are shown in Figure 6. Delivery ratio of DBR, EEDBR and EERBLC are 92%, 89% and 96%, respectively. The delivery ratio of DBR is higher than EEDBR, because DBR makes packets transmitted redundantly where multiple paths are followed to reach the sink node. For avoiding redundant data, EEDBR reduces the forwarding nodes. That leads to decline of delivery ratio of EEDBR. The data are aggregated effectively in EERBLC, and forwarding ratio is considered during the selection of forwarding node. That leads to more data packets received by sink nodes in EERBLC. So, EERBLC has the highest delivery ratio among the three protocols.

Energy consumption is the indicator to the network performance, and it reflects the status of network lifetime implicitly. Lower consumption energy causes longer lifetime of network. Figure 7 shows the simulation results of energy consumption. The energy consumption of DBR is higher than the other two protocols due to overmuch forwarding

nodes and redundant packets' transmissions. EEDBR selects forwarder based on the depth and the residual energy. It avoids the number of forwarding nodes. Moreover, in EEDBR, due to the priority assignment technique, repeat transmissions of the same data packets are reduced significantly. Hence, energy consumption of EEDBR is less than DBR. EERBLC uses the clustering technology to balance load of the whole network which results in balanced energy consumption and longer lifetime. In the network initial phase, clusters formation needs to consume more energy, therefore, the residual energy of EERBLC is less than EEDBR at first 300 s. However, the energy consumption is less than other two protocols after initial phase.

End-to-end delay is defined as the average time which data packets are received by sink nodes from source nodes. DBR, EEDBR and EERBLC are compared by simulations. The results are shown in Figure 8. Because a certain holding time is needed before forwarding, DBR has the longest end-to-end delay in the three protocols. Priority mechanism is adopted during the forwarding in EEDBR. The forwarder with highest priority transmits the packet immediately. Therefore, the delay is reduced. The data are aggregated effectively in EERBLC, so the reduction of data packets reduces the propagation time. Furthermore, link quality is considered during the selection of forwarding node, retransmissions are restricted effectively. So, EERBLC has the least end-to-end delay among the three protocols.

From the above simulation results, we can see that EERBLC has the longest network lifetime, highest delivery ratio and least end-to-end delay with the lowest energy consumption in the three protocols. It proves that EERBLC is an effective routing protocol for UWSNs.

5. Conclusions

It is hard to replace the batteries of underwater sensor nodes, therefor, energy efficiency is one of research hotspots in underwater wireless sensor networks. In this paper, a

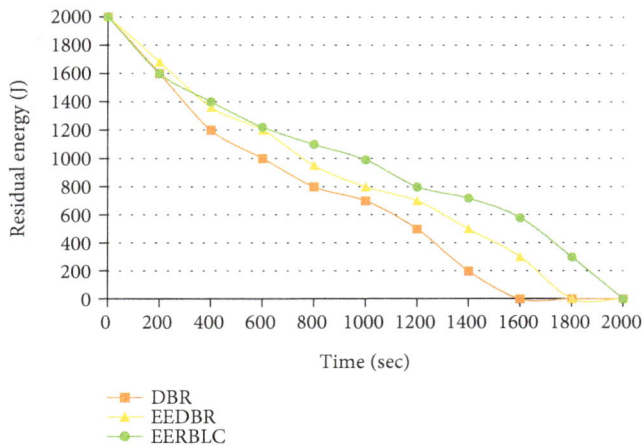

FIGURE 7: Comparison of energy consumption.

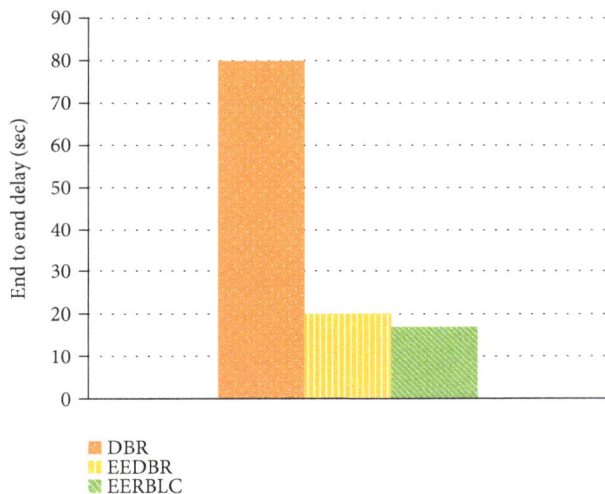

FIGURE 8: Comparison of end-to-end delay.

localization-free routing protocol, named energy efficient routing protocol based on layers and clusters (EERBLC) is proposed. This protocol aims to solve problems of high energy consumption, long end-to-end delay and high error rate. In this protocol, underwater monitoring area is layered, and then the sensor nodes are clustered at each layer. EERBLC includes three phases: cluster formation, transmission routing, maintenance and update of clusters. In the cluster formation phase, a new unequal cluster method suitable for UWSNs is proposed to solve "hotspot" problem. And, EERBLC improves the cluster head selection algorithm. A multi-objective optimization technique is introduced to calculate the cost value for each candidate head. In the transmission routing phase, EERBLC utilizes the forwarding ratio and the residual energy of sensor nodes as routing metrics. Finally, Intra and inter cluster updating method is presented. Through simulations, the EERBLC protocol is compared to DBR and EEDBR routing protocols in UWSNs. The results prove that EERBLC can effectively improve the performance in terms of network lifetime, energy consumption, delivery ratio, throughput and end-to-end delay.

Conflicts of Interest

The authors declare that there is no conflict of interest regarding the publication of this paper.

Acknowledgments

This work is supported by the Fundamental Research Funds for the Central Universities under Grant No. N172304027.

References

[1] J. Heidemann, M. Stojanovic, and M. Zorzi, "Underwater sensor networks: applications, advances and challenges," *Philosophical Transactions*, vol. 370, no. 1958, pp. 158–175, 2012.

[2] G. Tuna and V. C. Gungor, "A survey on deployment techniques, localization algorithms, and research challenges for underwater acoustic sensor networks," *International Journal of Communication Systems*, vol. 30, no. 17, article e3350, 2017.

[3] E. Felemban, F. K. Shaikh, U. M. Qureshi, A. A. Sheikh, and S. B. Qaisar, "Underwater sensor network applications: a comprehensive survey," *International Journal of Distributed Sensor Networks*, vol. 11, no. 11, Article ID 896832, 2015.

[4] H. Mythrehee and A. Julian, "A cross layer UWSN architecture for marine environment monitoring," in *2015 Global Conference on Communication Technologies (GCCT)*, pp. 211–216, Thuckalay, India, April 2015.

[5] M. R. Ahmed, M. Aseeri, M. S. Kaiser, N. Z. Zenia, and Z. I. Chowdhury, "A novel algorithm for malicious attack detection in UWSN," in *2015 International Conference on Electrical Engineering and Information Communication Technology (ICEEICT)*, pp. 1–6, Dhaka, Bangladesh, May 2015.

[6] M. R. Ahmed, S. M. Tahsien, M. Aseeri, and M. Shamim Kaiser, "Malicious attack detection in underwater wireless sensor network," in *2015 IEEE International Conference on Telecommunications and Photonics (ICTP)*, pp. 1–5, Dhaka, Bangladesh, December 2016.

[7] S. Srinivas, P. Ranjitha, R. Ramya, and G. Kumar Narendra, "Investigation of oceanic environment using large-scale UWSN and UANETs," in *2012 8th International Conference on Wireless Communications, Networking and Mobile Computing*, pp. 1–5, Shanghai, China, September 2013.

[8] C. C. Kao, Y. S. Lin, G. D. Wu, and C. J. Huang, "A comprehensive study on the internet of underwater things: applications, challenges, and channel models," *Sensors*, vol. 17, no. 7, p. 1477, 2017.

[9] M. Li, X. Du, K. Huang, S. Hou, and X. Liu, "A routing protocol based on received signal strength for underwater wireless sensor networks (UWSNs)," *Information*, vol. 8, no. 4, p. 153, 2017.

[10] P. Jiang, Y. Feng, F. Wu, S. Yu, and H. Xu, "Dynamic layered dual-cluster heads routing algorithm based on krill herd optimization in UWSNs," *Sensors*, vol. 16, no. 9, p. 1379, 2016.

[11] M. Jouhari, K. Ibrahimi, and M. Benattou, "New greedy forwarding strategy for UWSNs geographic routing protocols," in *2016 International Wireless Communications and Mobile Computing Conference (IWCMC)*, pp. 388–393, Paphos, Cyprus, September 2016.

[12] Z. Zhou, B. Yao, R. Xing, L. Shu, and S. Bu, "E-CARP: an energy efficient routing protocol for UWSNs in the internet of underwater things," *IEEE Sensors Journal*, vol. 16, no. 11, pp. 4072–4082, 2016.

[13] M. Aslam, F. Wang, Z. Lv et al., "Energy efficient cubical layered path planning algorithm (EECPPA) for acoustic UWSNs," in *2017 IEEE Pacific Rim Conference on Communications, Computers and Signal Processing (PACRIM)*, pp. 1–6, Victoria, BC, Canada, August 2017.

[14] P. Xie, J. H. Cui, and L. Lao, "VBF: Vector-Based Forwarding Protocol for Underwater Sensor Networks," in *NETWORKING 2006. Networking Technologies, Services, and Protocols; Performance of Computer and Communication Networks; Mobile and Wireless Communications Systems. NETWORKING 2006. Lecture Notes in Computer Science, vol 3976*, F. Boavida, T. Plagemann, B. Stiller, C. Westphal, and E. Monteiro, Eds., pp. 1216–1221, Springer, Berlin, Heidelberg, 2006.

[15] N. Nicolaou, A. See, and P. Xie, "Improving the robustness of location-based routing for underwater sensor networks," in *OCEANS 2007 – Europe*, pp. 1–6, Aberdeen, UK, June 2007.

[16] S. M. Mazinani, H. Yousefi, and M. Mirzaie, "A vector-based routing protocol in underwater wireless sensor networks," *Wireless Personal Communications*, vol. 100, no. 4, pp. 1569–1583, 2018.

[17] M. R. Khosravi, H. Basri, and H. Rostami, "Efficient routing for dense UWSNs with high-speed mobile nodes using spherical divisions," *Journal of Supercomputing*, vol. 74, no. 2, pp. 696–716, 2018.

[18] J. M. Jornet, M. Stojanovic, and M. Zorzi, "Focused beam routing protocol for underwater acoustic networks," in *Proceedings of the 3rd ACM International Workshop on Wireless Network Testbeds, Experimental Evaluation and Characterization (WuWNeT'08)*, pp. 75–82, San Francisco, CA, USA, September 2008.

[19] F. A. Salti, N. Alzeidi, and B. R. Arafeh, "EMGGR: an energy-efficient multipath grid-based geographic routing protocol for underwater wireless sensor networks," *Wireless Networks*, vol. 23, no. 4, pp. 1301–1314, 2017.

[20] J. Jiang, G. Han, H. Guo, L. Shu, and J. J. P. C. Rodrigues, "Geographic multipath routing based on geospatial division in duty-cycled underwater wireless sensor networks," *Journal of Network and Computer Applications*, vol. 59, pp. 4–13, 2016.

[21] R. W. L. Coutinho, L. F. M. Vieira, and A. A. F. Loureiro, "DCR: depth-controlled routing protocol for underwater sensor networks," in *2013 IEEE Symposium on Computers and Communications (ISCC)*, pp. 453–458, Split, Croatia, July 2013.

[22] H. Yan, Z. Shi, and J. Cui, "DBR: depth-based routing for underwater sensor networks," in *NETWORKING 2008 Ad Hoc and Sensor Networks, Wireless Networks, Next Generation Internet. NETWORKING 2008. Lecture Notes in Computer Science, vol 4982*, A. Das, H. K. Pung, F. B. S. Lee, and L. W. C. Wong, Eds., pp. 16–1221, Springer, Berlin, Heidelberg, 2008.

[23] A. Wahid and D. Kim, "An energy efficient localization-free routing protocol for underwater wireless sensor networks," *International Journal of Distributed Sensor Networks*, vol. 8, no. 4, Article ID 307246, 2012.

[24] M. Ayaz and A. Abdullah, "Hop-by-hop dynamic addressing based (H2-DAB) routing protocol for underwater wireless sensor networks," in *2009 International Conference on Information and Multimedia Technology*, pp. 436–441, Jeju Island, South Korea, December 2009.

[25] S. Basagni, C. Petrioli, R. Petroccia, and D. Spaccini, "CARP: a channel-aware routing protocol for underwater acoustic wireless networks," *Ad Hoc Networks*, vol. 34, pp. 92–104, 2015.

[26] Z. ying, S. Hongliang, and Y. Jiancheng, "Clustered routing protocol based on improved K-means algorithm for underwater wireless sensor networks," in *2015 IEEE International Conference on Cyber Technology in Automation, Control, and Intelligent Systems (CYBER)*, pp. 1304–1309, Shenyang, China, June 2015.

[27] A. Majid, I. Azam, A. Waheed et al., "An energy efficient and balanced energy consumption cluster based routing protocol for underwater wireless sensor networks," in *2016 IEEE 30th International Conference on Advanced Information Networking and Applications (AINA)*, pp. 324–333, Crans-Montana, Switzerland, March 2016.

[28] A. Khan, N. Javaid, I. Ali et al., "An energy efficient interference-aware routing protocol for underwater WSNs," *KSII Transactions on Internet and Information Systems*, vol. 11, no. 10, pp. 4844–4864, 2017.

[29] A. Khan, I. Ali, and H. Mahmood, "A localization-free variable transmit power routing protocol for underwater wireless sensor networks," in *Advances in Network-Based Information Systems. NBiS 2017. Lecture Notes on Data Engineering and Communications Technologies, vol 7*, L. Barolli, T. Enokido, and M. Takizawa, Eds., pp. 136–147, Springer, Cham, 2018.

[30] F. Hidoussi, H. Toral-Cruz, D. E. Boubiche et al., "PEAL: power efficient and adaptive latency hierarchical routing protocol for cluster-based WSN," *Wireless Personal Communications*, vol. 96, no. 4, pp. 4929–4945, 2017.

[31] M. Khanafer, I. Al-Anbagi, and H. T. Mouftah, "An optimized cluster-based WSN design for latency-critical applications," in *2017 13th International Wireless Communications and Mobile Computing Conference (IWCMC)*, pp. 969–973, Valencia, Spain, June 2017.

[32] S. Krishnamoorthy, "Enhanced adaptive clustering mechanism for effective cluster formation in WSN," *International Journal of Current Engineering and Scientific Research*, vol. 4, no. 9, p. 8, 2017.

[33] S. Mondal, P. Dutta, S. Ghosh et al., "Energy efficient rough fuzzy set based clustering and cluster head selection for WSN," in *2016 2nd International Conference on Next Generation Computing Technologies (NGCT)*, pp. 439–444, Dehradun, India, October 2017.

[34] G. Tuna, "Clustering-based energy-efficient routing approach for underwater wireless sensor networks," *International Journal of Sensor Networks*, vol. 27, no. 1, pp. 26–36, 2018.

[35] S. Rani, S. H. Ahmed, J. Malhotra, and R. Talwar, "Energy efficient chain based routing protocol for underwater wireless sensor networks," *Journal of Network and Computer Applications*, vol. 92, pp. 42–50, 2017.

[36] N. Goyal, M. Dave, and A. K. Verma, "Improved data aggregation for cluster based underwater wireless sensor networks," *Proceedings of the National Academy of Sciences, India Section A: Physical Sciences*, vol. 87, no. 2, pp. 235–245, 2017.

[37] D. Das and P. M. Ameer, "Energy efficient geographic clustered multi-hop routing for underwater sensor networks," in *TENCON 2017 - 2017 IEEE Region 10 Conference*, pp. 409–414, Penang, Malaysia, November 2017.

[38] P. Jiang, J. Liu, and F. Wu, "Node non-uniform deployment based on clustering algorithm for underwater sensor networks," *Sensors*, vol. 15, no. 12, pp. 29997–30010, 2015.

11

Systematic Development of a Wireless Sensor Network for Piezo-Based Sensing

Jian Chen ⓘ,[1] Peng Li,[2] Gangbing Song ⓘ,[2] Yu Tan,[1] Yongjun Zheng,[1] and Yu Han ⓘ[3]

[1]*College of Engineering, China Agricultural University, 17 Tsinghua East Rd, Beijing 100083, China*
[2]*Department of Mechanical Engineering, University of Houston, 4800 Calhoun Rd, Houston, TX 77204, USA*
[3]*College of Water Resources & Civil Engineering, China Agricultural University, 17 Tsinghua East Rd, Beijing 100083, China*

Correspondence should be addressed to Yu Han; yh916@uowmail.edu.au

Academic Editor: Christos Riziotis

A low-power wireless sensor/actuator network was specially developed and optimized for piezoceramic transducer-based active sensing applications. Wireless sensor network promises increased system flexibility, lower system cost, and increased robustness through decentralization. Piezoceramic signal conditioning circuit, actuating circuit, power management, and wireless microcontroller were integrated in the hardware design. IEEE 802.15.4 wireless stack protocol was implemented on the hardware, and user input/output management together with a shell provided easier debugging and configuring interface. The designed system provides a low-power wireless solution towards many applications such as wireless structural health monitoring and wireless structural vibration control.

1. Introduction

During the last two decades, a large amount of research work has focused on the wireless sensor networks (WSNs) and its potential application due to the increased flexibility and lower costs they promise to provide compared with wired installations [1–3]. Among them, many efforts have been made on the design and development of wireless sensor network platforms. The design of a WSN is highly system oriented. The physical and logical structure of the system together with other environmental requirements constrains the design of the wireless network and the wireless sensor devices [4]. The diverse constraints from different applications make it necessary to have a platform-based development approach where much of the components can be reused.

On the hardware side, many companies and universities have proposed their WSN platforms. From the power consumption and system performance point of view, current sensor network platforms fall into one of the two main categories: low-power consumption, application-specific platforms and high-power consumption, full-featured high-performance platforms. The former generally consists of a power-efficient microprocessor and a few application-specific resources, while the latter generally has a general purpose processor and a full complement of resources available via a general-purpose input/output (GPIO) interface [5]. The best known low-power system is the open-source hardware platform, the MICA series [6], designed by UC Berkeley [7] and manufactured by Crossbow [8]. MICA combined sensing, communication, and computing into a complete architecture, while consuming only a fraction of a watt of power. In the MICA platform, Atmel's eight-bit AVR microprocessor [9] was used as the central processing unit, and Texas Instrument's 2.4 GHz wireless transceiver [10, 11] was used as the wireless component. Different extension sensor boards can be connected to the MICA platform to realize the sensing of light, temperature, acceleration, or acoustic. Another popular sensor node developed also by UC Berkeley is the Telos. Telos used TI's 16-bit microcontroller MPS430 as a CPU and TI's CC2420 radio for wireless communication. Compared with Mica series, Telos boasts better performance on both power consumption and computation capacity [12].

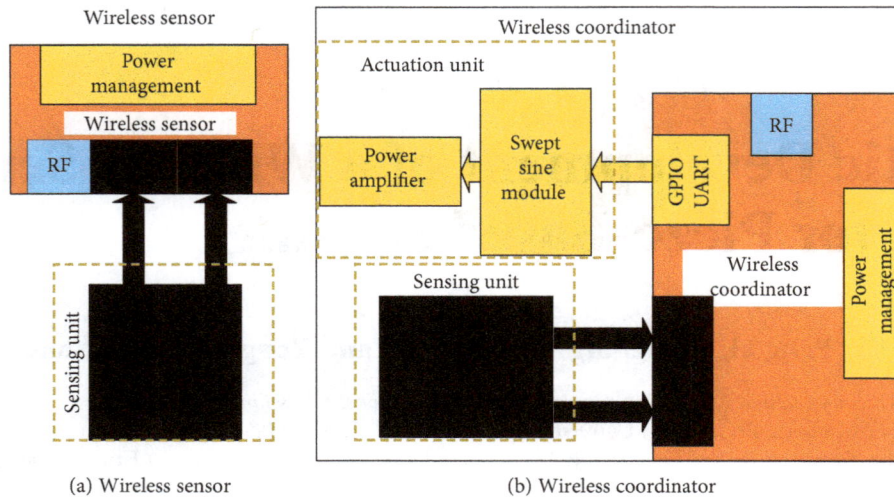

FIGURE 1: Architecture of wireless sensor and wireless coordinator.

On the high-performance side, Yale University designed the XYZ platform with an ARM7TDMI CPU and the TI CC2420 wireless transceiver [13]. The ARM core processor provided much larger RAM and much higher system performance; however, the power consumption was also greater. Another well-known platform is the Imote2 platform from Intel [14]. With the XScale PXA271 processor, Imote2 can run up to several hundred MHz and can afford complex processing algorithms. However, the power consumption was too much in a high-speed mode that it was not suitable for long-term battery-powered monitoring applications. There are some similar hardware platforms from several different companies: Ember Inc. [15], Dust Inc. [16], and Sentilla [17, 18], to name a few. Sandia National Lab proposed a modular platform, the MASS [5]. Instead of the traditional architecture where the resources are controlled by a single microcontroller, each board of MASS had its own specific processing unit. Communication between the modules was handled by the Inter-Integrated Circuit (I^2C) bus. This design technique was very modular for both software and hardware and is a simple architecture for lab experiments. However, the induced complexities from the multiple processing units impose a large amount of coding overhead for a simple sensor node, which makes it unfeasible for real application.

It should be noted that the above short review is just partial and does not include all available solutions. Although many WSN platforms have been produced, most platforms target general purpose applications and are not optimized for specific applications. There is not much work on a WSN platform optimized for a piezoceramic-based sensor network. Piezo-based active sensing WSN has special requirements, such as relatively high sampling rate (at a few thousand Hz), incorporation of an amplifier for the piezo-ceramic element for actuation, and low energy consumption to realize active sensing. Moreover, a low-power cost design is needed for the battery-based WSN. In this paper, a wireless network was specially designed and optimized for piezoceramic transducers, and a platform was used to

realize impact detection, active sensing for structural health monitoring (SHM) [2] and wireless structural vibration control [3].

2. Hardware Design

Designs of the wireless coordinator and the wireless sensor were presented for piezo-based wireless sensing. The functions of the wireless coordinator include building and managing the wireless network, controlling the performance of all the wireless sensor nodes, realizing the routing of the information, and transferring data from WSN to the wired station. The wireless sensor is used to measure the impacts, record the vibration, perform data acquisition, and transfer the data to a wireless coordinator.

The wireless coordinator is composed of four parts: the sensing unit, the actuation unit, the wireless microcontroller, and the power management module. The wireless sensor nodes have a similar structure as the wireless coordinator, except for the simplified power management module and the elimination of the actuation unit to reduce the power consumption. Figure 1 illustrates the architecture of the wireless sensor nodes (Figure 1(a)) and wireless coordinator (Figure 1(b)). Details of the wireless sensing unit and wireless coordinator will be presented in the following subsections.

2.1. Sensing Unit. The sensing unit is comprised of two parts: the impact detection circuit and the signal conditioning circuit. The principle behind piezomaterial-based sensing is piezoelectricity, where electrical charges are generated from the force exerted on the piezoelectric sensor. The function of the sensing unit is to sense and regulate the charge signal to an appropriate voltage signal which can be sampled by the analogue to digital (ADC) converter in the microcontroller system. The piezoelectric sensing circuit is different from ordinary signal conditioning circuits as the piezoelectric sensor is a capacitive load and has a large output resistance. A low-power piezo-sensing unit has specially been designed for the piezomaterial-based wireless sensor network.

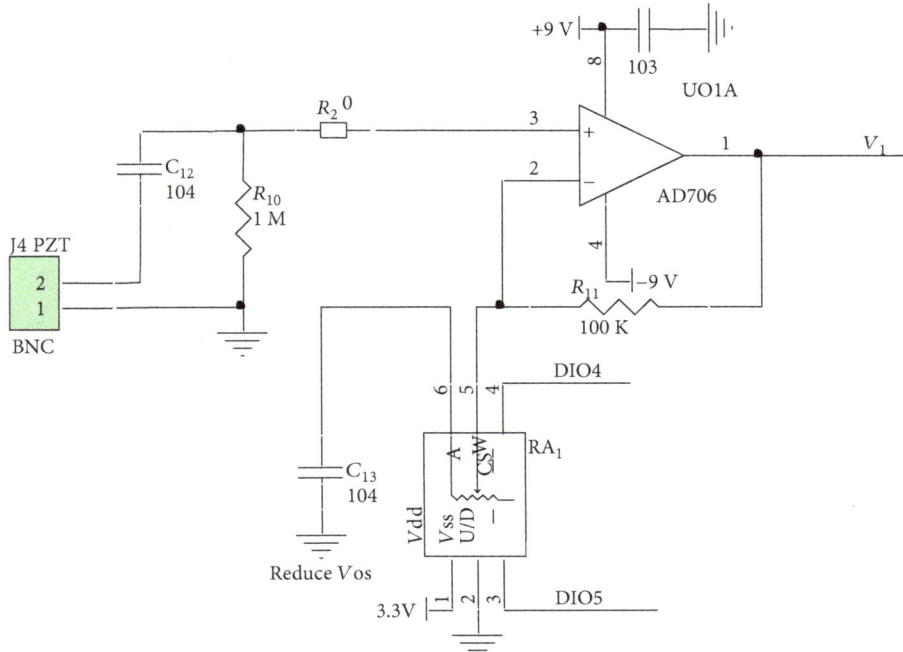

FIGURE 2: First-stage amplifying circuit.

2.1.1. Signal Conditioning Circuit. The amplitude of the charge and therefore the current generated by piezoelectric sensor are proportional to the magnitude of the vibration on it. The piezoelectric sensor is often used to measure dynamic signals, such as vibrations; therefore, this design does not require the circuit to hold the charge and the charge leakage will not be an issue. There are several ways to measure the charge signal. One common method is to use the integration circuit to convert charge into current and then to a voltage signal. This method can solve the problem of the high-impedance feature of a piezoelectric sensor. However, it causes another major problem, namely, the saturation of the decoupling capacitor in the circuit. This requires the circuit parameters to be tuned for a certain range of inputs. If the amount of charge has dramatic changes, such as in our case, it is not suitable to use the integration circuit. To optimize the signal conditioning for PZT, the method of using a large resistor to transform the charge signal to the voltage signal is adopted in this design. Although the impedance of the piezoelectric sensor is large, it does not affect the charge converted to voltage on the large resistor. And with a proper selection of an op-amp, this circuit works much better than the integration circuit. The tradeoff in this design is that a high-value resistor could produce a larger signal and small noises can be easily amplified through high-value resistors. Experimental results show that a 1 M resistor is the best choice for usage.

In Figure 2, J4 is the connector to the piezoceramic transducer, and C_{12} (the negative capacitor is included in the charge pump circuit) eliminates the DC value from the PZT while the 1 M ohm resistor R_{10} converts the charge signal to a voltage signal. Resistor R_2 is an optional resistor which can be tuned to match different load requirements.

Op-amp UO1A is employed to amplify the voltage with its gain adjustable by resister RA_1. The gain of this stage can be represented as

$$\text{Gain} = \frac{R_{11} + RA_1}{R_{11}}. \tag{1}$$

The signal generated by the piezoceramic material is bipolar; however, the ADC on the microcontroller is singularly polar. The first-stage op-amp only scaled the signal into a proper range, and an additional op-amp is needed to add offset to the signal. Figure 3 shows the offset circuit.

This circuit further amplifies the input signal and adds a 1.25 volts offset to the output, so the output can be a range from 0 to 2.5 volts. The input-output relation can be represented as

$$V_2 = \left(1 + \frac{R_{13}}{R_{12}}\right) - \frac{R_{13}}{R_{12}} V_1 = 1.25 - \frac{R_{13}}{R_{12}} V_1. \tag{2}$$

The last part of the signal conditioning circuit is a sallen-key low pass filter for the anti-aliasing purpose required by ADC sampling, as shown in Figure 4. The default sampling speed is 250 Hz, and the R_5 is an adjustable resistor. The cutoff frequency is represented by

$$f_c = \frac{1}{2\pi\sqrt{R_5 R_6 C_5 C_6}} \approx 500 \text{ Hz}, \tag{3}$$

when R_5 is 10 K. The larger the value of R_5, the lower the cutoff frequency is. SHDN (shutdown) is a power down control logic input that provides a way to shut the circuit down when not in use. The antialiasing low pass filter is very important in digital signal processing since the

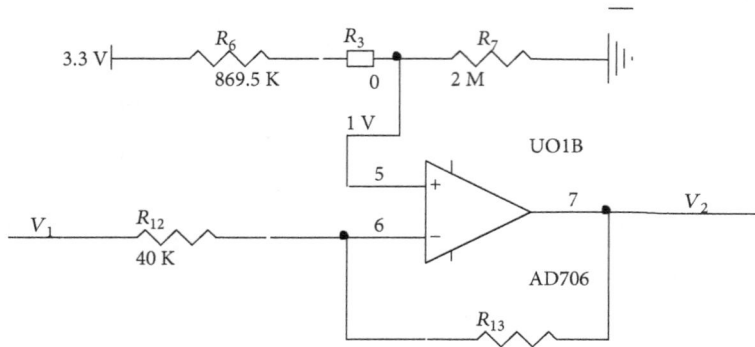

FIGURE 3: Offset amplifying circuit.

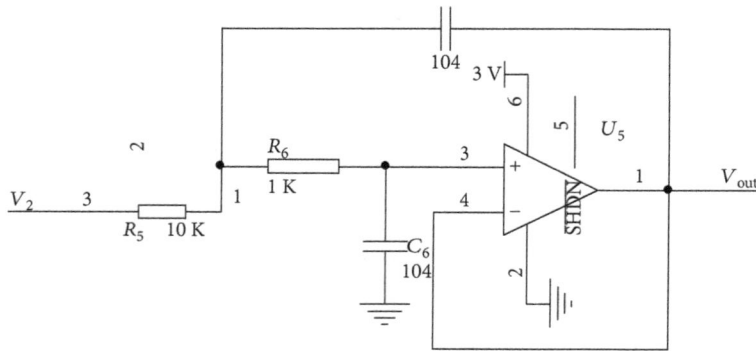

FIGURE 4: Sallen-key low pass filter.

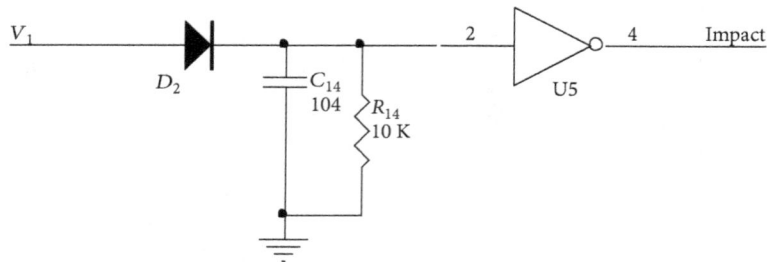

FIGURE 5: Impact detection circuit.

frequency above the Nyquist frequency is folded back into the low-frequency range. High-frequency noise is commonly seen in the microcontroller system due to the high speed switching happening in the digital circuit. These frequencies are mostly blocked by the antialiasing. The antialiasing filter used before the ADC sampler will attenuate the high frequencies larger than the Nyquist frequency and keeps the aliasing devices from being sampled.

2.1.2. Impact Detection Circuit. The purpose of the impact detection circuit is to sense the impact on the piezoelectric sensor and generate an interrupt signal to the microcontroller to inform the impact event. This feature can be very useful for low-power battery-based monitoring systems, where the interruption is used as a wake-up signal for the microcontroller. The threshold of the impact is

configurable in the signal conditioning circuit by changing the resistance of the op-amp circuit. Figure 5 shows the impact detection circuit.

The impact detection circuit takes the output from the first-stage amplifier, and the RC circuit and the diode are mainly peak detectors.

2.2. Actuation Unit. The piezoceramic material will generate an electric charge when it is subjected to a stress or strain (the direct piezoelectric effect); the piezoceramic material will also mechanically stress or strain when an electric field is applied to the piezoelectric material in its poled direction (the converse piezoelectric effect). The piezoelectricity property of piezoelectric material enables the active sensing ability, which allows a piezoelectric material to act as either a sensor or an actuator for structural health monitoring

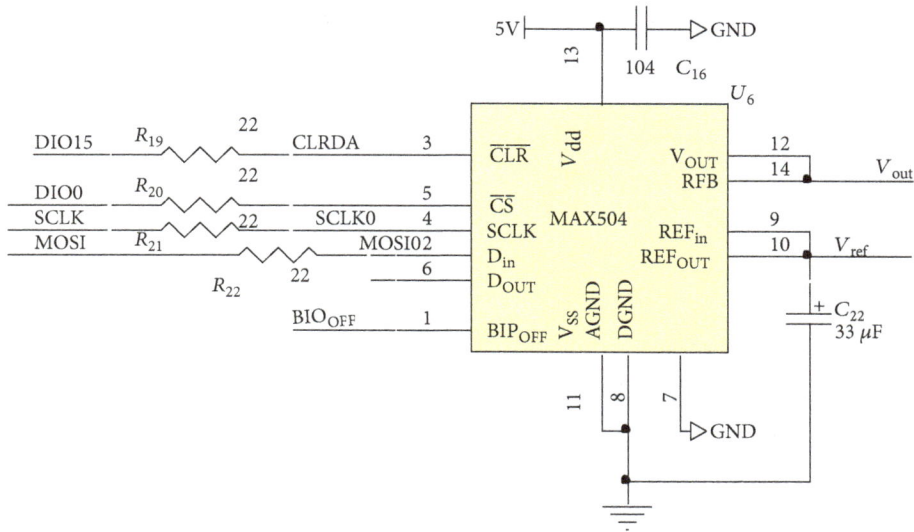

FIGURE 6: DAC circuit.

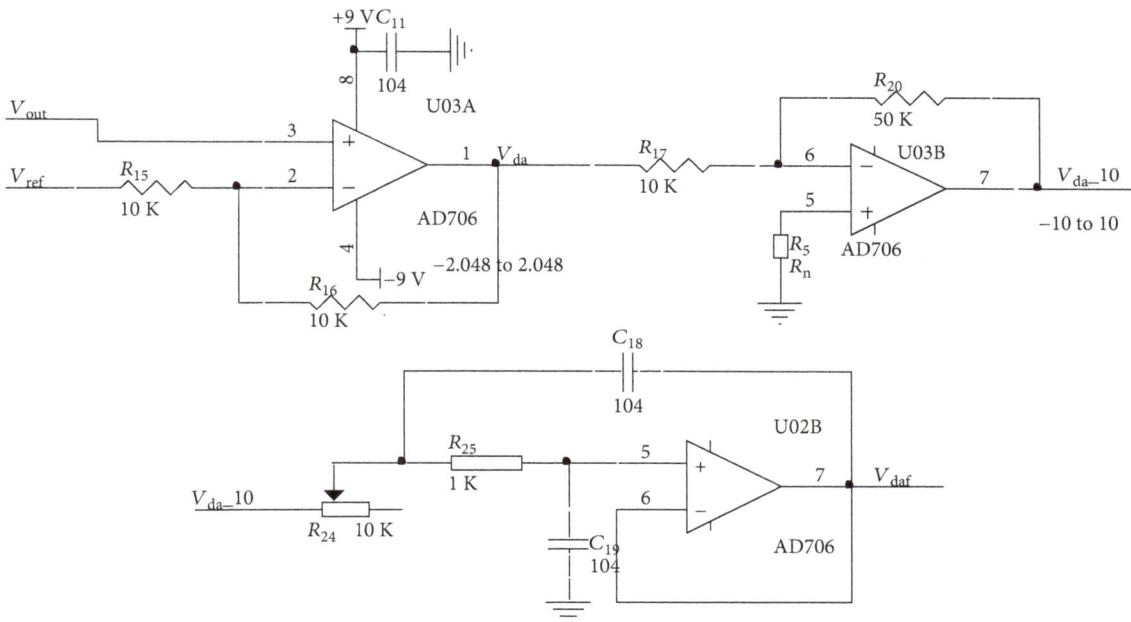

FIGURE 7: DAC amplifying and filtering circuit.

purposes, reducing the number of transducers and the system complexity. Thus, the actuation unit is used to generate an actuation signal for control and excitations. For example, in the structural health monitoring applications, a swept sine signal produced by the actuation unit can be used as the excitation source owing to the active sensing property and generates vibration waves to propagate inside a structure. The health state is evaluated from the vibration response.

A lower power MAX504 is utilized. The 10-bit digital to analog converter Max504 is powered by a single +5 volts; thus, the digital logic one on the SPI bus is also 5 volts. However, the microcontroller operates in the 3.3-volt range.

Therefore, a 22-ohm serial resistor is put on the SPI bus to protect the microcontroller from being damaged. Figure 6 illustrates the DAC circuit.

The output range can be configured to be 0 to 2.048 volts or 0 to 4.096 volts. The range of 0 to 2.048 volts is used in the design, and it can work with another amplifying circuit to adjust the voltage to −10 to +10 volts. The output from the DAC is discrete, inducing high-frequency signal. A low-pass filter is employed to eliminate these high frequencies generated by the DAC changes. Figure 7 shows the amplifying and the low pass filter circuit. Utilizing the 2.048-volt voltage reference V_{ref} from MAX504, op-amp U03A changes the actuation signal from a single polar (0- to 2.048-volt

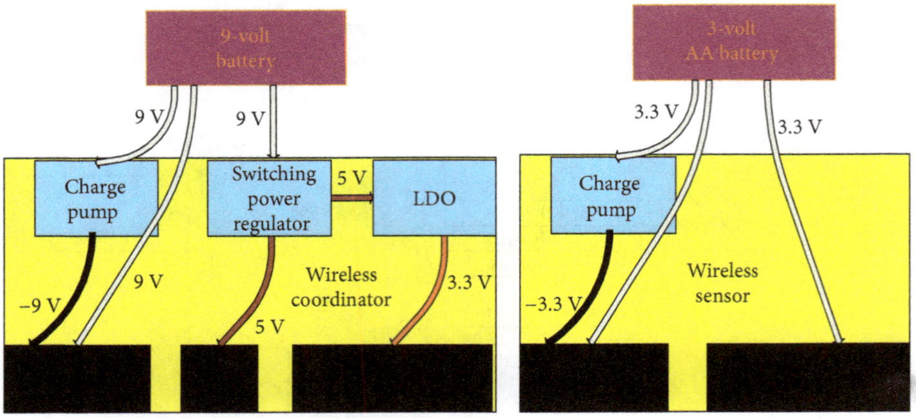

FIGURE 8: Power system design for the wireless coordinator and sensor.

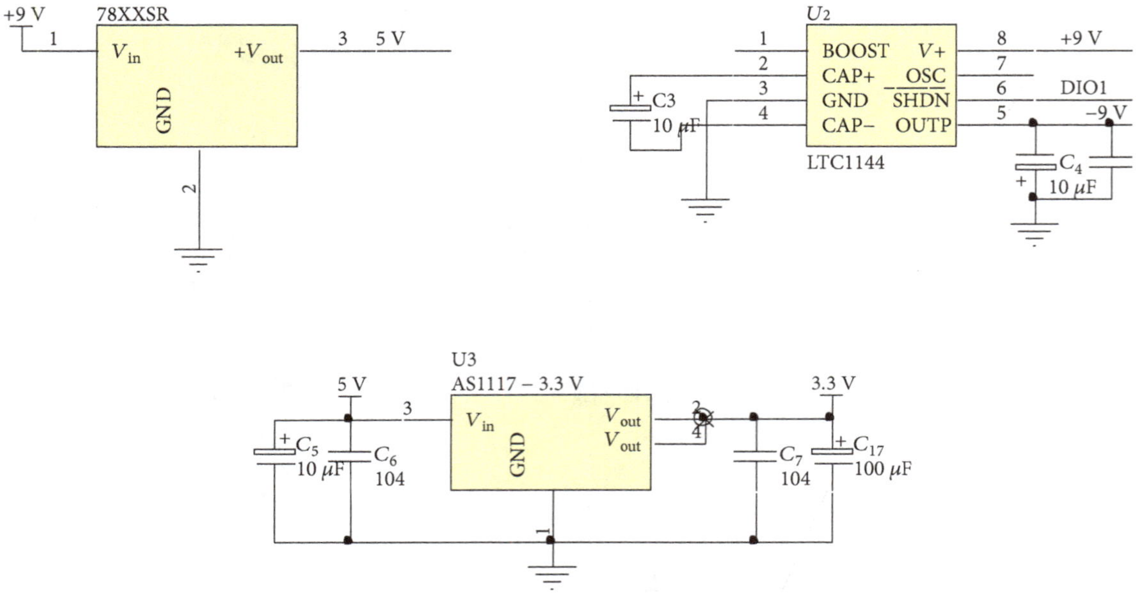

FIGURE 9: Power system circuit design.

range) to a bipolar (−2.048- to +2.048-volt range). The input-output relationship is shown in

$$V_{da} = 2V_{out} - V_{ref}. \qquad (4)$$

The op-amp U03B is used to amplify V_{da} five times, so that it reaches the −10- to 10-volt range. The low pass filter U02B is a sallen-key low pass filter, which is the same as the antialiasing filter in the sensing unit.

2.3. Power Management Unit. A power management unit directly determines performances of power saving. The WSN targeting piezo-based application needs dual voltage supplies to complete the sensing and actuating tasks. Several voltage regulators are used to adjust the battery voltage to ideal values. During voltage conversion, an efficient circuit is able to optimize the power efficiency to a maximum level.

Figure 8 illustrates the optimized power management units of the wireless coordinator (left) and the wireless

sensor (right). And Figure 9 shows the circuit design of the power system.

The 78XXSR is the switching power supply that regulates 9 Volts to 5 Volts for DAC, U2 is the charge pump that generates the negative power supply, and U3 is the LDO (Low-dropout regulator) that provides clean power supply to the MCU (Microcontroller Unit).

2.4. Wireless Microcontroller. The wireless coordinator is the key device of the whole system. Figure 10 shows the system block diagram of the JN5139 microcontroller. The microcontroller consists of four parts: the RISC CPU core with memory, different on-chip peripherals, 2.4 GHz radio components, and the power management.

The piezoceramic-based active sensing WSN differs from other WSNs in that a higher sampling rate, computational power, and lower power cost are needed. The JN5139 microcontroller is a good solution. The JN5139 component can work at 32 MHz and transfer 32 MIPS. In order to

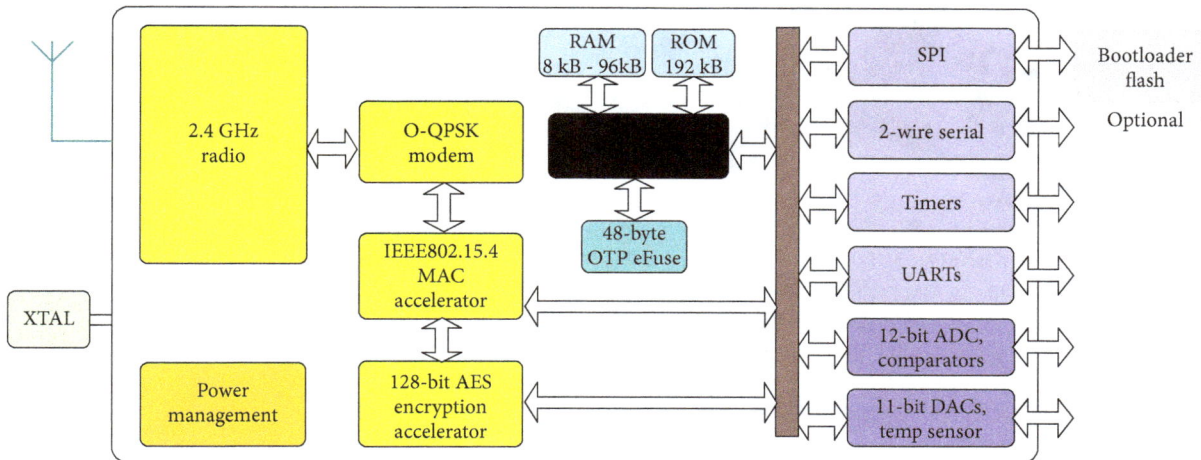

FIGURE 10: System block diagram for JN5139.

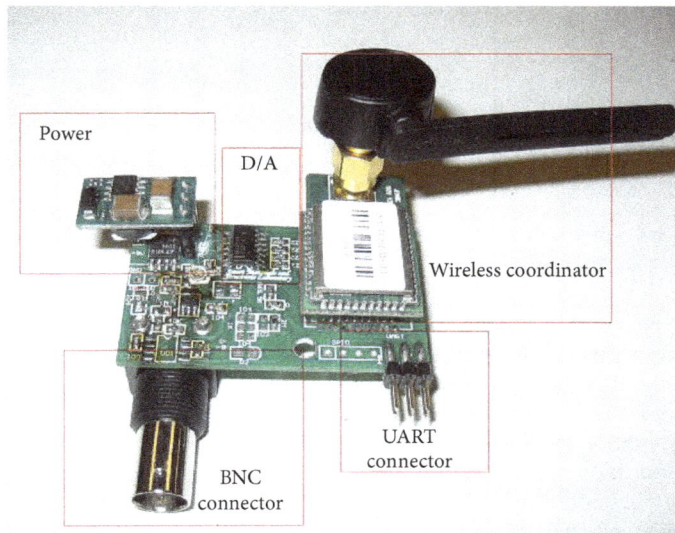

FIGURE 11: Photo of the hardware PCB (front view).

substantially increase the battery life, three working modes operated by the JN5139 guarantee that the power cost of the whole system is constrained.

2.5. Fabricated Hardware System. Photos of the wireless coordinator are shown in Figures 11 and 12. Figure 11 offers a top view, while Figure 12 provides a bottom view. Note that the DAC, the microcontroller, and part of the power management are placed on the top of the PCB, while the MCU peripheral, the BNC connector, and the rest of the power management are placed on the bottom. The universal asynchronous receiver/transmitter (UART) connector is placed on the side of the PCB.

Figure 13 shows the performance comparison between Crossbow Imote2, MICA, Telos, and the designed Jennic WSN system. Although the Imote2 provides a much higher speed and more resources, both the price and the power consumption are higher, which disqualify it in large-scale SHM applications. The MICA has relatively low power

FIGURE 12: Photo of the hardware PCB (rear view).

Characteristic	Crossbow Imote2 ($550/node)	Crossbow MICA ($100/node)	Telos ($200/node)	Jennic ($50/node)
Model (price)	Crossbow Imote2 ($550/node)	Crossbow MICA ($100/node)	Telos ($200/node)	Jennic ($50/node)
Processor	Intel XScale (416 MHz)	Atmel ATmega 128l (4 MHz)	TI MPS430 (16 MHz)	JN5139 RISC (16 MHz)
Memory	256 k SPRAM, 32M FLASH, 32M SDRAM	4 k SRAM, 4 k EEPROM, 128 k FLASH, 4M Serial FLASH	10 KB RAM, 48 KB FLASH 16 k EEPROM	96 k RAM, 192 k ROM
Max current draw	500 mA @ 5 V	17.5 mA @ 3 V	23 mA @3 V	37 mA @ 3.6 V
ADC	4ch, 12-bit	8ch, 10-bit	8ch, 12-bit	4ch, 12-bit
DAC	None	None	12-bit	2× 1ch, 11-bit
Max sample rate	20+ kHz	1 kHz	20 kHz	20 kHz
OD and programming	Linux, TinyOS, or Microsoft .NET/C, C++, or .NET	TinyOS/C or C++	TinyOS/C	NO OS/C
Programming environment	Generic, open-source programming environment			
Compiler	GCC on Linux or Cygwin			Propiatary
Programming interface	Onboard USB interface		USB interface	UART interface

FIGURE 13: Comparison of Crossbow Imote2, MICA, and the designed Jennic system.

consumption and a lower cost; however, the limited RAM size and low speed make it impossible to meet the high sampling rate requirement from the PZT-based applications. Although Telos has a higher speed, the limited RAM size is an issue. The designed Jennic WSN platform provides specially designed piezo signal conditioning circuit and low power actuating DAC circuit. In addition, the power management unit provides a high efficient bipolar power supply that is needed in piezo-based sensing. It can be seen that compared with other commercially available WSN platforms, the designed Jennic WSN system is optimized for piezoceramic transducer-based applications because it provides a relatively high sampling frequency, low power consumption, large RAM for data storage, and abundant on-chip resources.

3. Embedded Software Platform Design

Together with the hardware system, the embedded software system provides a platform for different wireless applications. The software platform has the following functions: managing access to hardware peripherals, handling various interrupt requests, realizing wireless sensor network software stack, performing user inputs and outputs, and providing services to the application layer. Those functions can be categorized into several different layers. Figure 14 illustrates the software platform of the system.

3.1. Hardware Abstraction Layer (HAL). The management of the hardware peripherals is realized through drivers.

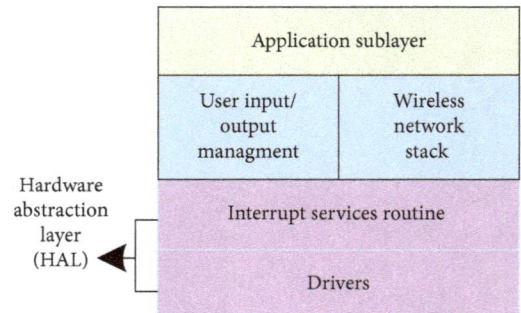

FIGURE 14: Software architecture.

Hardware drivers include timer drivers, ADC drivers, Serial Peripheral Interface (SPI) bus drivers, DAC drivers, radio drivers, GPIO drivers, and UART drivers. A timer driver provides counting and timing services to both the wireless stack and other applications. An ADC driver realizes the control and reading of ADC on the microcontroller. A SPI bus driver and a DAC driver together realize the function of sending out a voltage. A radio driver is responsible for handling input wireless packages and sending wireless messages out. A GPIO driver enables applications to read and write a logic level on a GPIO pin. A UART supplies function interfaces such as sending and receiving characters. The interrupts are handled in the interrupt service routines (ISR). This layer works closely with the hardware driver layer to produce different requests in the application layer. The driver layer together with the ISR layer forms the so-called HAL.

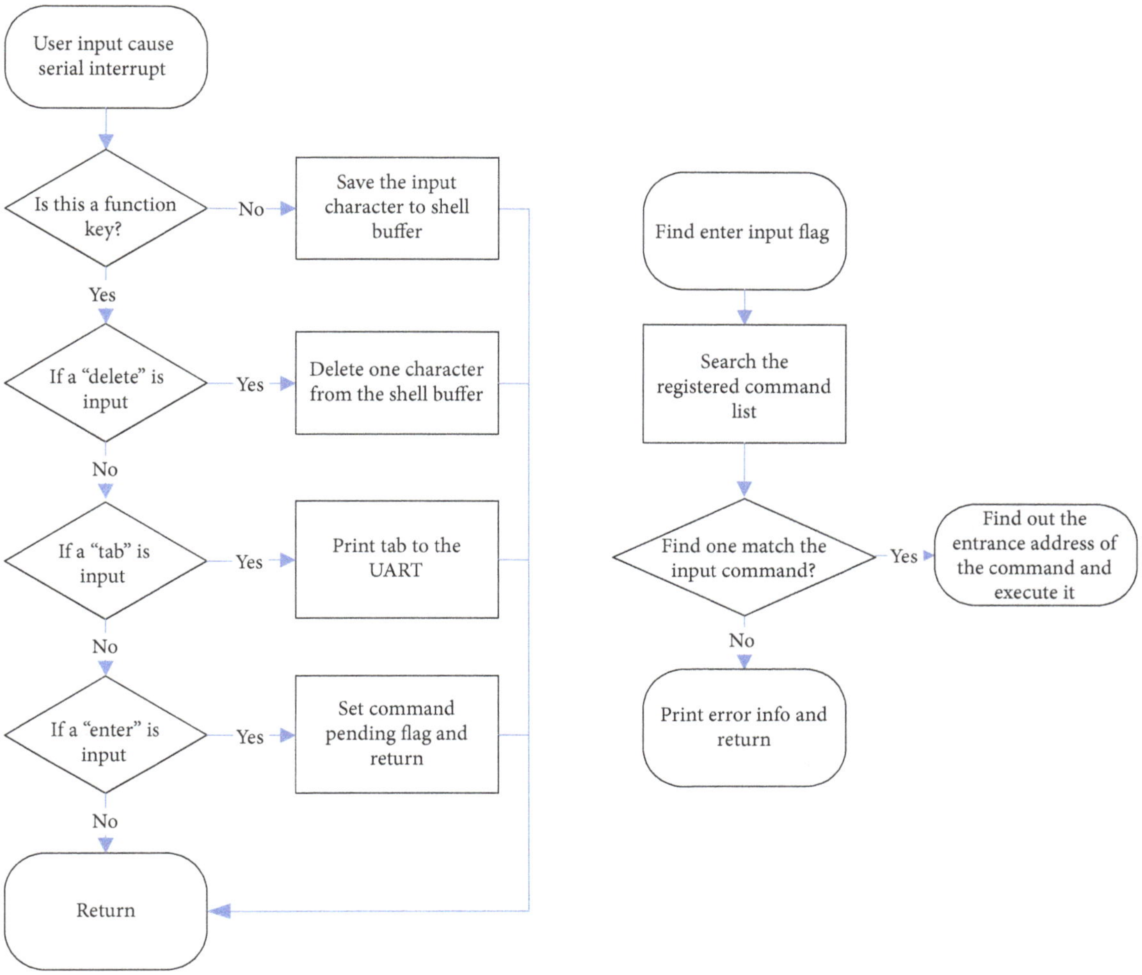

FIGURE 15: Flow chart for shell processing.

3.2. *User Input/Output Layer.* The user input/output was designed through the UART. A shell program was constructed, so that both system debugging and system configuration can be realized.

The shell program utilizes the UART driver and retrieves and parses the user input from the UART serial register. The input character is then either appended to the end of the shell buffer or recognized as a function key, such as "backspace," "delete," or "tab." For the latter case, the corresponding operation is performed and determined by the input function key. "Delete" will trigger the pointer and counter of the shell buffer to reduce by one. "Tab" will send out a tab key through the UART to the console; however, it has no effect on the shell buffer. "Enter" is the flag indicating an input is completed. Figure 15 shows the flow chart for the shell processing.

The shell on the UART is used to test and configure the system functions and manage the wireless network. In fact, by operating and using shell commands through a computer with any serial-port debugging program, such as Putty, any process can be coded. Figure 16 shows a snapshot of the shell running with Putty. In the figure, the "help" command prints out all the registered commands in the system.

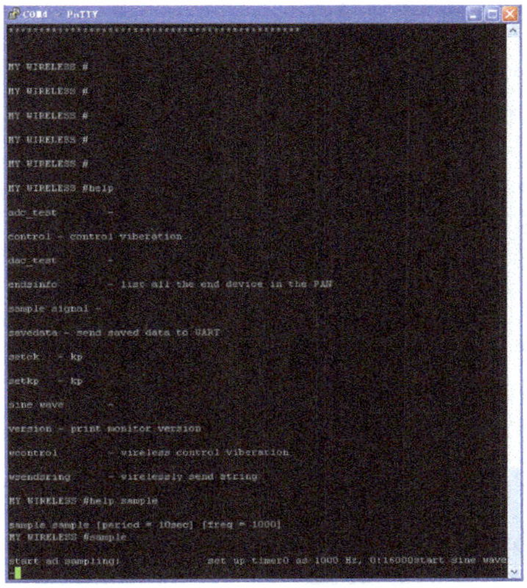

FIGURE 16: Snapshot of shell on Putty.

3.3. *Wireless Sensor Network Stack.* The wireless sensor network stack is a critical software component for WSN applications. The JN5139 wireless microcontroller integrated

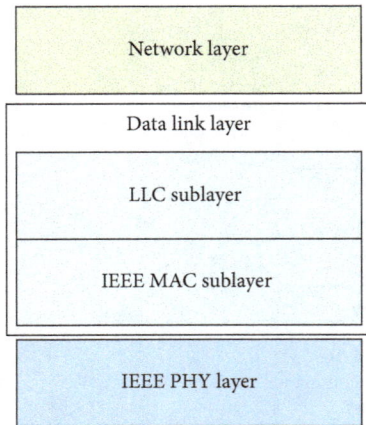

FIGURE 17: 802.15.4 network stack.

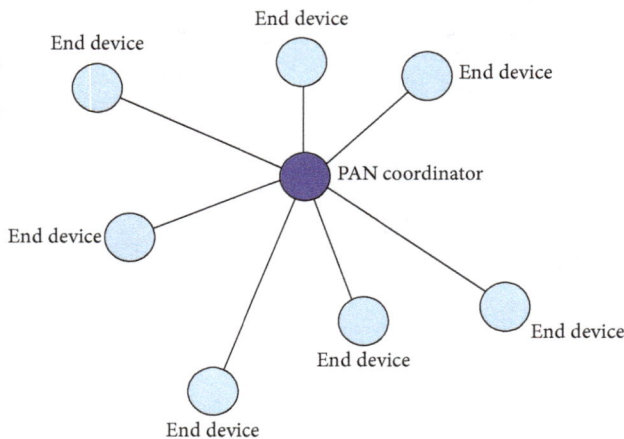

FIGURE 19: Tree topology.

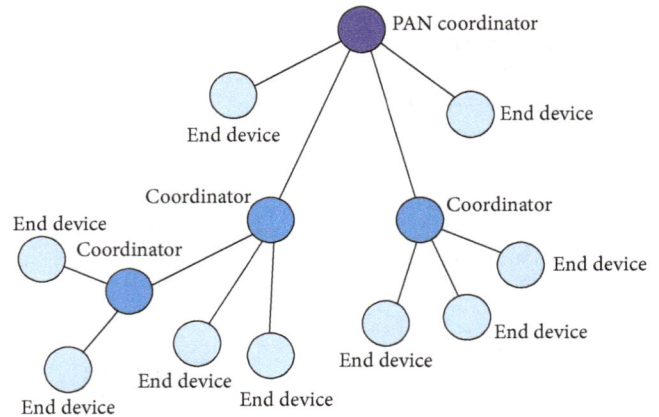

FIGURE 18: Star topology.

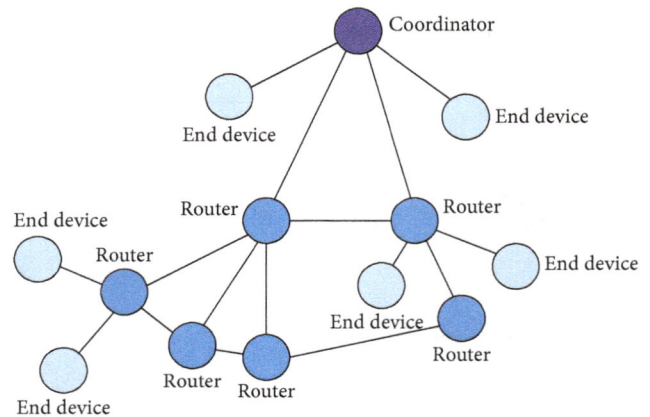

FIGURE 20: Mesh topology.

a wireless transceiver that is fully compatible with 2.4 GHz IEEE 802.15.4 protocol, and the network protocol used in the system is based on IEEE 802.15.4. The network stack is organized into three layers as shown in Figure 17. A MAC sublayer and an LLC sublayer together constitute the data link layer.

The physical (PHY) layer interacts directly with the radio driver and handles physical transmission as well as data bit exchanging. It realizes functions such as channel assessment, bit modulation, packet synchronization, and radio control. The media access control (MAC) layer provides services for associating and disassociating, beacon generation, guaranteed timeslot generation management, and access control to shared channels.

3.4. Network Topology. IEEE 802.15.4 supports the following types of network topologies: the star topology, the tree topology, and the mesh topology. The tree topology is configured as a single central PAN coordinator surrounded by many other sensor nodes. Each of these nodes can communicate only with the PAN coordinator. Therefore, in order for one node to send a message to another node, the message must be relayed through the PAN coordinator. The advantage of

this is the simplicity of the configuration and management, whereas the disadvantage is the limited communication range and the lack of an alternative route in case of the breakdown of one path. Figure 18 illustrates the star topology.

The tree topology is based on the parent and child structure and has more additional nodes—router nodes—in the system. Still, there is only one PAN coordinator in the system, and all end nodes and router nodes have a parent. Each node can only communicate with its children or parent. The tree topology is more advanced than the start topology since it can cover a much larger area; however, the management is also more complex. Figure 19 illustrates the tree topology.

A similar network topology is the mesh topology, where communication rules are more flexible in that router nodes within the range of each other can communicate directly. The advantage of the mesh topology is the increased alternative routes. Higher robustness can be achieved in a mesh topology, since one broken route can easily be replaced by another. Figure 20 shows the mesh topology.

3.5. Message Propagation. Message propagates through the sensor network depending on the network topology adopted. For star topology, both of the addresses will be the PAN coordinator. For tree topology, the next hop address is the parent of the sending node. For mesh topology, the next hop address will be the first node if the target node is out of range.

The implementation of one specific topology is determined by the program on the router nodes and the PAN coordinator. It is realized in the data link and the network layer. Star topology is the simplest implementation and do not require a network layer. Under this topology, the PAN coordinator maintains an end node table and manages the joining and detaching of each sensor node. There is no routing table in this case, since every route is just the direct path to the coordinator. Sensor nodes only need to know the network address of the PAN coordinator, which appears in the input source address and output destination addresses.

For tree and mesh topologies, a routing table is maintained in each router node. Either a fixed routing scheme or a route discovery process is needed in the network layer for building the routing table. The fixed routing topology is much more rigid and is only suitable for situations where sensor positions are fixed. Route discovery strategy provides a more flexible configuration. During the route discovery, one router broadcasts the source address of the initiating end node, which contains the destination network address of the targeting end node.

4. Conclusions

In this paper, a piezoceramic-based wireless embedded system has been designed and tested. The hardware system was composed of four parts: signal conditioning, actuating, power management, and wireless microcontroller. The software platform system consisted of a HAL layer, a network stack layer, and a user input/output management layer. The shell provided easier debugging and configuration interface to the users. The network stack was based on the IEEE 802.15.4 protocol. With the hardware and software platforms ready, different SHM and control applications can be developed on the system.

Conflicts of Interest

The authors declare that they have no conflicts of interest.

Acknowledgments

This work is supported by the National Key R&D Program of China under Grant nos. 2017YFD0701003, 2016YFD0200702, 2016YFC0400207, and 2017YFC0403203; National Natural Science Foundation of China under Grant no. 51509248; Jilin Province Key R&D Plan Project under Grant no. 20180201036SF; and Chinese Universities Scientific Fund under Grant nos. 2018QC128 and 2018SY007.

References

[1] J. H. Taylor and H. M. S. Ibrahim, "A new, practical approach to maintaining an efficient yet acceptably-performing wireless networked control system," in *2010 International Conference on System Science and Engineering*, pp. 269–274, Taipei, Taiwan, July 2010.

[2] J. Chen, P. Li, G. Song, and Z. Ren, "Piezo-based wireless sensor network for early-age concrete strength monitoring," *Optik - International Journal for Light and Electron Optics*, vol. 127, no. 5, pp. 2983–2987, 2016.

[3] J. Chen, "Model based compensation for overcoming sensor breakdown in a piezo-based wireless sensor network," *Optik - International Journal for Light and Electron Optics*, vol. 127, no. 5, pp. 3138–3142, 2016.

[4] F. Linnarsson, P. Cheng, and B. Oelmann, "SENTIO: a hardware platform for rapid prototyping of wireless sensor networks," in *IECON 2006 - 32nd Annual Conference on IEEE Industrial Electronics*, pp. 3002–3006, Paris, France, November 2006.

[5] N. Edmonds, D. Stark, and J. Davis, "MASS: modular architecture for sensor systems," in *IPSN 2005. Fourth International Symposium on Information Processing in Sensor Networks*, pp. 393–397, Boise, ID, USA, April 2005.

[6] J. L. Hill and D. E. Culler, "Mica: a wireless platform for deeply embedded networks," *IEEE Micro*, vol. 22, no. 6, pp. 12–24, 2002.

[7] B. Webs, "Wireless embedded systems," *Tiny OS Tutorial Index*, vol. 1, pp. 1-2, 2003.

[8] P. Trenkamp, M. Becker, and C. Goerg, "Wireless sensor network platforms — datasheets versus measurements," in *2011 IEEE 36th Conference on Local Computer Networks*, pp. 966–973, Bonn, Germany, October 2011.

[9] R. Barnett, S. Cox, and L. O'Cull, *Embedded C Programming and the Atmel AVR*, Cengage Learning, Clifton Park, NY, USA, 2006.

[10] P. Le-Huy and S. Roy, "Low-power 2.4 GHz wake-up radio for wireless sensor networks," in *2008 IEEE International Conference on Wireless and Mobile Computing, Networking and Communications*, pp. 13–18, Avignon, France, October 2008.

[11] S. Tennina, M. Di Renzo, F. Graziosi, and F. Santucci, "Locating zigbee nodes using the ti's cc 2431 location engine: a testbed platform and new solutions for positioning estimation of WSNs in dynamic indoor environments," in *Proceedings of the ACM International Workshop on Mobile Entity Localization and Tracking in GPS-less Environments*, pp. 37–42, San Francisco, USA, September 2008.

[12] J. Polastre, R. Szewczyk, and D. Culler, "Telos: enabling ultra-low power wireless research," in *IPSN 2005. Fourth International Symposium on Information Processing in Sensor Networks*, pp. 364–369, Boise, ID, USA, April 2005.

[13] D. Lymberopoulos and A. Savvides, "XYZ: a motion-enabled, power aware sensor node platform for distributed sensor network applications," in *IPSN 2005. Fourth International Symposium on Information Processing in Sensor Networks*, pp. 449–454, Boise, ID, USA, April 2005.

[14] L. Nachman, J. Huang, J. Shahabdeen, R. Adler, and R. Kling, "IMOTE2: serious computation at the edge," in *2008 International Wireless Communications and Mobile Computing Conference*, pp. 1118–1123, Crete Island, Grace, August 2008.

[15] A. Wheeler, "Commercial applications of wireless sensor networks using zigbee," *IEEE Communications Magazine*, vol. 45, no. 4, pp. 70–77, 2007.

[16] K. S. J. Pister, "Smart dust-hardware limits to wireless sensor networks," in *23rd International Conference on Distributed*

Computing Systems, 2003, pp. 2–2, Providence, RI, USA, May 2003.

[17] A. Fernandez-Montes, L. Gonzalez-Abril, J. Ortega, and F. Morente, "A study on saving energy in artificial lighting by making smart use of wireless sensor networks and actuators," *IEEE Network*, vol. 23, no. 6, pp. 16–20, 2009.

[18] M. Johnson, M. Healy, P. van de Ven et al., "A comparative review of wireless sensor network mote technologies," in *2009 IEEE Sensors*, pp. 1439–1442, Christchurch, New Zealand, October 2009.

12

A Biologically Inspired Energy-Efficient Duty Cycle Design Method for Wireless Sensor Networks

Jie Zhou

College of Information Science and Technology, Shihezi University, Shihezi, China

Correspondence should be addressed to Jie Zhou; jiezhou@shzu.edu.cn

Academic Editor: Gabriele Cazzulani

The recent success of emerging wireless sensor networks technology has encouraged researchers to develop new energy-efficient duty cycle design algorithm in this field. The energy-efficient duty cycle design problem is a typical NP-hard combinatorial optimization problem. In this paper, we investigate an improved elite immune evolutionary algorithm (IEIEA) strategy to optimize energy-efficient duty cycle design scheme and monitored area jointly to enhance the network lifetimes. Simulation results show that the network lifetime of the proposed IEIEA method increased compared to the other two methods, which means that the proposed method improves the full coverage constraints.

1. Introduction

Recent technological advances in sensing, nanosystems technologies, and communication have made it possible to equip inexpensive small, low-cost, vulnerable, and fast response sensing units [1]. Wireless sensor networks (WSNs) are composed of some sensors having limited sensing, computing, communication, and self-organizing abilities [2]. Each sensor node consists of five modules: the computation module, the data acquisition module, the RF module, the data storage module, and the power module [3]. Wireless sensor networks have applications in many areas such as military surveillance, traffic avoidance, intelligent family, preventing forest fire loss, building monitoring and control, and advanced health care delivery [4].

The energy-efficient duty cycle design problem has recently attracted the attention of many researchers in the field of wireless sensor networks [5]. Limited by their size, small wireless sensors are equipped with restricted sensing capacity [6]. To have a long network lifetime, energy-efficient duty cycle scheme should be designed properly. However, finding the ideal energy-efficient duty cycle design is an NP-hard problem. For large-scale wireless sensor networks, the exhaustive search cannot be used to get the ideal duty cycle design in real time [7].

Due to its computational complexity, many heuristics are proposed to get near-optimal solutions in reasonable time. In [8], an energy-efficient duty cycle design technique that enables trade-offs between computation cost and network lifetime is investigated. It provides a wider search space by randomly selecting energy-efficient duty cycle design solutions when each one in the population gets updated. However, the reference does not take into consideration factors such as sensing radius. A research effort to the energy-efficient duty cycle design for the wireless sensor networks based on the genetic algorithm (GA) can be found in [9]. The work therein focused on an energy-efficient duty cycle design optimization with many constraints. But it suffers from premature convergence and low convergence rate when the number of nodes is high. A particle swarm optimization (PSO) is developed by employing a refined fitness evaluation technique [10]. In the work therein, the authors maximize the network lifetime without considering the fact that the energy of a battery is limited. PSO approach has a problem of algorithm convergence and complexity.

Elite computing and immune theory have always attracted attention of the scholars of artificial intelligence [11]. Besides that, the superior performance of evolutionary theory for combinatorial optimization problems is demonstrated in [12, 13]. In particular, in this paper, an improved elite immune evolutionary algorithm (IEIEA) is investigated to effectively represent the individuals to explore the search space with a small group and to exploit the global optimal solution in the search space within a few iterations, respectively.

We first formulate an optimization problem as integer programming that is proven to be NP-complete. Then advanced evolutionary algorithm is used to solve the problem. Moreover, immune clone operator is adopted to enhance global search ability. The immune clone strategy helps to avoid local optima, and the improved stopping criterion automatically finds the best solution. Extensive simulations are conducted which compare IEIEA with SA and PSO. Simulation results show that the proposed IEIEA algorithm outperforms, regarding network lifetime, SA and PSO with the same computational complexity. Furthermore, IEIEA has better robustness.

The rest of this paper is briefed as follows. In Section 2, we describe the full mathematical model of energy-efficient duty cycle design. In Section 3, IEIEA for energy-efficient duty cycle design as well as its detailed implementation is presented. In Section 4, we evaluate the performance of the proposed algorithm and show the simulation results and comparisons. We also analyze the characteristics of IEIEA. Finally, in Section 5, concluding remarks are presented.

2. System Model

In this section, we build the system model to demonstrate the energy-efficient duty cycle design problem concerning the constraints of the energy of a battery and sensing radius. In [14], the authors propose a set of inequalities for modeling the energy efficiency. Later in [15], the authors showed a similar model by a fewer number of sensors and targets. The model proposed in [16] is more flexible than the model proposed in [14]. In this work, a similar set of inequalities for maximizing the network lifetime is employed to model the energy-efficient duty cycle design in WSNs. Consider a WSN system with F sensors and E targets in the monitoring area. The monitoring relationship can be shown as

$$Q = \begin{bmatrix} q_{1,1} & q_{1,2} & \cdots & q_{1,E-1} & q_{1,E} \\ q_{2,1} & q_{2,2} & \cdots & q_{2,E-1} & q_{2,E} \\ \vdots & & q_{f,e} & & \vdots \\ q_{F-1,1} & q_{F-1,2} & \cdots & q_{F-1,E-1} & q_{F-1,E} \\ q_{F,1} & q_{F,2} & \cdots & q_{F,E-1} & q_{F,E} \end{bmatrix} \quad (1)$$

$$\left(q_{f,e} \in \{0,1\} \right),$$

where $q_{f,e}$ is the monitoring relationship between the e_{th} target and f_{th} sensor; $q_{f,e} = 1$ means that the e_{th} target is within the monitoring range of the f_{th} sensor and $q_{f,e} = 0$ otherwise. Due to the limited battery capacity of the sensor

nodes, the maximum length of work time is assumed to K rounds. The duty cycle of the WSN can be shown as

$$D = \begin{bmatrix} d_{1,1} & d_{1,2} & \cdots & d_{1,F-1} & d_{1,F} \\ d_{2,1} & d_{2,2} & \cdots & d_{2,F-1} & d_{2,F} \\ \vdots & & d_{i,f} & & \vdots \\ d_{FK-1,1} & d_{FK-1,2} & \cdots & d_{FK-1,F-1} & d_{FK-1,F} \\ d_{FK,1} & d_{FK,2} & \cdots & d_{FK,F-1} & d_{FK,F} \end{bmatrix} \quad (2)$$

$$\left(d_{i,f} \in \{0,1\} \right),$$

where $d_{i,f} = 1$ means that the i_{th} sensor node is active in round i and $d_{i,f} = 0$ otherwise. The monitoring cycle of the whole network can be shown as

$$DQ$$

$$= \begin{bmatrix} \sum_{f=1}^{F} d_{1,f} q_{f,1} & \cdots & \sum_{f=1}^{F} d_{1,f} q_{f,E-1} & \sum_{f=1}^{F} d_{1,f} q_{f,E} \\ \sum_{f=1}^{F} d_{2,f} q_{f,1} & \cdots & \sum_{f=1}^{F} d_{2,f} q_{f,E-1} & \sum_{f=1}^{F} d_{2,f} q_{f,E} \\ \vdots & \ddots & \vdots & \vdots \\ \sum_{f=1}^{F} d_{FK-1,f} q_{f,1} & \cdots & \sum_{f=1}^{F} d_{FK-1,f} q_{f,E-1} & \sum_{f=1}^{F} d_{FK-1,f} q_{f,E} \\ \sum_{f=1}^{F} d_{FK,f} q_{f,1} & \cdots & \sum_{f=1}^{F} d_{FK,f} q_{f,E-1} & \sum_{f=1}^{F} d_{FK,f} q_{f,E} \end{bmatrix}, \quad (3)$$

where the element $\sum_{f=1}^{F} d_{j,f} q_{f,e} = 1$ is positive and represents the fact that the e_{th} target is monitored by at least one sensor node in round j and $\sum_{f=1}^{F} d_{j,f} q_{f,e} = 0$ otherwise. When the WSN maintains a full coverage in the j_{th} round, all the elements in the j_{th} row must be positive. Thus, from top to bottom, the previous row number of the first zero element is the network lifetime. Thus, the mathematical model of the duty cycle design problem can be shown as follows:

Objective is

$$f(D) = \text{zero\_row}(DQ) - 1 \quad (4)$$

subject to

$$\sum_{i=1}^{FK} d_{i,f} \leq K, \quad f = 1 \cdots F, \quad (5)$$

where zero_row is the row number of the first zero element which occurs from row 1 to row FK in matrix. The constraint $\sum_{i=1}^{FK} d_{i,f} \leq K$ means that the maximum lifetime of node is K.

3. Energy-Efficient Duty Cycle Design Based on IEIEA for Wireless Sensor Networks

An evolutionary algorithm (EA) is a search method that mimics the mechanism of Darwinian's evolutionary theory such as genetic recombination and survival of the fittest. EA is used for a large number of multiobjective optimization problems which is confirmed to be suitable for locating the best and near-optimal answers. The EA has good searching ability. However, the algorithm is often not suitable for large-scale complex problems. In many cases, the algorithm falls into local optima.

In recent times, elite computing and immune theory impressed strategic methods are getting evolved. They are confirmed to be very successful while looking for better solutions, targeting not only more efficient convergence but also solving combinatorial optimization problems. In this paper, we propose an improved elite immune evolutionary algorithm (IEIEA), which is a modification of EA, for energy-efficient duty cycle design in wireless sensor networks. In IEIEA, based on the concept and principles of elite computing and immune theory, new evolutionary theory techniques are getting developed, which can achieve an increased balance in exploitation of the solution space and receive better results compared with the standard EA. In such a hybrid technique, search ability of elite evolutionary strategy and immune clone operator are involved to generate faster and efficient convergence. It operates parallel exploration in complicated search regions.

Throughout each generation, IEIEA gets a different population out of the current population by using genetic operators, such as selection, crossover, and mutation. Additionally, suggested elite evolutionary strategy is using the elite idea, and the provided immune clone operator is using the immune idea to improve the global search capability of IEIEA.

3.1. Initialization.
In IEIEA, a solution is represented by a chromosome (individual). We select a string as a chromosome to symbolize a solution to the energy-efficient duty cycle design problem. Each chromosome within the population can signify a group of randomly selected duty cycles. The energy-efficient duty cycle design problem is acquiring a chromosome within prospective solution regions making use of IEIEA. Therefore, if matrix (2) is arranged in rows, a feasible solution could be presented as a binary sequence. So we use the way of binary coding for each chromosome. Each gene in the chromosome is mapped to a binary number, and the gene value means that the sensor is in active or sleep mode. In other words, the binary chromosome is made up of Boolean variables implying whether the sensor is active or not.

The number of binary solution genes matches the number of sensor nodes and the maximum rounds. As an example, consider a system with three sensor nodes and four rounds. Binary coding representation with 36 binary numbers is accurate and reliable as it covers up to the entire solution space, and, besides, the string size is the number of sensor nodes square times the maximum rounds of the duty cycle. For ten sensor nodes with the highest lifetime of 20 rounds, the total string length is 2000. Here, the search space is the space of all matrices that satisfy (5). The total amount of all possible solutions is very large.

IEIEA solves optimizing challenges by using a population of a preset number, named the population size, of solutions. IEIEA maintains a number of such solutions. The

initialization design applied by this algorithm is as follows. The algorithm starts out by constructing an initial population of random candidate solutions. The original population was built randomly expecting to search globally to find some solutions. In each solution, a previously chosen number of random bits in the chromosome are set to 1. The rest bits in the chromosome are set to 0. By doing this, the initial population contains randomly produced chromosomes. Each solution in the population will be initialized to a binary number randomly selected from the uniform distribution over the set $[0, 1]$. A large number of low fitness solutions are constructed via utilizing the discrete random values in the initialization step. If a solution owns a low fitness value, random bits are determined from the set $[0, 1]$ to renew the solution. After obtaining the genes for each chromosome, the initial population is generated. The process is applied over and over unless all the chromosomes in the population get feasible. There are P_size individuals in the population of IEIEA. The population in the initialization process carries on randomly to ensure that the diversity is well kept.

3.2. Selection and Crossover.
For every generation, children are generated by mixing the characteristics of two parent individuals. The selection operator picks a solution coming from the most recent population for the next population with possibility proportional to its fitness value.

Solutions in the current population are evaluated for their merit to survive in the following population. Every solution inside population is associated with a figure of benefit as well as a fitness value. Every time a set of two parent solutions are to be selected from a current population for developing an offspring by a crossover operation. The fitness value of each solution within the current population is calculated as the objective function by (4).

Picking out parents is based on random and selective approaches. At each iteration, couples of parents are selected by fitness, whereby every parent is chosen as the best of those randomly selected among the best candidates. In the selection step, fitter individuals are more regularly used to generate offspring than less fit individuals, which tends to raise the average fitness as the algorithm continues. Because the objective is to maximize the network lifetime, a solution string with a practically significant number of cycles must have comparatively high fitness value. So the solutions that are better will have a greater chance to be chosen. The selection operation mimics relatively neutral selection in real life.

The RWPS (roulette wheel proportionate selection) is used for choosing the solutions from the latest population. Better solutions have a relatively higher possibility to be selected for reproduction making use of the RWPS. The idea of RWPS is simple. The selection possibility of each solution is defined by the roulette wheel selection utilizing the linear scaling. By using the scaling, the probability of choosing the relevant solution for reproduction is proportional to the amount of scaled fitness.

In each step, two parents are picked out first according to the roulette wheel rule. This process will likely be applied repeatedly except when the whole set of child individuals in the population turns into feasible. In this way, all child strings in the population were split into several pairs, with every pair establishing two children in the crossover operation. When all the parents were selected, their properties will be passed to their offspring. The previously mentioned selection process permits efficient constraint raising the quality of solutions by selecting the child chromosomes from high fitness districts.

In a crossover, the genes of the parent are replicated to its offspring. It signifies the mixing up of information from both parents to construct the children. We use a crossover operator to each of the selected pairs of parent solutions. In crossover operation, a couple of offspring are produced by the selected pair of parent solutions. In this way, two new children are manufactured by the random mixture of the properties of their parents. The crossover procedure is done between two chromosomes in their binary form. In a binary crossover, two new child chromosomes are created while using the interval between the two parents.

Standard crossover operator switches the segments of selected strings through the crossover points with a probability. The crossover place is randomly selected. One-point crossover specifies the point of crossover of the two parent chromosomes. Therefore, in single-point crossover, one crossover suggests that a child gets the left side of the binary string coming from the first parent and the right side coming from the second parent. For the second child, the crossover is accomplished by combining other parts of child chromosomes of the parents. The second child from this union obtains the complementary substances not taken by the first child, that is, receives the right section from the first parent and the left side from the second parent, having the same crossover location. Two-point crossover indicates that the first child gets both right and left sections from the first parent. Meanwhile, the first child also gets the middle section from the second parent. The crossover point is randomly chosen.

It is often known that uniform crossover works more efficiently at solving combinatorial optimization issues than just one-point or two-point crossover. In this case, we present a binary execution of uniform crossover. The crossover operation is done in the mating pool to build children. In uniform crossover, every one of these couples produces two children utilizing an adaptive number of crossovers. Binary strings are changed between the two parents to generate two children. Despite establishing the crossover point in the conventional crossover, we use the random binary number as the mask to form the binary uniform crossover.

In binary uniform crossover operation, a probability term is set to manage the rate of crossover. We generate a random decimal between 0 and 1. Then choose each gene with an equal chance from two child strings and recombine the genes to create a new offspring. Exchange positions in the two chromosomes are selected arbitrarily. Once the decimal is over the probability term, no change takes place, and the corresponding bit in solutions one and two is passed on to child one and child two, respectively. Otherwise, the crossover operation is carried out. If the bit in the mask is equal to one, the related bit in solutions one and two is distributed to child two and child one, respectively, if the crossover opportunity term is below the probability term. For this reason, in

a binary uniform crossover, each pair of parents has a probability term chance of using uniform crossover operator. This process should be implemented continually except in cases where all the child strings become feasible. All elements of genes handed down from the parents ought to keep the order as they are available in their parents. So, there have been child individuals soon after the crossover has been carried out.

Observe that since binary encoding is applied, the tiniest unit that is exchanged in binary uniform crossover could be a bit. It allows developing the offspring positioned according to the parents. Crossover creates a designed yet randomized exchange of genetic elements between strings. Hence the offspring contains the properties received from each of its parents. Moreover, using uniform crossover can mix the genes from two solutions with a more consistent way.

3.3. Mutation and Recombination. The content of this portion is kept in the mutation procedure. In IEIEA, random mutation is conducted to the offspring to prevent the premature convergence. This is executed by using random adjustments to a ratio of the children generated. In this paper, mutation operation is established by randomly picking out any chromosomes utilizing prespecified possibilities. In this way, plenty of random mutations are exercised to add some diversity to the population. For the duty cycle design problem, this means replacing some genes with new random figures. Selected genes in a single individual are randomly chosen to carry out the mutation.

In IEIEA, thinking of a sufficiently large, diverse initial population where zeros and ones are available for every bit within the chromosomes, full convergence could conceivably be obtained without having mutations at all because any practical solutions can be manufactured from some blend of the initial population. Nevertheless, the initial population dimension is limited, and the tiniest gene that may be changed in a crossover is binary. Recognize that mutations are an essential method for IEIEA; we need to apply a mutation operator to develop new solutions.

Subsequent mutations of the offspring apply diversity to the population and investigate different locations of the searching space. Offspring experience a constraint amount of mutations completed, and so the offspring are then scored for fitness. Each and every gene in chromosomes is mutated as mentioned above applying a probability, labeled as mutation probability. The chances that a certain gene is mutated are operated by the mutation rate. In this paper, an initial mutation rate of 5% means that each gene has a 5% possibility of being replaced with a new value. We employ reverse mutation in IEIEA. The reverse mutation is an important operation for preserving diversity in the population to be able to stay away from local optima. It will be shown in the simulation section that this mutation assists in reaching more efficient solutions.

For stopping the premature convergence of the algorithm, we employed dynamic mutation operators to diversify the population to be able to result in making the strings in the population dissimilar to each other. Dynamic mutation function is an iteration based function that determines the mutation rate. Smaller mutation levels indicate a more local search, while larger mutation levels are more efficient for introducing diversity. In case the constructed variables converge at the low objective value, the degree of mutation can be higher. If a solution has a superior fitness value, a lower mutation rate is picked. In this way, the convergence length of IEIEA is reduced.

As in natural reproduction, there exists a likelihood of illegal mutation during the mutation process. The solutions that comply with constraints (5) are classified as legal strings; otherwise, they are recognized as illegal chromosomes. In illegal strings, a fully new random legal individual will be created to replace the illegal one.

Following using the crossover operator and the mutation operator, there have already been P_size chromosomes in the population that consisted of P_size parent strings and P_size offspring solutions. More advanced than the execution in EA, the selection and recombination processes in IEIEA are blended to manufacture a much better population. The recombination technique is a sort of uniform crossover which selects one offspring from two strings. After recombination, the new offspring fitness will be estimated depending on the fitness function.

3.4. Evaluation and Elite Clone Operation. In the examination procedure, the values of fitness for each string are assessed. Each chromosome in the population is linked to a figure of merit (fitness value) following the value of the function to be optimized. These individuals are scored, with the best performers probably going to be parents for the next iteration. Individuals with the greatest scores are implemented to be parents. In this paper, the aim is to increase the network lifetime. Within the current context, the fitness value of a chromosome would depend upon the entire duty cycle. The function values (fitness values) linked to the duty cycle are then assessed. The duty cycle design problem can be adapted swiftly to suit the full coverage constraints by mapping the duty cycle to the chromosomes and defining a fitness function to evaluate the potential fitness computation function for solution strings by the chromosome.

Here, we intend to identify the formulation of examining the level of chromosomes. To be able to calculate the level of each solution string in the population, formula (4) associates a fitness function. We determine fitness as the amount of network lifetime that can be shown by a combination of duty cycles; thus, fitness is an integer value from 0 to maximum rounds. The function value denoted by network lifetime is calculated by delivering binary individuals into the objective function. The solutions are then decoded into their matrix equivalent and examined for constraints violation. The fitness function is going to be scaled to avoid premature convergence. Through this definition, the fitness value is more effective when it is larger.

This procedure will be used for Max_gen times except for the solution that becomes feasible before Max_gen times. In this way, the algorithm ends till the fitness of the highest quality string converges to the maximum rounds. If it does not reach this value, the final chromosome distribution is adopted as the fittest individual (demonstrating the optimized network lifetime) following the Max_gen generations.

In this way, the IEIEA is terminated, and the fittest string in the population is taken as the final solution.

The individuals in the elite collection demonstrate the promising zones to seek out the suitable solution. The string in the elite group was harvested over all the generations. The population of the next generation includes the certain percentage of solutions from the elite group and most from offspring. The percentage between 2 kinds of individuals inside the new generation is defined by the probabilities of the alternative. The activity repeats for $Elite\_num$ times, and, as a consequence, a new population is constructed.

The IEIEA commences with the building of an initial population, normally by spreading random binary numbers within investigated space. First, all individuals in a population are rated in lowering the order of fitness. The first 10% points are selected for cloning. Individuals with higher fitness will have more clones. The clones, not the original individual, then undergo the clonal mutation progress. The clonal mutation only performs on the cloned chromosomes to conserve the information. By doing this, the benefits of using the clonal mutation provide a neighboring exploration throughout the original chromosome, while the utilization of the chaotic mutation provides a global survey surrounding the string. Following that, strings are evaluated on the fitness function, and merely the best of each clone can successfully pass to the population, preserving an equal scale of the population. With this type of alternative, the diversity is conserved, and new fields of looking region tend to be quite possibly explained.

4. Simulation and Discussion

In this section, we compare the proposed IEIEA method with the simulated annealing algorithm (SA) and PSO method for energy-efficient duty cycle design in WSN. Simulations are performed to validate and test the performance of the proposed IEIEA approach to the energy-efficient duty cycle design problem in WSN. To examine the applicability for practical implementation, we evaluate the performance of the schemes on a PC with Intel Core i5, 4 G RAM, WIN-10 OS, and MATLAB software. The fitness function in Section 2 is used to evaluate the IEIEA efficiency. Then we generate the position of sensors and targets randomly specified within a 500×500 m square area. We assume that each sensor node has the same energy consumption per unit time. We placed 70, 80, 90, and 100 sensors, respectively. The sensing range of all sensor nodes is 300 meters. IEIEA has been used to solve this problem, and the generations obtained have been compared with SA and PSO. The maximum number of generations of IEIEA, SA, and PSO is 100. In all three methods, a population size of 100 individuals (solutions) is used. In SA, the initial temperature is set to 500 degrees and the annealing temperature coefficient is set to 0.98. In IEIEA, crossover and mutation probabilities are 0.9 and 0.05, respectively. In PSO, the dynamic range of the particle has been set to 0.4, the maximum velocity is 4, and the cognitive and social parameters are $c1 = c2 = 0.5$.

Figure 1 presents the network lifetime obtained by IEIEA, SA, and PSO on the energy-efficient duty cycle design problem with 100 sensors and 10, 20, 30, and 40 targets,

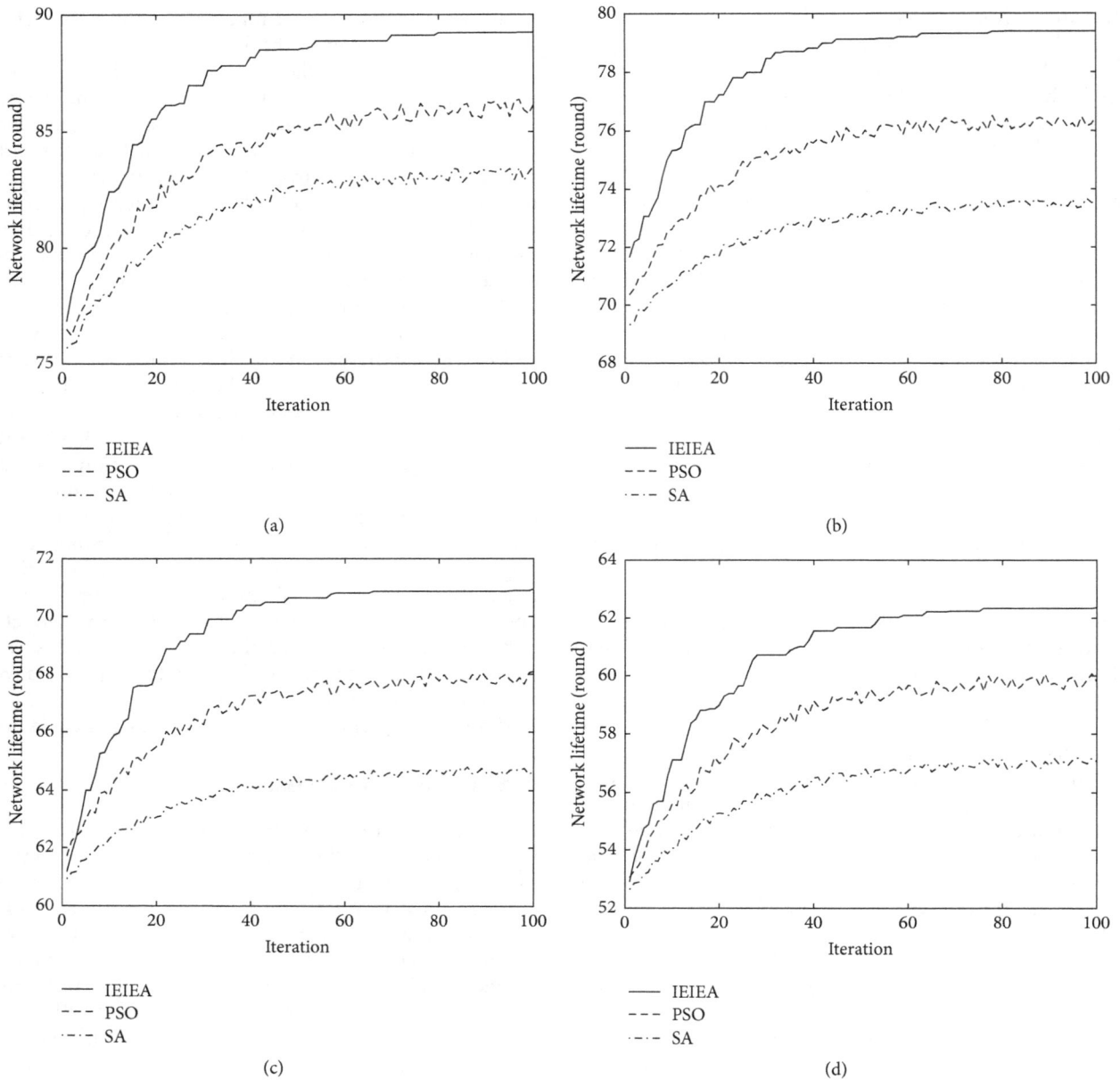

FIGURE 1: Network lifetime with 100 sensors and 10 targets (a), 20 targets (b), 30 targets (c), and 40 targets (d).

respectively. For each method, the fitness of the best individual at each generation is recorded. The results are the average of 30 time runs with three different methods to show the convergence difference. So there is a certain difference between the result of Figure 1 and the average value in Table 1.

From Figure 1, we can see that the network lifetime of IEIEA is better than those obtained by SA and PSO methods and it provides improvement of network lifetime after the 100 iterations. In the beginning, the value of network lifetime increased with the growth of the generations, as shown in the figure. After 50 iterations, IEIEA still has a fast convergence rate, but PSO and SA can only slowly converge to a lower value than IEIEA. As a result, diverse population leading to better solutions is maintained, so IEIEA is prevented from

TABLE 1: Network lifetime with different targets and 100 sensors.

Number of targets	IEIEA	PSO	SA
10	89.32	86.49	83.58
20	80.24	77.39	74.48
30	71.75	68.89	65.54
40	63.06	60.19	57.28

premature convergence. Over 100 iterations, IEIEA provides a longer network lifetime than PSO and SA, and IEIEA converges with a faster rate. Similar results can be obtained with 90 sensors in Figure 2.

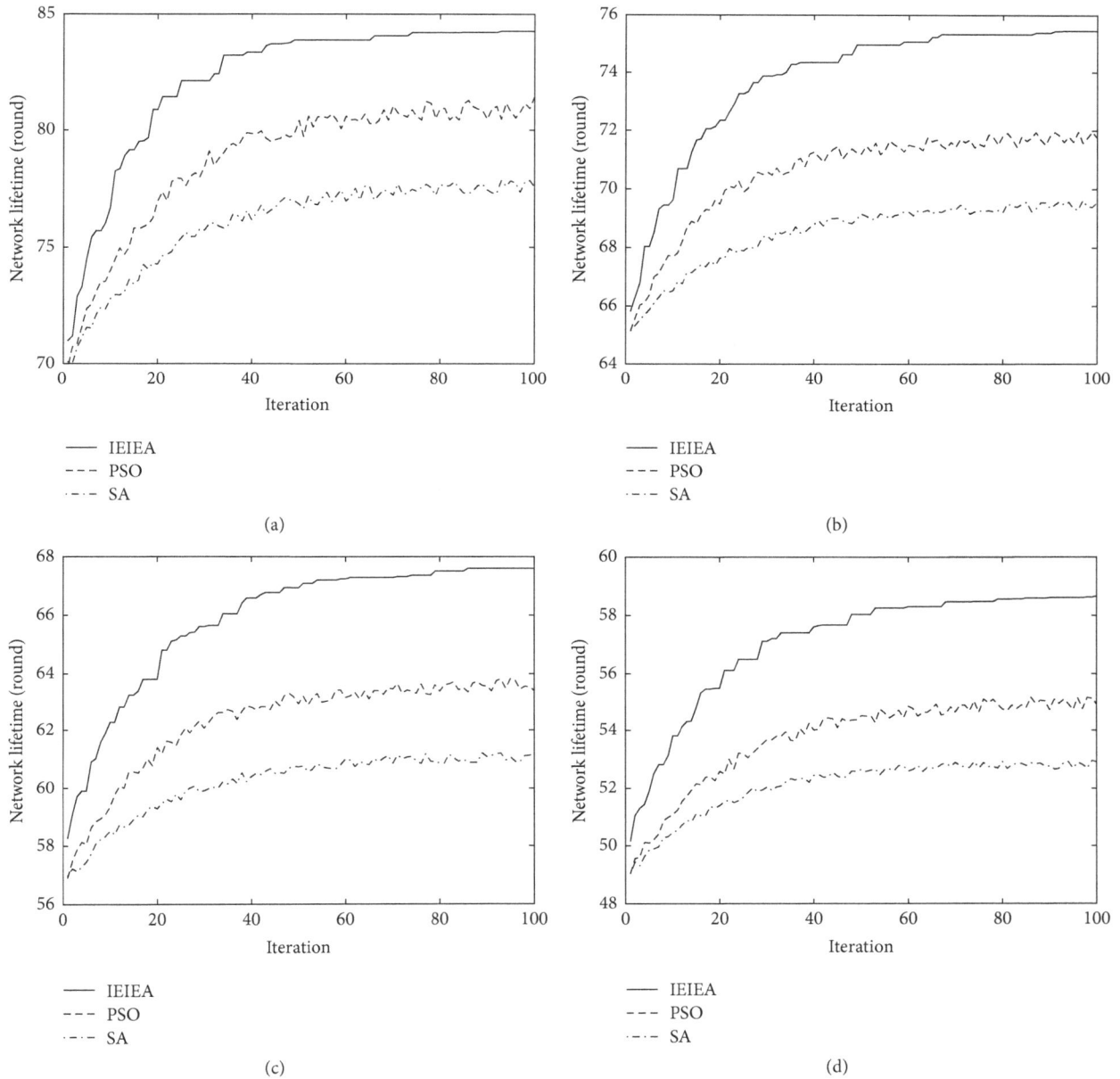

FIGURE 2: Network lifetime with 90 sensors and 10 targets (a), 20 targets (b), 30 targets (c), and 40 targets (d).

Figure 3 presents the obtained results of IEIEA, SA, and PSO approaches after 100 iterations with 100, 90, 80, and 70 sensors and 10, 20, 30, and 40 targets, respectively. All the results were averaged over 100 runs. Table 1 shows the results of Figure 3(a) with 100 sensor nodes and 10, 20, 30, and 40 targets, respectively. It can be seen that, after 100 iterations, the average network lifetime of IEIEA is 89.32, 80.24, 71.75, and 63.06 rounds, respectively. However, PSO cannot always obtain the optimal solution within the predefined iterations and the network lifetimes obtained by the PSO are 86.49, 77.39, 68.89, and 60.19, respectively. On the other hand, the SA shows quite slower convergence and the network lifetimes obtained by the SA are 83.58, 74.48, 65.54, and 57.28,

respectively, hence proving the superior reliability of IEIEA. Similar results can be obtained with 90, 80, and 70 sensors.

5. Conclusion

In this paper, we use an improved elite immune evolutionary algorithm (IEIEA) to solve the energy-efficient duty cycle design problem in WSN. In this work, we first propose a novel formulation of the objective function to maximize the network lifetime to satisfy the full coverage constraints. Simulations are conducted by using the IEIEA and the energy-efficient duty cycle design methods based on SA, PSO, and GA. Simulation results show that the proposed IEIEA

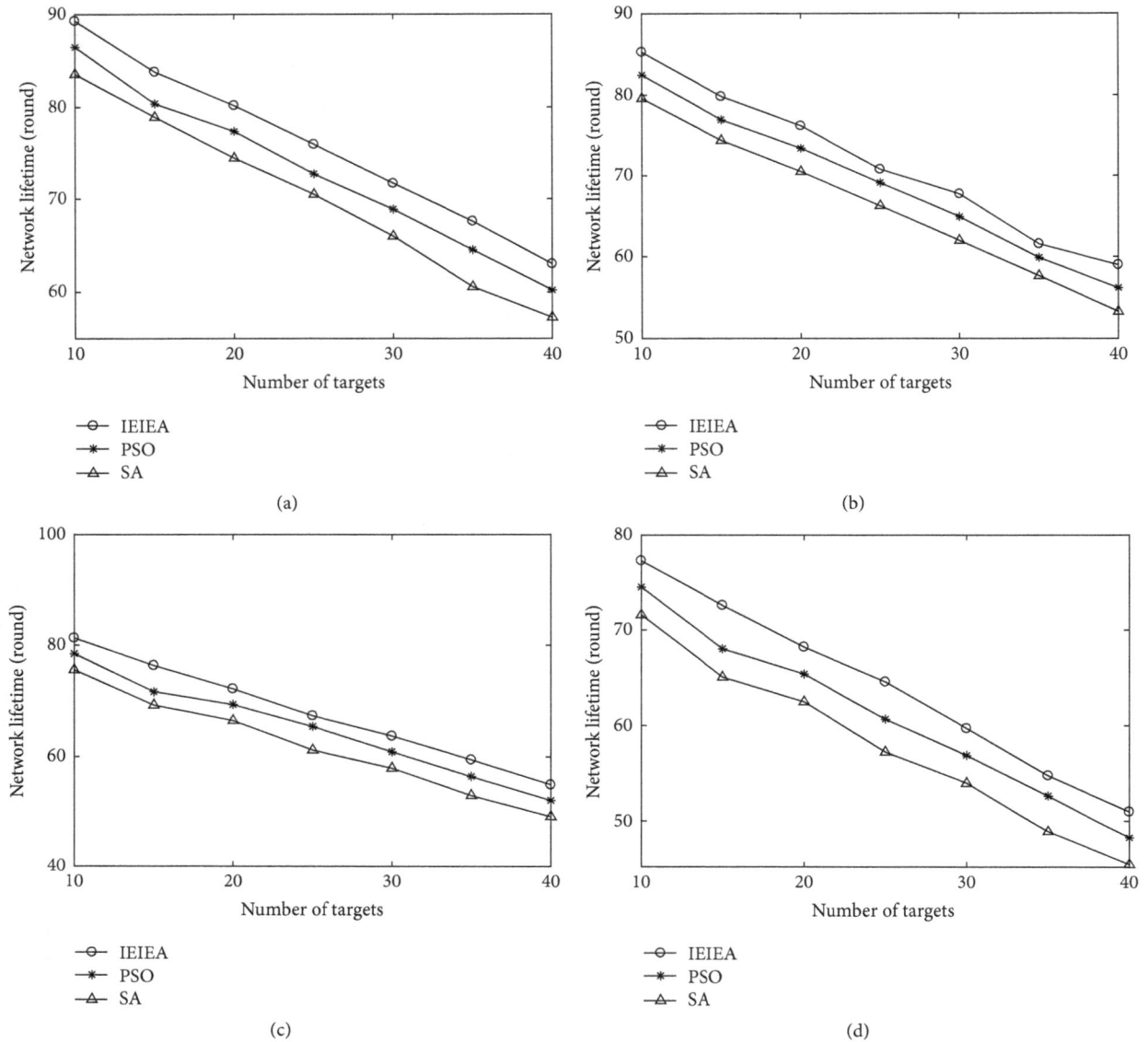

FIGURE 3: Network lifetime with different targets and 100 sensors (a), 90 sensors (b), 80 sensors (c), and 70 sensors (d).

provides longer network lifetime compared with previous SA, PSO, and GA. Besides that, the network lifetime of the proposed IEIEA is much higher than those of GA and SA. So the proposed method significantly enhanced the monitoring results.

Conflicts of Interest

The author declares that there are no conflicts of interest regarding the publication of this paper.

Acknowledgments

This paper is supported by the National Natural Science Foundation of China (no. 61662063) and High-Level Talent Research Project of Shihezi University (no. RCZX201530).

References

[1] G. Y. Li, Z. Xu, C. Xiong et al., "Energy-efficient wireless communications: tutorial, survey, and open issues," *IEEE Wireless Communications*, vol. 18, no. 6, pp. 28–34, 2011.

[2] X. Zhao, Y. Zhuang, and J. Wang, "Local adaptive transmit power assignment strategy for wireless sensor networks," *Journal of Central South University*, vol. 19, no. 7, pp. 1909–1920, 2012.

[3] N. Sun, Y.-S. Jeong, and S.-H. Lee, "Energy efficient mechanism using flexible medium access control protocol for hybrid wireless sensor networks," *Journal of Central South University*, vol. 20, no. 8, pp. 2165–2174, 2013.

[4] D. Feng, C. Jiang, G. Lim, L. J. Cimini Jr., G. Feng, and G. Y. Li, "A survey of energy-efficient wireless communications," *IEEE Communications Surveys & Tutorials*, vol. 15, no. 1, pp. 167–178, 2013.

[5] A. A. Kumar Somappa, K. Øvsthus, and L. M. Kristensen, "An industrial perspective on wireless sensor networks—a survey of requirements, protocols, and challenges," *IEEE Communications Surveys & Tutorials*, vol. 16, no. 3, pp. 1391–1412, 2014.

[6] D. Dong, X. Liao, K. Liu, Y. Liu, and W. Xu, "Distributed coverage in wireless ad hoc and sensor networks by topological graph approaches," *Institute of Electrical and Electronics Engineers. Transactions on Computers*, vol. 61, no. 10, pp. 1417–1428, 2012.

[7] W.-C. Ke, B.-H. Liu, and M.-J. Tsai, "Constructing a wireless sensor network to fully cover critical grids by deploying minimum sensors on grid points is NP-complete," *Institute of Electrical and Electronics Engineers. Transactions on Computers*, vol. 56, no. 5, pp. 710–715, 2007.

[8] K. Han, J. Luo, L. Xiang, M. Xiao, and L. Huang, "Achieving energy efficiency and reliability for data dissemination in duty-cycled WSNs," *IEEE/ACM Transactions on Networking*, vol. 23, no. 4, pp. 1041–1052, 2015.

[9] X. Hu, J. Zhang, Y. Yu et al., "Hybrid genetic algorithm using a forward encoding scheme for lifetime maximization of wireless sensor networks," *IEEE Transactions on Evolutionary Computation*, vol. 14, no. 5, pp. 766–781, 2010.

[10] E. R. Dosciatti, W. Godoy Junior, and A. Foronda, "TQ/PSO - A new scheduler to optimize the time frame with PSO in WiMAX networks," *IEEE Latin America Transactions*, vol. 13, no. 1, pp. 365–376, 2015.

[11] T. Back, U. Hammel, and H.-P. Schwefel, "Evolutionary computation: Comments on the history and current state," *IEEE Transactions on Evolutionary Computation*, vol. 1, no. 1, pp. 3–17, 1997.

[12] Z.-J. Zhang, J. Huang, and Y. Wei, "Frequent item sets mining from high-dimensional dataset based on a novel binary particle swarm optimization," *Journal of Central South University*, vol. 23, no. 7, pp. 1700–1708, 2016.

[13] S. Sarafijanović and J.-Y. Le Boudec, "An artificial immune system approach with secondary response for misbehavior detection in mobile ad hoc networks," *IEEE Transactions on Neural Networks*, vol. 16, no. 5, pp. 1076–1087, 2005.

[14] Z. Lu, W. W. Li, and M. Pan, "Maximum lifetime scheduling for target coverage and data collection in wireless sensor networks," *IEEE Transactions on Vehicular Technology*, vol. 64, no. 2, pp. 714–727, 2015.

[15] O. Demigha, W. Hidouci, and T. Ahmed, "On Energy efficiency in collaborative target tracking in wireless sensor network: a review," *IEEE Communications Surveys and Tutorials*, vol. 15, no. 3, pp. 1210–1222, 2013.

[16] C.-L. Yang and K.-W. Chin, "Novel algorithms for complete targets coverage in energy harvesting wireless sensor networks," *IEEE Communications Letters*, vol. 18, no. 1, pp. 118–121, 2014.

13

Electromagnetic Bridge Energy Harvester Utilizing Bridge's Vibrations and Ambient Wind for Wireless Sensor Node Application

Farid Ullah Khan⑩ **and Muhammad Iqbal**

Institute of Mechatronics Engineering, University of Engineering and Technology, Peshawar, Pakistan

Correspondence should be addressed to Farid Ullah Khan; dr_farid_khan@uetpeshawar.edu.pk

Academic Editor: Jesus Corres

This paper presents novel electromagnetic bridge energy harvesters (BEHs) utilizing bridge vibrations and ambient wind surges to power wireless sensor nodes used for bridges' health monitoring. The developed BEHs are cantilever-type and are comprised of a wound coil, permanent magnet, an airfoil, cantilever beam, and a support. Harvesters are characterized in-lab under different vibration levels and are subjected to variable speed air surges. The harvesters exhibit multiresonant frequencies; prototype I has resonant frequencies of 3.6, 14.9, and 17.6 Hz. However, 7.6, 33, and 45 Hz are the resonant frequencies for prototype II. Under vibration testing, prototype I produced a maximum voltage of 206 mV and an optimum power of 354.51 μW at a frequency of 3.6 Hz and 0.4g acceleration. However, at a frequency of 7.6 Hz and 0.6g acceleration, prototype II showed the capability of generating a maximum voltage of 430 mV and an optimum power of 2214.32 μW. Moreover, when BEHs are characterized under variable speed air surges, prototype I generated a load voltage of 19 mV and a power of 7.84 μW at an air speed of 9 m/s; however, 22 mV and 9.14 μW load voltage and power, respectively, are developed by prototype II at 6 m/s air speed.

1. Introduction

Energy harvesting from ambient vibration is a nondestructive and maintenance-free solution for powering remote, integrated, abandoned, and embedded sensing systems used for health monitoring of machines (such as reciprocating engines, compressors, pumps, turbines, electrical generators, and motors) and civil structures (such as bridges, flyovers, and high risers). In civil infrastructure, health monitoring is significant to avoid catastrophic failure, to save human lives, and also to alleviate detours and traffic nonuniformities [1]. In reality, due to environmental hazards, fatigue, improper usage, material aging, and earthquakes, the condition of civil infrastructure continues to degrade with the passage of time [2]. For deterioration, damage recognition, and risk investigation, wireless sensor node- (WSN-) based health monitoring systems are normally used to interrogate continuously and remotely the condition of bridges [3]. The schematic of

a typical WSN is shown in Figure 1. For bridge's health monitoring, either accelerometer or strain sensor is used to provide the real-time signals of bridge's vibration or strain levels. The sensor's signals are processed (for analog to digital conversion, noise filtration, and amplification) with condition and processing part of the WSN. The processed data is then supplied to the transceiver and to the on-board memory. The power management part on the WSN is responsible for the power distribution from the power source (battery or super capacitor) to all the components on the WSN platform. The programmable microcontroller manages all the activities (sleep mode, measuring mode, and transmission mode) which are programmed for the WSN.

Usually, WSNs deployed for health monitoring require batteries for their operation. However, batteries are of limited life span and have to be either replaced or recharged repeatedly which makes WSNs prohibitively inconvenient and in some cases, expensive. Additionally, interrupted

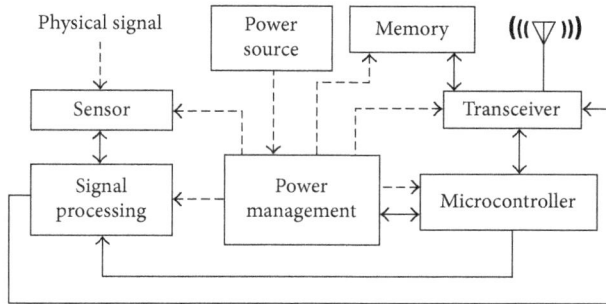

FIGURE 1: Schematic diagram of a wireless sensor node.

TABLE 1: Bridges' vibration data.

Bridge	Frequency (Hz)	Acceleration (g)	Ref.
RT11 (New York, USA)	1	0.55	[19]
Ferrite (Sweden)	14-15	0.02	[20]
Barrel Springs	3.003	0.09	[21]
Grove Street (Michigan, USA)	2–30	0.01–0.035	[22]
—	2–8	0.1	[23]
New Arsta (Sweden)	1–5	0.3–1.5	[24]
Komtur (Berlin)	2–2.6	0–0.006	[25]
New Carquinez (California, USA)	2–30	0.01–0.13	[26]
(North, France)	2	0.05	[27]
NC (USA)	1–40	0.01–0.1	[28]
Huanghe cable-stayed bridge (China)	1-2	0.015	[29]
Golden Gate (San Francisco, USA)	0–1.5	0–0.061	[30]
IH-35N over Medina River (Texas)	3.1	0.15	[31]
California (USA)	10–20	0.0002	[32]
Box girder (Austin, USA)	1–15	0.12	[33]
Iwate Prefecture (Japan)	0–25	0.005	[34]
Ypsilanti (Michigan, USA)	2–30	0.01–0.035	[35]
Seohae Grand Bridge (South Korea)	1	0.0125	[36]
Voigt bridge	4.8	0.0025	[37]
Seohae Grand (South Korea)	1	0.02	[38]
Iriri	—	0.278	[39]
Ferrite (Sweden)	4.1	0.02	[40]

functionality (due to lack of battery power) of a WSN is a burden on the network and may result in network congestion. Although during operational mode, the power need of commercial ultralow power MEMS sensors is in the range of 1.8 to $324\,\mu$W for a continuous current supply of 1.7 to $2.5\,\mu$A [4] only. However, in WSNs, more power is required for data transmission (0.09 to 128 mW) [5] rather than sensing. Furthermore, due to the rapid developments in low power sensors, microcontrollers, conditioning circuits [6], power management circuits [7], and transmission module and because of efficient wireless sensor networks [8], the power requirement of WSNs is on a sharp decline. The energy harvesting technique [9–13] that is developed two decades ago has the tendency and capability to power these low power WSNs [14]. The energies, those are present in bridge's environment and can be taken into the account for energy harvesting, are vibration (vehicle-induced vibrations), acoustic (vehicle noise), wind (naturally blowing wind and air surges produced due to vehicle motion), and solar. However, bridge's vibration [15–17] and ambient wind [18] are more reliable to be utilized for the harvesting energy on bridges.

Usually, the traffic and wind-induced bridge's vibrations exhibit low frequency and low acceleration amplitude. The vibration data of various bridges is summarized in Table 1. The maximum acceleration of bridge vibration reported by Sazonov et al. [19] is $0.55g$; however, the maximum frequency (40 Hz) is recoded for the bridge reported by [20]. Short and medium span bridges vibrate with frequency typically ranging from 2 to 8 Hz and acceleration levels less than $0.1g$ [21]. Some bridges exhibit random vibrations with low frequency (2 to 30 Hz) and low excitations of 0.01 to $0.05g$ [22]. However, at other bridge structures, the vibrations are comparatively more severe with a frequency range of 1 to 5 Hz and acceleration levels from 0.3 to $1.5g$. From Table 1, it can be seen that in the bridges' vibrations, the overall frequency content actually ranges from 0 to 40 Hz and the overall acceleration level is from 0 to $0.55g$.

The bridges' vibration levels provided in Table 1 are quite enough to drive the vibration-based energy harvesters [41, 42] to produce electrical power for the operation of WSNs mounted for the heath monitoring of bridges. In vibration-based energy harvesters, there has been a substantial research and development and to extract the bridge's excitations, vibration-based piezoelectric [43],

electromagnetic [44–46], and electrostatic [47, 48] energy harvesters can be utilized. Vibration-based piezoelectric energy harvester usually consists of a piezoelectric membrane or beam. When the harvester is exposed to base excitation, the membrane or beam also starts oscillations and deformation or strain is produced in the piezoelectric material. That strain actually induces the voltage (polarization) in the piezoelectric material [49–51]. Electromagnetic energy harvester comprised of a beam, coil, magnet, and base support. When the harvester is excited, the beam starts vibrating and the magnet attached to the beam also starts moving relative to the coil, and due to the changing magnetic flux, a voltage is induced in the coil [52–54]. However, electrostatic energy harvester is made of two conductive plates which are parallel to each other. The plates are separated by vacuum or air and initially, an electric field is provided between the plates (with the help of a battery). When external vibration is applied to an electrostatic energy harvester, the plates of charged capacitor separate and the mechanical energy is converted into electrical energy [55, 56].

For the health monitoring system of bridge structures, several BEHs are devised and reported in the literature. The reported BEHs are mainly either electromagnetic type or piezoelectric type. An electromagnetic BEH [19] produced with wound coil, permanent magnets, and spring has a resonant frequency of 3.1 Hz. It is reported to generate 10 V voltage and $1000\,\mu W$ power at 3 mm displacement and 3.1 Hz frequency. Moreover, a power development of $12500\,\mu W$ is also reported when the energy harvester is subjected to 10 mm displacement amplitude and resonant frequency.

In BEH [33], a rod having a number of cylindrical magnets is enclosed in a nylon casing and is allowed to oscillate vertically (with the aid of the end repulsive magnets located in the end caps) to generate energy from vibrations. Various parts (such as caps, frame, and casing) of the BEH are fabricated with rapid prototyping machine. When tested under sinusoidal excitation, the harvester is reported to produce a maximum power of 26 mW at $0.08g$ and 2.2 Hz.

Neodymium magnets, wound coil, and a pipe (acrylic) are used to develop an electromagnetic BEH [36]. End magnets in the outer caps are utilized to help the moving magnets to levitate in the pipe (having the wound coil) during operation. When mounted on the bridge structure, the developed BEH generated 10 mV voltage and $2\,\mu W$ power from the excitations at 8 mg acceleration and frequency of 14 Hz.

A rod (steel) having stacked magnets and core (steel) discs is kept stationary while wound coils attached to helical springs are allowed to vibrate in response to external oscillation in BEH developed by [40]. A voltage and power production of 0.71 V and 0.12 mW, respectively, is obtained from the developed BEH when vibrated at a frequency of 4.1 Hz and base acceleration of 25 mg.

An electromagnetic BEH comprised of permanent magnet, wound coil, and microfabricated planar spring is reported for low frequency vibration applications [57]. For the developed harvester, a power generation of $163\,\mu W$ (at load resistance = $220\,\Omega$) is reported at 10 Hz frequency and $1g$ base acceleration.

Membrane (latex), wound coils, PCB-planar coils, neodymium magnets, and spacers (Teflon) are used to develop a BEH [58]. In the harvester, the magnets are placed on the membrane and are allowed to oscillate; however, the wound coils and planar coils are all kept fixed. Under sinusoidal excitation at 27 Hz and $3g$, single wound coil is reported to produce 11.05 mV voltage and $2.1\,\mu W$ power; however, under same operation condition, a power of $1.8\,\mu W$ and voltage of 15.5 mV are generated by a single planar coil.

A piezoelectric cantilever-type BEH [25] comprised three cantilever beams. Frequency upconversion technique is utilized for the harvester's operation. In the harvester, the main beam has a proof mass at the free tip; moreover, two beams (piezoelectric bimorph) are located on top of the main beam just above the proof mass. The frequency of the main beam is kept low (2 Hz); however, the resonant frequencies of the piezoelectric beams are comparatively high. During operation when the proof mass hit the stopper, it excites the piezoelectric beams which then vibrate relative at higher frequencies. For the single piezoelectric beam, a load power of $64\,\mu W$ (at $70\,k\Omega$ optimum load) at 2 Hz and $0.1g$ vibration is reported.

A piezoelectric cantilever-type BEH [27] is devised for narrow band vibrations. In the harvester, two piezoelectric patches are bonded (near the clamped end of the beam) to the top and bottom surfaces of the steel cantilever beam; however, a proof mass is bonded to the free end of the beam to tune the BEH at the resonant frequency of 14.5 Hz. When tested at the bridge's real vibration in which the frequency content is less than 15 Hz, the BEH is reported for the voltage generation of 1.8 to 3.6 V and a power production of $30\,\mu W$.

A cantilever-type piezoelectric BEH [59] is developed with tunable capability. A couple of helical springs is attached in between the proof mass (at beam tip) and frame (at beam fixed end) to keep the piezoelectric cantilever beam in compression. The preload technique is utilized to tune the piezoelectric beam. The screws provided in the frame are used to set the preload on the beam. The harvester exhibits the frequencies 44.5 Hz and 40 Hz at preload of 2 N and 4 N, respectively, and it is reported to generate a load voltage of 27 and 28 V when connected to a load of $10\,M\Omega$ and vibrated at $1g$.

For low frequency operation, a cantilever-type BEH [59] is produced. Multi-impact phenomenon and frequency upconversion technique are utilized for the basic working of the harvester. In the BEH, there are two vertically oriented piezoelectric contained cantilever beams. Moreover, protruding knobs are made on the free end of piezoelectric beams. In between the piezoelectric beams, a proof mass having rollers is suspended with the help of a coil spring. The resonant frequency of the spring-mass component is kept low (2.71 Hz); however, the resonant frequencies of the beams are relatively high (120 Hz). During operation, due to vertical vibration of the proof mass, the sudden impact of the roller beams' knobs causes the piezoelectric beams to vibrate horizontally at its own resonant frequency. When the devised BEH is applied to the sinusoidal vibration, it generated a mean load power of 7.7 mW (at optimum load = $9.7\,k\Omega$) and average power of 9.4 mW at $0.29g$ and $4.4g$, respectively. Moreover, under simulated bridge's oscillations, the harvester delivered 2.8 mW mean power to the optimum load.

Cantilever-type piezoelectric BEHs [59] used piezoelectric bimorph material to extract the energy for bridge's health monitoring system. Three prototypes are reported in this work. Prototype I which has a resonant frequency of 117.1 Hz is a bimorph cantilever beam having no proof mass. In prototype II, proof mass is added due to which its resonant frequency is lowered down to 65.2 Hz. However, to extract the energy from the comparatively broader band of excitation an array type, prototype III is developed that have six piezoelectric beams. The beams in prototype III are tuned to frequencies of 63.25, 76.63, 71.5, 66.25, 63.13, 58.88, and 55.38 Hz. Under optimum load condition, prototype I and prototype II are reported to generate power levels of $197\,\mu W$ (at $11.8\,k\Omega$) and $657\,\mu W$ (at $14.9\,k\Omega$), respectively, at $0.21g$ acceleration. Moreover, for prototype III, an aggregate power production of 1.73 mW is also reported when it is subjected to $0.2g$ acceleration excitation.

A piezoelectric BEH [60] is devised to extract the energy from the bridge's bearing vibrations. Six 2 mm thick piezoelectric plates are adhesively bonded to the steel plate with conductive epoxy to form a composite sheet. By sandwiching rubber layers in between five composite sheets, the reported BEH is developed. Under a dynamic load of 17.8 kN amplitude and 2 Hz frequency, the harvester is reported to generate

650 mV voltage. Moreover, at a forcing frequency of 1.5 Hz, the energy harvester delivered a power of 83.5 µW to the optimum load of 480 Ω.

Two linear and one nonlinear piezoelectric BEHs [61] are produced with the same material and have the identical cantilever beam architectures; moreover, the piezoelectric layers and sizes of these are also kept the same. Linear BEH-1 has the tip mass of 36 grams and its resonant frequency is 28.2 Hz; however, in contrast, linear BEH-2 has a resonant frequency of 80.4 Hz and has no mass at the tip of the cantilever beam. With the nonlinear BEH-3, one magnet is attached to the tip mass and the other is fixed near to moving magnet in repulsive configuration. The excitation of the bridge is simulated with the vibration shaker in order to characterize (at 10 kΩ load resistance) the developed BEHs. The BEHs are tested for the bridge's vibration at the entrance, mid span, and exit. At the entrance and mid span, the performance of BEH-3 is reported to be better than the other energy harvesters. A maximum voltage of 0.64 V and power of 41.1 µW are produced by BEH-3 at the bridge's entrance $(0.364g)$; however, at the mid span $(0.27g)$, it generated a voltage of 0.46 V and a power of 21.3 µW. However, at the bridge's exit $(0.157g)$, BEH-1 is reported to perform well and an optimum voltage of 0.31 V and a power of 9.7 µW are reported for this linear BEH-1.

An energy harvester for the train-induced vibration in bridge structure is reported in [62]. Finite element modeling is used to predict the bridge's excitation and power production from the piezoelectric BEH (located at the underside bridge's surface) for various types of trains. It is reported that the power production BEH with PZT as transduction material is much better than that of PVDF (power generation is 52% of PZT). The developed BEHs are tested on a single-span, composite (steel-concrete) bridge (36 m long and 6.7 m wide), and at a train's speed of 120 km/hr, it is reported to generate a maximum power of 1.6 µW (with PZT-based harvester) and 0.82 µW (with the PVDF-based harvester).

Modeling and simulation for a linear electromagnetic BEH driven by the lateral vibration (wind induced) of the bridge structure are described in [63]. Two such mass-tuned BEHs are installed on the bridge's girder and are modeled as single degree of freedom (lumped parameter model) mechanical oscillators. At a wind speed of 30 m/s, the simulation results predicted a mass displacement of 0.5 m for the harvester mass, when the BEH is installed at the mid span of the bridge; moreover, it is reported to produce an optimized power of 2400 W at the same wind speed.

Lumped parameter model for the bridge and an electromagnetic BEH is utilized for the analytical modeling of the coupled system [64]. Single degree of freedom and two degree of freedom-tuned mass energy harvesters are modeled and simulated for the bridge's excitations. A nondimensional steady state power of 4.8 and 5 is reported for a single and two degree of freedom BEH at equivalent damping ratio of 0.01 and 2, respectively.

In this research, work BEHs are developed to power the bridge monitoring system. The work is the extension of the BEH developed by the authors in [65]. The energy harvesters are capable to produce power simultaneously from the

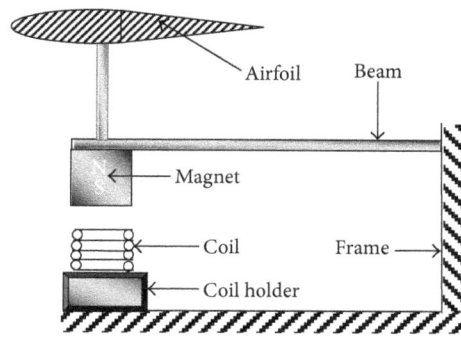

FIGURE 2: Architecture of the bridge energy harvester, prototype I.

vibrations and wind available at bridge's structure. Two low frequency, electromagnetic, cantilever-type, vibration-based and wind-based harvesters, prototype I, and prototype II are produced. For the prototypes, novel device architecture is utilized in which an airfoil is mounted at the free end of the cantilever beam and is used to also extract the wind energy. Moreover, the airfoil and the relevant attachment are also exploited to have an extra device's resonant frequency. Prototype II has dual cantilever beams: one beam contains the magnet; however, the other lower beam has the wound coil. The wound coil beam actually adds an additional resonant frequency to the harvester. Furthermore, most of the reported BEHs are monoresonant and exhibit narrow bandwidth; however, in this work, the multiresonant nature of the developed energy harvesters is exploited to increase the bandwidth of the devices and to harvest the energy from the real narrow band bridge's vibration environment.

2. Architecture and Working Principle of Prototypes

The architecture of bridge energy harvester (BEH), prototypes developed in this work, is shown in Figures 2 and 3. The BEHs are comprised of a wound coil, a cylindrical permanent magnet, an airfoil, cantilever beams, and a frame. In BEHs, the magnet and airfoil are attached to the free end of the cantilever beams; however, the wound coils are fixed just underneath the magnets. Moreover, a gap is provided between the magnet and coil to allow free vibration of the magnets over the coils. Prototype I consists of a single beam; however, two cantilever beams are provided in prototype II. The cantilever beam-II, in prototype II, will actually add an extra resonant frequency to the energy harvester. Moreover, a slightly long rod is used with the attached airfoil, in order that the airfoil assembly behaves like an inverted pendulum with a flexible base (beam) and can add another resonant frequency to both the prototypes. The coil holder, in prototype I, can be adjustable in the vertical direction and is used to fix the gap between the magnet and coil according to the level of base vibration. However, in prototype II, provision is provided to move beam-II in vertical direction as well as its length can be adjusted with the help of beam holder slots allowed in the frame. Since prototype II is comprised of two beams, therefore it is capable of extracting the energy from the real random (narrow band) bridge's vibration at both

---

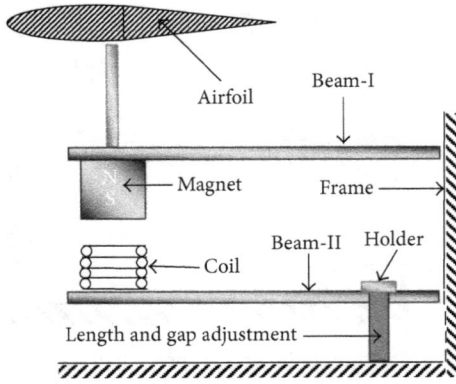

FIGURE 3: Architecture and working principle of prototype II.

resonant frequencies at once. Moreover, in prototype II, the beams are designed stiffer and are more suitable for high acceleration level bridge's vibrations. Furthermore, in accordance to bridge's vibrations, the lower beam in prototype II can easily be tuned to adjust its resonant frequency and alter the bandwidth of the harvester for better performance. When the BEHs are subjected to vibrations, the cantilever beam and the magnet attached to it start oscillation over the coil and as a result of which, a changing magnetic flux density is experienced by the coil and an EMF is induced in the coil according to Faraday's law of electromagnetic induction. Similarly, when the ambient wind surges flow over the airfoil, it produces an upward lift force; however, usually the natural wind flow and even the air surges due to moving traffic are normally fluctuating in nature; therefore, due to these wind surges, the cantilever beam and magnet attached to the airfoil will start oscillating vertically and induce an EMF in the coil. The EMF induced in the coil of the developed BEHs depends on the number of coil turns, magnetic field strength, level of base acceleration, wind speed, and relative velocity between the magnet and the coil.

3. Modeling of Prototypes

3.1. Harvester's Beam Design. With lumped parameter model, the undamped fundamental frequency

$$f_n = \frac{1}{2\pi}\sqrt{\frac{k}{m_e}} \qquad (1)$$

of the cantilever-type energy harvester depends on the beam's stiffness k and equivalent mass m_e.

The cantilever beam's stiffness (lumped) [66]

$$k = \frac{3EI}{L^3} \qquad (2)$$

is a function of beam's length L, modulus of elasticity E, and moment of inertia [66]

$$I = \frac{bh^3}{12} \qquad (3)$$

of beam's cross-section that depends on the beam's thickness h and width b.

The equivalent mass [66]

$$m_e = \frac{33}{140}m_b + m \qquad (4)$$

of the uniform beam having mass (magnet and airfoil) at the tip can be obtained with beam's mass m_b and tip mass m.

Using (1), (2), (3), and (4), the length

$$L = \left(\frac{35Ebh^3}{4\pi^2 f_n^2(33m_b + 140m)}\right)^{1/3} \qquad (5)$$

of the beam can be obtained in terms of beam's parameters and fundamental frequency.

In order to estimate the beam's length for a certain fundamental frequency (1 to 40 Hz, the targeted frequency range based on bridge vibration data, Table 1) of the BEH and keeping in view the low frequency operation of the harvesters, the fundamental frequency of 3.6 Hz and 7.6 Hz has been selected for the developed prototype I and prototype II, respectively. Moreover, a beam's width and thickness of 4 cm and 0.4 mm, respectively, are taken for a hot dipped galvanized steel (modulus of elasticity = E = 200 GPa) beams. The mass (55.6 grams) of the magnet which is to be attached to the beam is measured; however, the mass (30.1 grams) of the airfoil assembly is estimated from the proposed dimensions. Furthermore, the mass of the beam (m_b) which is supposed to be less in comparison to the tip mass (m = 85.7 grams) is ignored during the simulation for the beam's length. Figure 4 shows the simulation result of (5), and for a fundamental frequency of 1 to 40 Hz, the beam's length can be adjusted from 335.7 mm to 28.7 mm. Moreover, the simulation predicts the beam's length of 142.9 mm and 86.8 mm for prototype I and prototype II, respectively.

The beam of cantilever-type energy harvester can also be designed based on the acceleration levels of the bridge's vibrations. By modeling the harvester as single degree of freedom system, the amplitude of the relative displacement [66]

$$Z = \frac{A(\omega/\omega_n)^2}{\omega^2\sqrt{\left[1 - (\omega/\omega_n)^2\right]^2 + [2\xi_T(\omega/\omega_n)]^2}} \qquad (6)$$

between the magnet and coil depends on the base acceleration's amplitude A, total damping ratio ξ_T, undamped natural frequency ω_n, and frequency of base excitation ω.

At resonance, the relative displacement amplitude

$$Z = \frac{A}{2\omega_n^2\xi_T} \qquad (7)$$

can be used to write the undamped natural frequency

$$\omega_n^2 = \frac{A}{2\xi_T Z} = \frac{k}{m_e} \qquad (8)$$

in terms of base acceleration A.

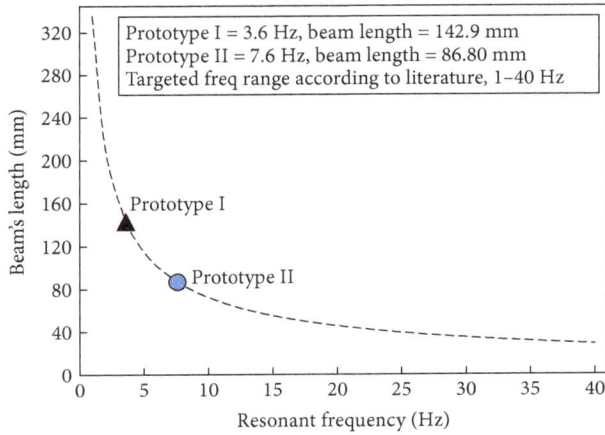

FIGURE 4: Beam's length as a function of the fundamental frequency of the BEH.

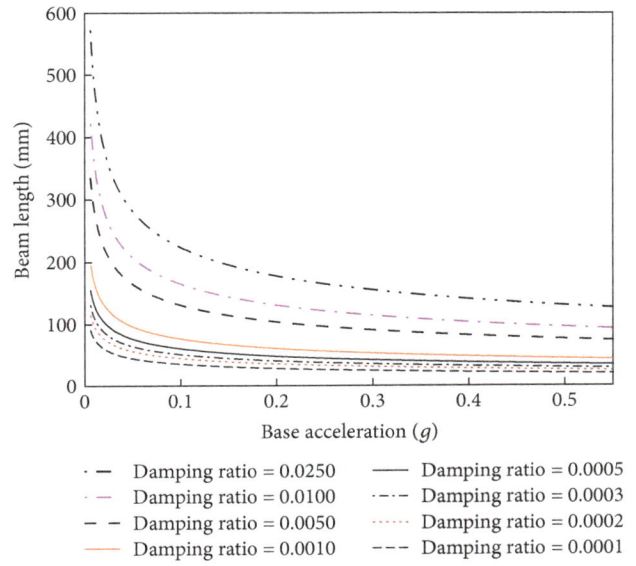

- · — Damping ratio = 0.0250   —— Damping ratio = 0.0005
- · — Damping ratio = 0.0100   -·-· Damping ratio = 0.0003
- – – Damping ratio = 0.0050   ········ Damping ratio = 0.0002
- —— Damping ratio = 0.0010   – – – Damping ratio = 0.0001

FIGURE 5: Harvester's beam length as a function of the bridge's base acceleration at different total damping ratios.

By utilizing (2), (3), and (8), the length

$$L = \left(\frac{\xi_T Z E b h^3}{2 A m_e} \right)^{1/3} \qquad (9)$$

of the cantilever beam can be expressed in energy harvester's parameters.

Figure 5 shows the simulation result of (9) for a constant gap between the magnet and coil ($Z = 1.5$ cm, amplitude of relative displacement). During simulation, the acceleration levels from 0.006 to $0.55g$ (bridges' vibration data, Table 1) are utilized to predict the length of hot dipped galvanized steel (modulus of elasticity $= E = 200$ GPa) beam with a width and thickness of 4 cm and 0.4 mm, respectively, and having a tip mass of 85.7 grams (mass of magnet and airfoil assembly). Moreover, the total damping ratio from 0.0001 to 0.025 is used for the simulation purpose. For high acceleration level vibrations, the stiffness of the harvester's beam needs to be also on higher side; however, to operate effectively in low acceleration vibration environment, the beam is required to be less stiff and the same is revealed by the simulation in Figure 5. As per acceleration level, the length of the beam is required to be adjusted, and long beam will be suitable for low acceleration vibration; however, for high acceleration vibration, the beam's length is kept to be short. Moreover, at a specific acceleration level, the beam's length also depends on the total damping ratio, while for harvesters with high damping, relatively long beams have to be adopted. At ultralower acceleration levels from $0.0006g$ to $0.031g$, the reduction in beam's length is about 41.5% irrespective of the total damping ratio of the harvester. A percentage reduction of 31.6 in the beam's length occurs for an acceleration range from $0.032g$ to $0.1g$. However, for relatively high acceleration levels from $0.102g$ to $0.55g$, the beam's length has to be reduced by 43%.

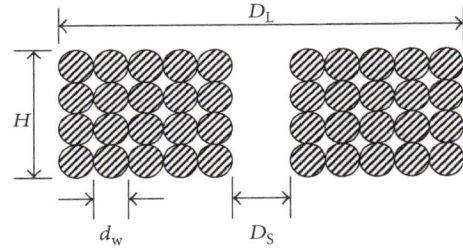

FIGURE 6: Cross-section of a wound coil of energy harvester.

3.2. Harvester's Wound Coil Design. For a wound coil (Figure 6) in an electromagnetic energy harvester, the number of turns

$$n = \frac{D_L - D_S}{2 d_w} \qquad (10)$$

in each layer and the number layers

$$N_1 = \frac{H}{d_w} \qquad (11)$$

in wound coil depend on coil's outer diameter D_L, inner diameter D_S, height H, and wire diameter d_w and can be used to compute the total number of coil turns

$$N_t = n N_1 = \frac{H}{d_w} \left(\frac{D_L - D_S}{2 d_w} \right). \qquad (12)$$

In electromagnetic energy harvesters, the coil's resistance

$$R_C = \frac{\rho_c L_t}{A_c} = \frac{4 \rho_c N_1 L_1}{\pi d_w^2} = \frac{\rho_c H L_1}{\pi d_w^3} = \frac{\rho_c H}{d_w^3} \sum_{i=1}^{n} [D_S + (2i - 1) d_w]$$

$$\qquad (13)$$

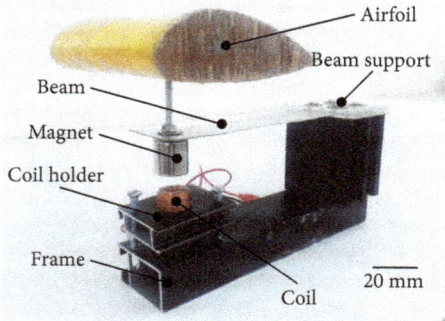

FIGURE 7: Photograph of the fabricated BEH, prototype I.

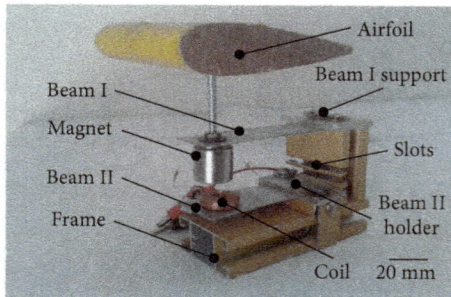

FIGURE 8: Photograph of the fabricated BEH, prototype II.

TABLE 2: Main features and dimensions of developed BEHs.

Feature	Prototype I	Prototype II
Coil's resistance	54.5 Ω	28 Ω
Magnetic flux density	1.32 T	1.32 T
Magnet's mass	55.6 grams	55.6 grams
Magnet's dimensions	20 mm × 20 mm	20 mm × 20 mm
Coil and magnet gap	1.5 cm	1 cm
Number of coil turns	965	400
Width of beam-I	4 cm	3 cm
Width of beam-II	—	2.9 cm
Thickness of beam-I	0.40 mm	0.40 mm
Length of beam-I	14 cm	8 cm
Thickness of beam-II	—	0.30 mm
Length of beam-II	—	5.7 cm
Airfoil dimensions	10 cm × 10 cm × 3 cm	8.5 cm × 8 cm × 2 cm
Device overall size	14 cm × 9 cm × 5 cm	8 cm × 6 cm × 4 cm

that is also the function of the coil's material and coil's parameters (dimensions) is important, since the optimum load is normally equal to the coil's resistance for mesoscale coils [45]. There are two ways to manufacture the wound coil in an electromagnetic BEH. By using the coil's dimensions as constraints, (12) can be utilized to compute the total number of coil's turns; however, by producing coil in this way has the disadvantage that the coil resistance is not a constraint.

However, by utilizing (13), the coil's resistance can also be kept as a constraint along the other coil's parameters and the number of coil's turns can be adjusted such that (13) is satisfied.

4. Fabrication of Prototypes

The developed BEHs, prototype I and prototype II, are shown in Figures 7 and 8. The base support (frame) of the BEHs are produced from commercially available aluminum sections (Forward Metals, UK). The cantilever beams holding the neodymium (NdFeB) magnets are fabricated from a galvanized steel sheet (Shanghai Metal Co., China). By using a manual winding apparatus, a wound coil is made from an enameled copper wire of 80 μm diameter. In prototype I, the produced wound coil has a thickness of 10 mm and diameter of 18 mm and contains 965 turns. However, a coil of 4 mm thickness and 20 mm diameter and having 400 turns is formed for prototype II. Moreover, for the prototypes, the airfoils are fabricated from a light weight Thermophore insulation sheet (Industrial Enterprises, Pakistan) and plywood. The fabricated airfoils are then attached to the free end of cantilever beams through a long bolt and a nut. A cylindrical permanent magnet is then attached to the bottom

side of the top cantilever beam. Just underneath the magnet, the wound coil, in prototype I, is bonded to the coil's holder located at the harvester's base and in prototype II, the coil is glued to the bottom cantilever beam-II. The coil holder of prototype I can move up and down with the help of screws to adjust the desired gap between the magnet and the coil. However, in prototype II, the beam-II support is kept movable and adjustable. This beam-II support is designed in such a way that with it, not only the magnet-coil gap is adjusted but also the beam holder is moved along the base, and the length of beam-II can be selected in order to tune beam-II accordingly. Furthermore, small size nuts and bolts and screws are utilized to securely clamp and assembled all the produced parts and components of the BEHs. In both prototypes, the gap between the magnet and coil is adjusted such that, these can be operated in low as well as in relatively high acceleration levels and can produce adequate voltage levels. If the gap is kept too small, the harvesters' performance will be better at low acceleration levels; however, at high acceleration levels, the magnet will touch the coil and alter the harvesters' operation. Moreover, for large gaps, the harvesters will generate high output voltage at high acceleration levels; however, minimum or no voltage will be produced when these are subjected to low accelerations. Dimensions and main parameters of the developed BEHs are provided in Table 2.

5. Experimentation and Results

5.1. Characterization of Prototypes under Harmonic Vibrations. With the experimental setup shown in Figure 9, the developed BEH prototypes have been characterized inside the laboratory. The setup composed of a 12/24 V, DC power supply (Universal Electronics, Pakistan), a power amplifier (Model: RM-AT2900, Rock Mars, United Arab Emirates), oscilloscope (Model: GOS 6112, GW Instek, New Taipei, Taiwan), a function generator (Model: GFG 8020H, GW Instek, New Taipei, Taiwan), digital multimeter (Model: UT81A/B, Uni-Trend Technology, Dongguan,

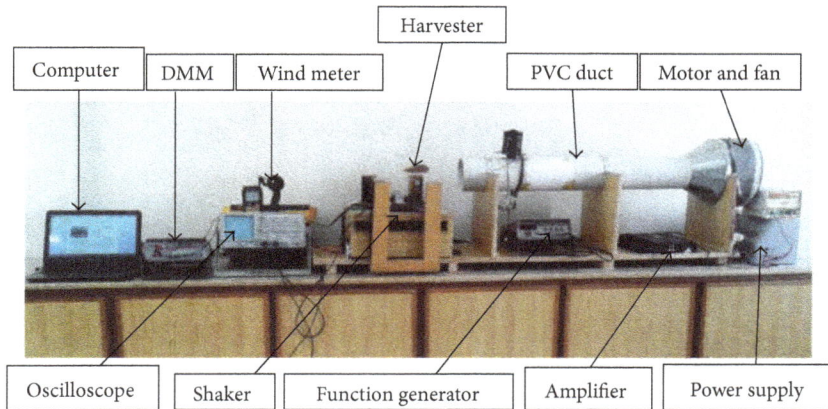

FIGURE 9: Experimental setup for characterization of BEHs.

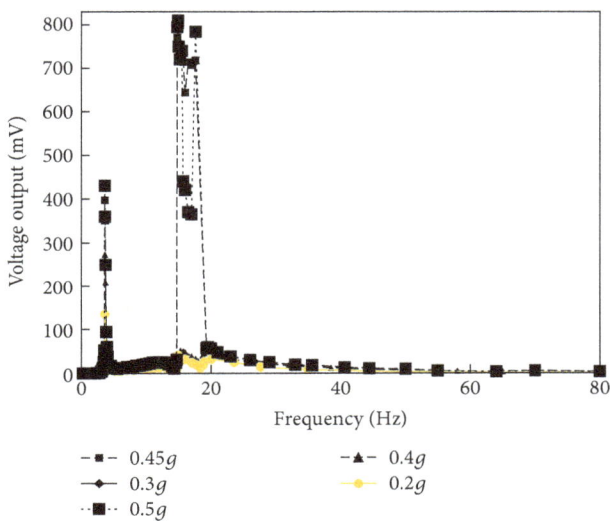

FIGURE 10: Open circuit RMS voltage produced as a function of frequency for BEH prototype I.

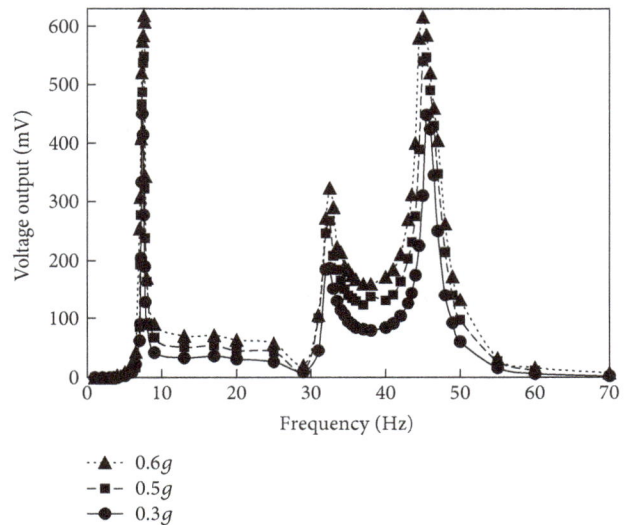

FIGURE 11: Open circuit RMS voltage produced as a function of frequency by BEH prototype II.

China), vibration shaker, and an accelerometer (Model: EVAL-ADXL335Z, Norwood, USA). Moreover, a variable speed electrical motor, a fan arrangement, and a PVC duct pipe are used to produce surges (variable speed) of air. In the setup, the function generator is used to generate a sinusoidal voltage signal of a desired frequency which is amplified up by the power amplifier and is then supplied to the vibration shaker. The BEH is mounted on a wooden block which is tightly fixed to the shaker's table, and an accelerometer is also attached to the block in order to monitor the acceleration levels to which the BEH is subjected. The BEH output voltage signals and the measurements from the accelerometer are measured and analyzed with the oscilloscope and multimeter.

Figures 10 and 11 show open circuit voltage levels of BEHs with respect to excitation frequency at different acceleration levels. The BEHs are subjected to a forward frequency sweep (FFS) from 0.2 Hz to 80 Hz and to acceleration levels from $0.2g$ to $0.6g$. As the acceleration level to which the BEHs are exposed is increased, the output voltage of these

devices also increases. In the tested frequency range, prototype I exhibits three resonant frequencies 3.6 (first resonance of the cantilever beam), 14.9 (resonance of the beam-airfoil-magnet assembly), and 17.6 Hz (second resonance of the cantilever beam, torsion mode). A maximum voltage of 810 mV is generated at a resonant frequency of 14.9 Hz and a base excitation of $0.5g$. In prototype I, at a base acceleration greater than $0.5g$, the magnet was touching the coil.

However, under the same testing conditions, the prototype II response is shown in Figure 11. Three resonant frequencies are exhibited by this BEH, 7.6 Hz (first resonance of the upper cantilever beam), 33 Hz (resonance of the beam-airfoil-magnet assembly), and 45 Hz (resonant frequency of the second beam upon which the wound coil is placed). At the first resonant frequency in which the magnet along and the upper beam are oscillating, the harvester produces greater output voltage. BEH prototype II is capable of generating a maximum voltage of 618 mV at a resonant frequency of 7.6 Hz and at base acceleration of $0.6g$. However, comparatively, less voltage levels are generated at the second

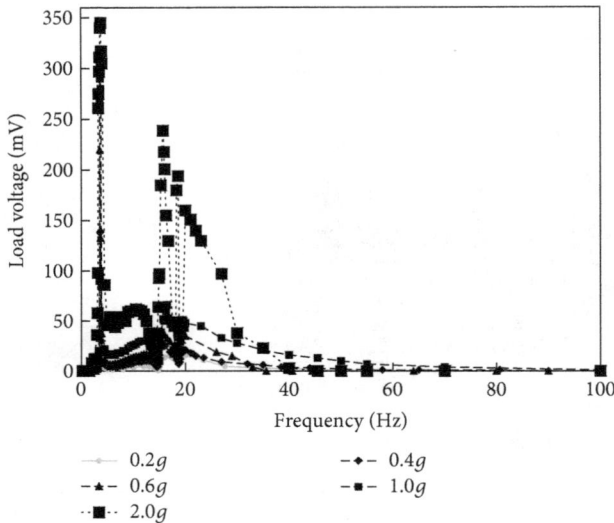

FIGURE 12: Load voltage as function of frequency for prototype I.

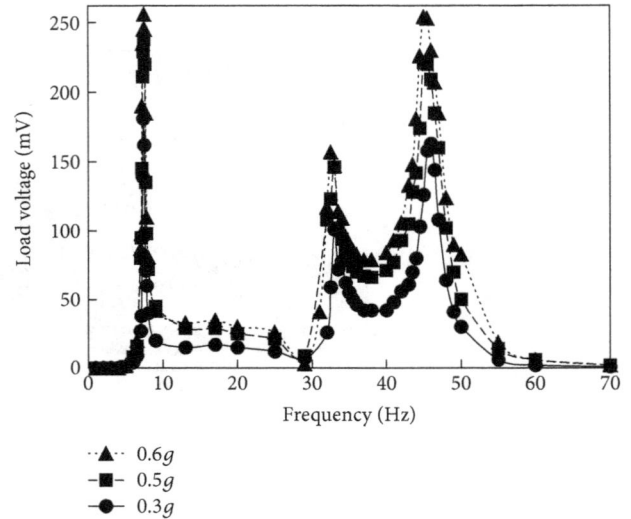

FIGURE 13: Load voltage as a function of frequency for prototype II.

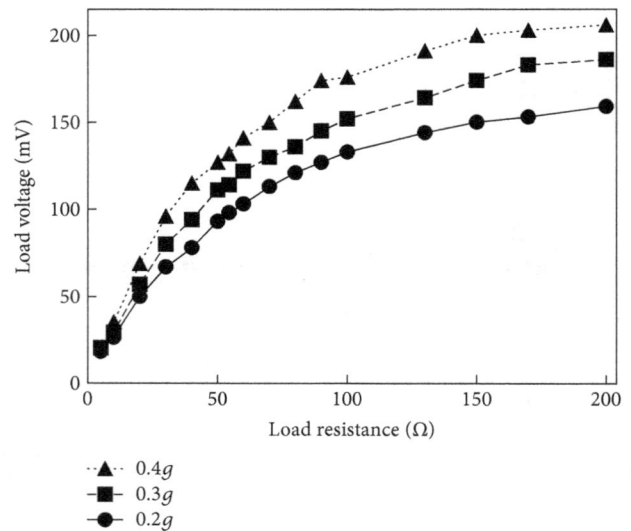

FIGURE 14: Load voltage with respect to load resistance at different g levels and at a resonant frequency of 3.6 Hz for prototype I.

resonant frequency; however, at the third resonance of the BEH prototype II, the voltage production is almost equal to that generated at the first resonant frequency. Since comparatively the beams of prototype II are stiffer, therefore, it is subjected to relatively high acceleration levels (up to 0.6g). The multiresonant behaviour of the prototypes is beneficial and is significant in a number of ways. For example, such characteristic increases the bandwidth of the harvester which is helpful during off resonance operation of the harvesters. The multiresonant prototype performance will be far better than the monoresonant energy harvesters when the operation frequency is slightly away from the resonance; moreover, with multiresonant frequencies, the developed BEH will be more capable to perform better under the real, narrow band bridge excitations. Since all the resonant frequencies of prototype I are associated with the upper beam only, therefore, it will not be able to harvest the energy from the bridge vibration at all frequencies at once; however, the resonant frequencies of prototype II are not associated with a single beam, therefore, comparatively, during operation, it will be capable to generate power at both beams' resonant frequencies at once.

In Figures 12 and 13, optimum load resistances of 54.5 Ω and 28 Ω are attached to prototype I and prototype II, respectively, and these are subjected to FFS at various acceleration levels. When connected to the optimum load, prototype I produced a maximum RMS load voltage of 345 mV at the resonant frequency of 3.6 Hz and at a base acceleration of 2g. On the other hand, prototype II produced a maximum RMS load voltage of 256 mV at a resonant frequency of 7.6 Hz and at base acceleration of 0.6g.

In Figures 14, 15, and 16, output load voltage from prototype I with respect to load resistance is depicted. In this experimentation, the BEH is excited at a resonant frequency of 3.6 Hz, 14.9 Hz, and 17.6 Hz and is subjected to acceleration levels of 0.2, 0.3, and 0.4g. Different load resistances were connected with the harvester, and output voltage of the prototype is measured. It is obvious from these figures

that, at greater load resistance, high load voltage levels are produced which are attributed to the low current that is flowing in the load. Furthermore, the load voltage is increased as the acceleration level to which the harvester is exposed is increased. At a base acceleration of 0.4g and load resistance of 200 Ω, maximum load voltage levels of 206 mV, 58 mV, and 25 mV are produced by the harvester when excited at a resonant frequency of 3.6 Hz, 14.9 Hz, and 17.6 Hz, respectively.

At the first, second, and third resonant frequencies of 7.6, 33, and 45 Hz, the output load voltage from the BEH prototype II with respect to load resistance is depicted in Figures 17, 18, and 19. The harvester is excited at 0.3g, 0.5g, and 0.6g acceleration levels and is characterized for a load resistance from 10 to 100 Ω. While attached to 100 Ω load, maximum load voltage levels of 430 mV, 250 mV, and

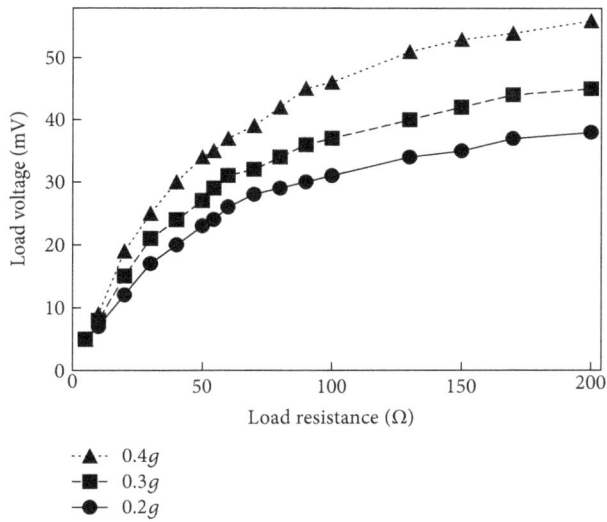

FIGURE 15: Load voltage versus load resistance at different g levels and at 14.9 Hz resonant frequency for prototype I.

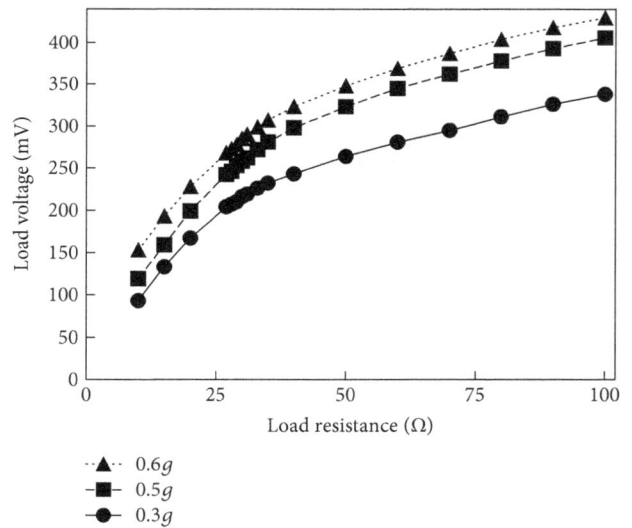

FIGURE 17: Load voltage with respect to load resistance at 7.6 Hz for prototype II.

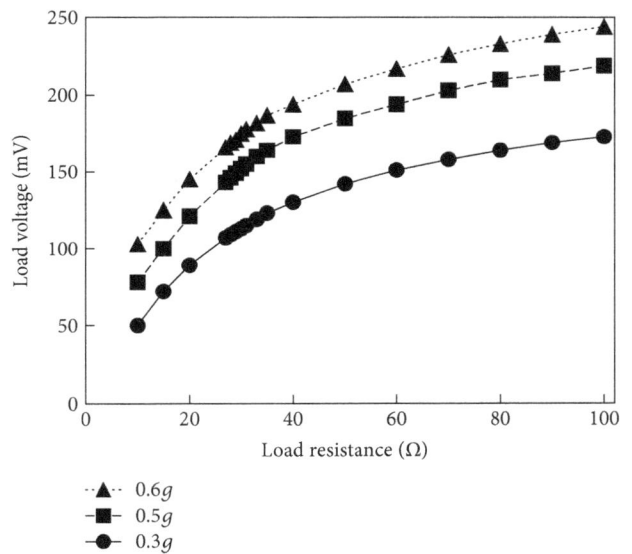

FIGURE 16: Load voltage against load resistance at different g levels and at 17.6 Hz resonant frequency for prototype I.

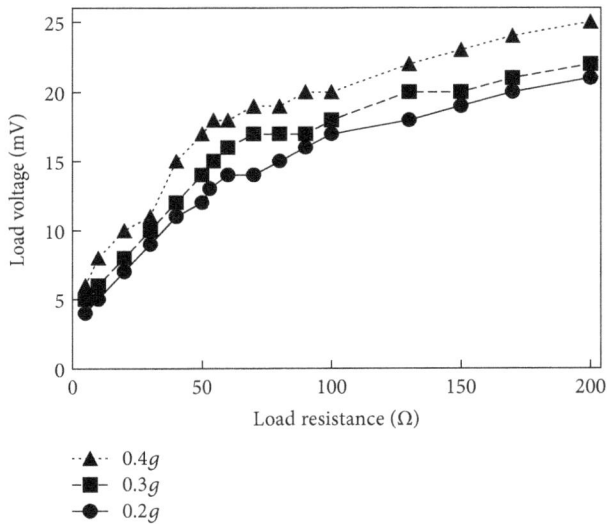

FIGURE 18: Load voltage versus load resistance at different g levels and at 33 Hz for prototype II.

390 mV are produced by the harvester when excited at a base acceleration of $0.6g$, $0.5g$, and $0.3g$, respectively.

At the resonant frequencies, the load power characteristics of prototype I as a function of the load resistance are shown in Figures 20, 21, and 22. These figures are obtained with the load voltage data depicted in Figures 15, 16, and 17. The average load power values are computed with the aid of measured load RMS voltage and the corresponding load resistance attached to prototype I. The load power characteristic lines correspond to the device's first (3.6 Hz), second (14.9 Hz), and third (17.6 Hz) resonant frequencies and at base acceleration levels of 0.2, 0.3, and $0.4g$. It is evident from the plots that as the base acceleration level is increased, the more power is generated by the harvester. Relatively, higher power levels are obtained at the first

resonant frequency of 3.6 Hz. Moreover, at all resonant frequencies, the average power delivered to the load is optimum at a condition (maximum power transfer theorem) where the attached load resistance is equal to the coil's resistance (54.5 Ω). Furthermore, the harvester when excited at $0.4g$ base acceleration level delivered an optimum power of 354.51 μW, 26.5 μW, and 7.33 μW at resonant frequencies of 3.6 Hz, 14.9 Hz, and 17.6 Hz, respectively, under optimum load condition.

The BEH prototype II performance in terms of load power (power characteristics) is also analyzed. Figures 23, 24, and 25 show the average load power produced by the harvester as a function of load resistance at different base accelerations. The device is characterized at three resonant states and under base acceleration levels of 0.3, 0.5, and $0.6g$. The

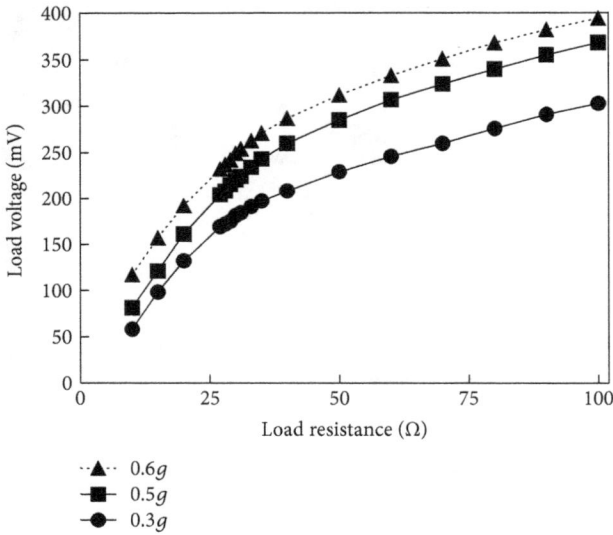

FIGURE 19: Load voltage against load resistance at 45 Hz for prototype II.

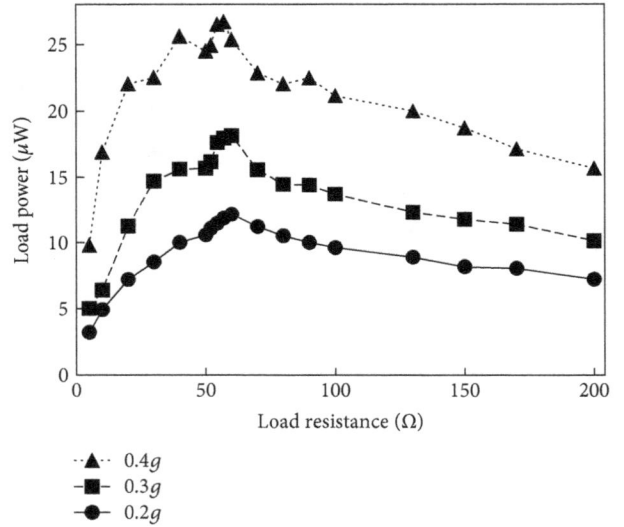

FIGURE 20: Load Power versus load resistance at different g levels and 3.6 Hz resonant frequency for prototype I.

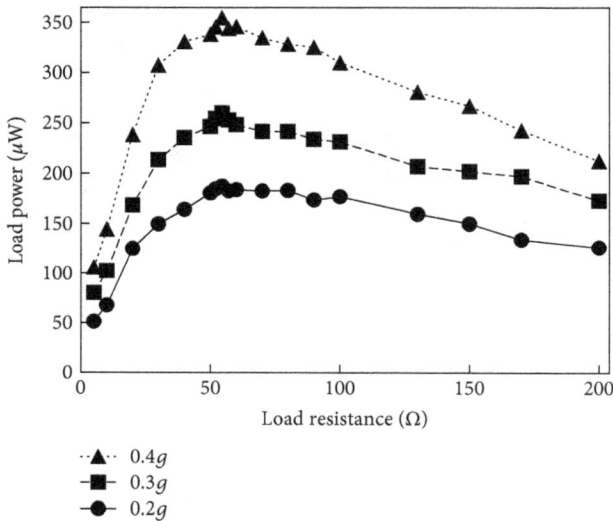

FIGURE 21: Load Power with respect to load resistance at different g levels and 14.9 Hz resonant frequencies for prototype II.

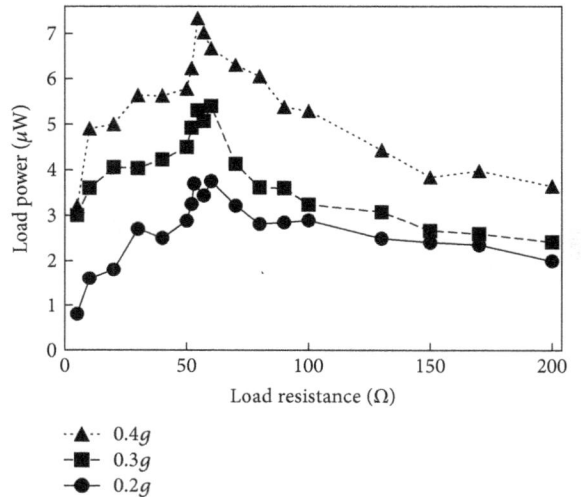

FIGURE 22: Load Power as a function of load resistance at different g levels and 17.6 Hz resonant frequencies for prototype I.

measured load RMS voltage levels obtained at the respective load resistance are utilized to compute the corresponding power levels ($P = V_{RMS}/R_{load}$) delivered to the load. Due to better design, optimized wound coil, optimum magnet-coil gap, and lower coil resistance, comparatively, higher power levels are obtained from prototype II. At optimal load resistance of 28 Ω (resistance of the coil), maximum power levels of 2214.32 μW, 836.03 μW, and 2178.89 μW are produced at 0.6g acceleration and at resonant frequency of 7.6 Hz, 35 Hz, and 45 Hz, respectively.

For a magnet and coil gap of 1.5 cm (prototype I) and 1 cm (prototype II), the developed prototypes have been characterized at acceleration levels from 0.2 to 0.6g. For a constant beam's length, too low acceleration levels will actually result in a very small relative displacement between

magnet and coil and hence the performance of the harvesters would be highly affected. If the gap between the coil and magnet is not altered, then in the harvester, the beam's length is required to be modified according to the specific low acceleration of the bridge's vibration. Based on vibration's acceleration level, the harvester's beam length can be adjusted as per simulation performed in Figure 5. For ultralow acceleration levels such as 0.0006g, the beams' length needs to be from 90.7 mm to 571.5 mm for the total damping ratio of 0.0001 and 0.025, respectively. However, for relatively high acceleration levels, like 0.03g, for the total damping ratio of 0.0001 and 0.025, the beams' length is required to be kept from 53 mm to 334.2 mm, respectively, for better operation. Moreover, for the same values of total damping ratio's and at a bridge's acceleration level of 0.1g, the beam's lengths of 35.5 mm and 223.7 mm, respectively, will be highly effective for the harvester's efficient operation.

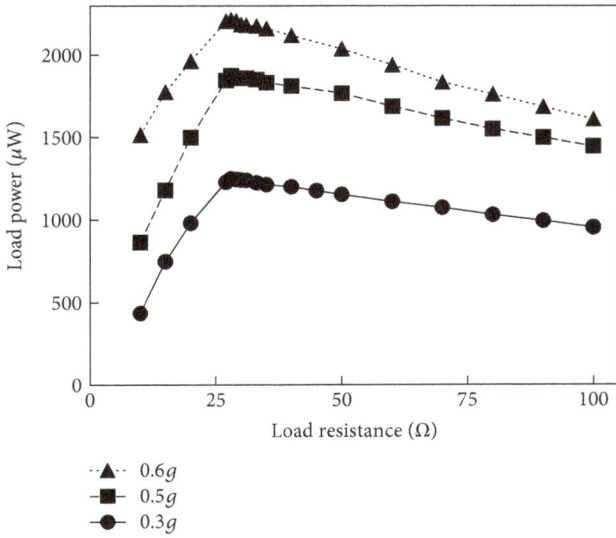

FIGURE 23: Load power as function of load resistance at 7.6 Hz for prototype II.

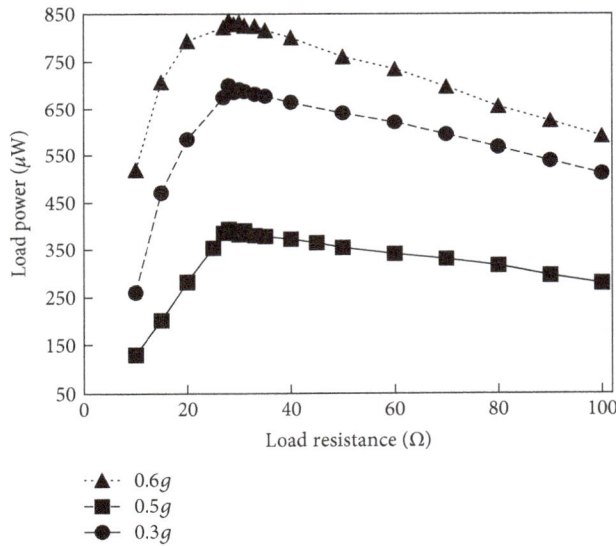

FIGURE 25: Load power as a function of load resistance at 45 Hz for prototype II.

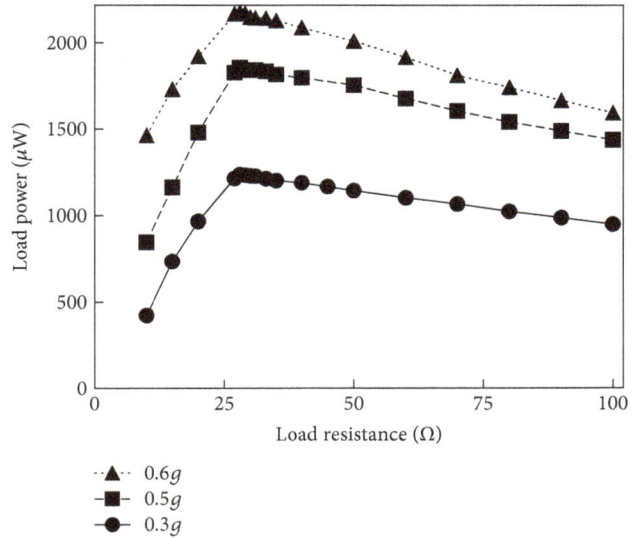

FIGURE 24: Load power with respect to load resistance at 35 Hz for prototype II.

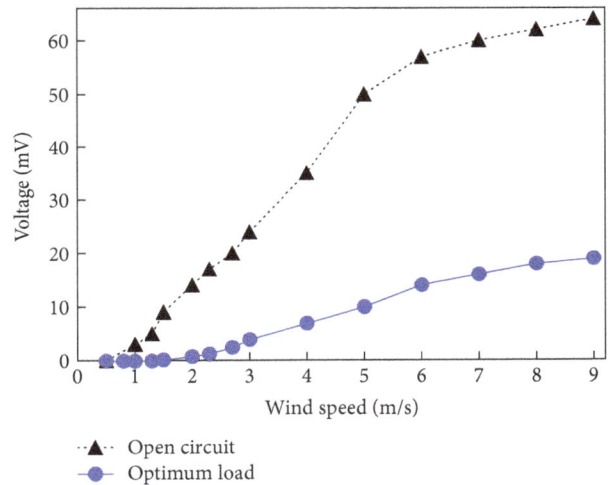

FIGURE 26: Generated voltage as a function of wind speed for prototype I.

However, in the developed prototypes, the provision of the gap adjustment is provided and with simple modification, these can be very easily utilized to operate efficiently and effectively under ultralow and low bridge's vibration. In the prototypes for the existing beams, the gap between the magnet and coil can be minimized (optimized) to significantly extract the energy from the ultralow and low levels of acceleration.

5.2. Characterization of BEH Prototypes under High Speed Air Surges. The harvesters are also characterized in the scenario of ambient high speed wind. In the testing rig, the wind blowing setup is utilized to perform experiments on energy harvesters. Figures 26 and 27 show the open circuit and load

voltage levels generated by prototype I and prototype II as a function of wind speed. In the measurements, the air speed is gradually increased by regulating the fan rotation and voltage levels are recorded at various air speeds. Moreover, for measuring the wind speed, a flow anemometer (Model AR-856, Intell Instruments™ Plus, China) is used. Prototype I is characterized for an air speed from 0.5 m/s to 9 m/s. It is clear from the graph that as the wind speed to which the harvesters are subjected is increased, the output voltage level increases. For prototype I, a maximum open circuit voltage of 64 mV and a load voltage of 19 mV (at optimum load resistance of 54.5 Ω) are produced at an air flow of 9 m/s. However, the performance of prototype II is determined under an air speed from 0.5 to 6 m/s. In prototype II, at an air speed higher than 6 m/s, the magnet was touching the coil; therefore, it was not subjected to air speed beyond 6 m/s. For prototype II, a maximum open circuit output voltage (84 mV) is obtained at an

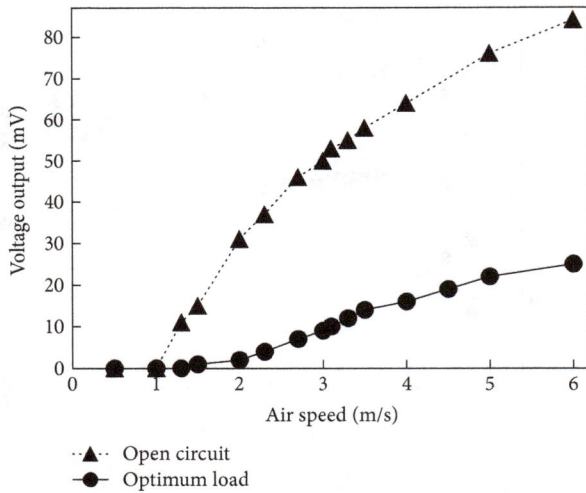

FIGURE 27: Voltage as a function of wind speed graph for prototype II.

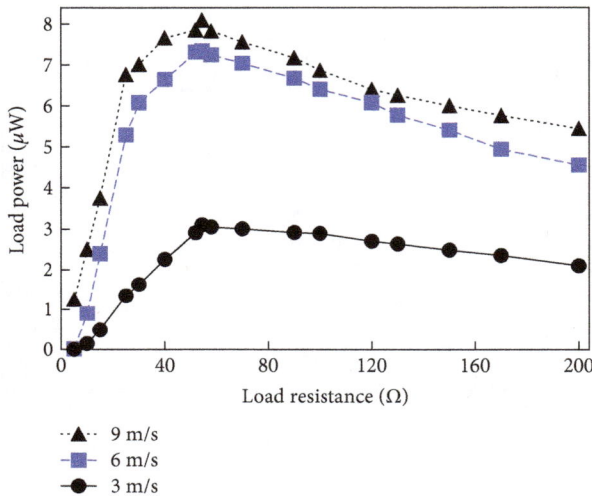

FIGURE 28: Load power with respect to load resistance at different wind speed for prototype I.

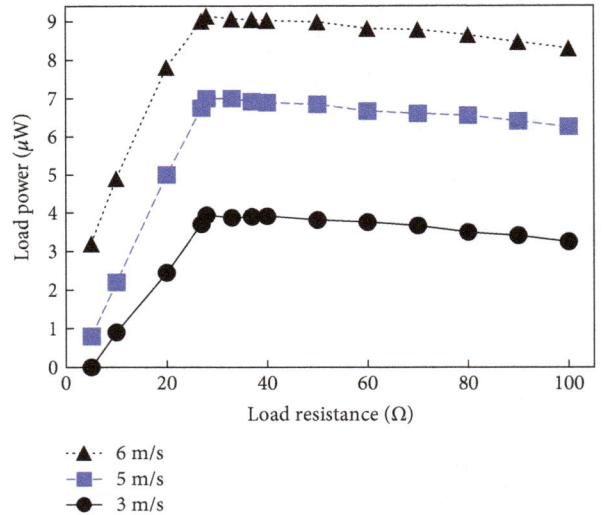

FIGURE 29: Load power with respect to load resistance at different wind speed for prototype II.

the harvester produced optimum power levels of $3\,\mu$W and $7.1\,\mu$W at 3 m/s and 6 m/s wind speed, respectively. However, the optimum power levels (Figure 29) obtained from prototype II while operating under optimum load condition ($28\,\Omega$ load resistance) are $9.14\,\mu$W, $7\,\mu$W, and $3.9\,\mu$W at 6 m/s, 5 m/s and 3 m/s wind speed, respectively. Similar to the operation of under vibration, prototype II showed the capability of generating more power than prototype I also when these are excited by variable speed air surges.

6. Comparison and Discussions

The developed BEHs are listed and compared with each other in Table 3. A number of factors can be utilized to compare these BEHs, such as device's energy transduction method, acceleration levels to which these are designed, harvesters' resonant frequency, number of device resonant frequencies, harvester's bandwidth, mechanism internal impedance, voltage, and power level generation. In all of the reported BEHs, either electromagnetic or piezoelectric energy transduction methods are used. Up until now, the electrostatic transduction method is not adopted for the BEHs and most probably it is due to the fact of the requirement of a battery for the initial charging of the harvester during operation. The resonant frequency range for the reported energy harvesters is 1–120 Hz. The overall acceleration levels to which the devised BEHs are subjected during characterization actually range from 0.008 to 0.6g. Comparatively, the voltage levels produced by the piezoelectric BEHs (350–28000 mV) are higher than those generated by the electromagnetic BEHs (10–10000 mV). Most of the electromagnetic BEHs are actually developing voltage levels well below 1 V. However, while relating these BEHs on the basis of power production, it is quite clear that the power generation ability of electromagnetic BEHs (0.7–26000 μW) is far better than the piezoelectric BEHs (0.6–7700 μW), and this advantage of electromagnetic BEHs in reality is attributed to the low internal impedance of electromagnetic BEHs (3.6–67 kΩ).

applied wind speed of 6 m/s. However, under load condition, when $28\,\Omega$ (optimum load) is attached to the harvester, a maximum load voltage of 22 mV is delivered. Just as under vibration testing, prototype II seems to produce high voltage levels than prototype I, when it is subjected to variable air speed flow.

Figures 28 and 29 show the output power generated by prototype I and prototype II as a function of load resistance at different wind speeds. In this experiment, prototype I is subjected to specific air speed (3, 6, and 9 m/s) and various load resistances are connected to the harvester and load voltage levels are measured. The recorded load voltage values are used to compute the corresponding power levels delivered to the load. It is clear from Figure 28 that at 9 m/s air speed, a maximum load power of 7.84 μW is obtained from prototype I, when connected to a load resistance of 54.5 Ω (identical to coil resistance). Moreover, under optimum load condition,

TABLE 3: Comparison of energy harvesters.

Harvester's type	Transduction mechanism	Acceleration (g)	Resonant frequency (Hz)	Impedance (Ω)	Voltage (mV)	Power (μW)	Ref.
Bridge energy harvesters	Electromagnetic	0.55	2	1.8	—	2.3	[26]
		0.08	2.2	—	—	26000	[25]
		0.38	3.1	67 k	10000	12500	[41]
		0.008	14	—	10	2	[42]
		0.025	4.1	290	710	120	[43]
		1	10	270	—	39.45	[45]
		1	10	240	—	163	
		0.05	2	240	—	57	[44]
		—	<1	—	—	0.7	
		1	10	0.22	0.1	13.6	[57]
		3	27	3.6	15.5	2.1	[58]
		0.2–0.4	3.6, 14.9, & 17.6	54.5	206	354.51	[This work]
		0.3–0.6	7.6, 33, & 45	28	430	2214.32	[This work]
	Piezoelectric	0.1	2	70 k	—	64	[25]
		0.02	14-15	100	1.8–3.6	30	[27]
		—	14.5	—	3600	30	[46]
		1	40	10 M	28000	78.4	[47]
		0.29	2.71 & 120	9.7 k	—	7700	[48]
		0.21	117.1	11.8 k	2000	197	[49]
		0.21	65.2	14.9 k	5000	657	
		—	1.5	480	650	83.5	[50]
		—	2	200 k	—	0.6	[52]
Multimode energy harvesters	Electromagnetic	0.76	369, 938, & 1184	0.8	3.2	3.2	[67]
		1	840, 1070, & 1490	626	3.7	0.0041	[68]
		50	4200-5000	580	10	0.4	[69]
		50	3300-3600	580	20	0.5	[70]
		1	368, 530, & 614	264	0.043	2.6	[71]
		0.1	10-100	100 k	1000	5	[72]
	Piezoelectric	0.1	17.5 & 21.7	100 k	22.5	500	[73]
		0.8	101.04, 108.16, & 134.4	—	3100	—	[74]
		0.05	387.1, 398.2, & 398.6	1 M	—	53	[75]
		1	14.2 & 25.4	71 M	13.2	24.5	[76]
		—	32 & 41.6	100 k	30500	1000	[77]
		0.2	71.8, 84.5, & 188.4	2 M	1000	0.136	[78]
	Hybrid	0.1	65-95	116	—	2160	[79]
		0.2	22.8 & 25.6	240	—	4200	[80]

The internal impedance of piezoelectric BEHs is relatively on higher side (480–10 MΩ). The overall power production (0.6–26000 μW) of the developed BEHs is fairly enough to operate the WSNs utilized for the health monitoring of bridges.

In comparison, the BEHs, prototype I and prototype II developed in this work, are the only BEHs, those exhibit multiresonant frequencies and therefore are capable of harvesting the energy from relatively wider band of input bridge's vibrations. The relatively broader bandwidth of the developed prototypes and multiresonant frequencies actually ensure better performance for these harvesters during off resonance operation in comparison to monoresonant energy harvesters whose performance drastically degrade when these are operating at off resonance frequencies. Furthermore, in contrast, the developed harvesters will perform far better in narrow band random vibration environments (such as the bridge's vibration). Moreover, the internal impedance of the harvesters reported in this work is lower than all the reported BEHs except [58]. Among electromagnetic BEHs, the voltage production of the harvesters [this work] is also adequate, and only the harvesters [41, 43] are capable of generating more power than these (due to large number of turns in the coil). As far as power generation is concerned, only the BEHs [25, 41, 48] have the tendency to produce larger power than the harvesters developed in this work. Furthermore, the BEHs in this work are the only harvests which have the ability to extract the energy simultaneously from the bridge's vibration and wind surges (natural or traffic induced) and therefore this feature of the developed BEHs makes these exceptional than the other reported BEHs.

Moreover, the developed BEHs are also compared with the reported multimode energy harvesters listed in Table 3. In comparison, the resonant frequencies of multimode energy harvesters are relatively higher than those of the developed BEHs. There are only few multimode energy harvesters [73, 76, 77, 80], for which the resonant frequencies are reported less than 42 Hz. Due to their higher resonant frequencies, most of the multimode energy harvesters are incapable to perform well under bridges' vibration (1–40 Hz). Moreover, most of the multimode energy harvesters are subjected to the acceleration levels higher than that at which the developed BEHs are characterized. Only multimode energy harvesters [72, 73, 75, 79] are reported to be tested at acceleration levels less than $0.2g$. By comparing with respect to internal impedance, only the multimode energy harvester [67] has lower impedance than that of the developed BEHs. Furthermore, the voltage generation of the BEHs developed in this work is greater than all of the electromagnetic multimode energy harvesters and the piezoelectric multimode energy harvesters reported in [73, 76]. Among the unhybrid multimode energy harvesters, the developed BEH, prototype II, is capable of producing the highest power level (2214.32 μW); however, the unhybrid multimode energy harvesters [73, 77] are reported to generate power levels higher than the developed BEH, prototype I. Due to dual transduction mechanisms, the performance of the hybrid multimode energy harvesters [79, 80] is better and these are the only reported multimode energy harvesters which produce comparable or higher power levels than the developed BEHs.

7. Conclusions

Electromagnetic-based bridge energy harvesters (BEHs) have been developed in this work. The reported energy harvesters utilize the available energies on bridge, such as bridge's vibration and ambient wind to produce power for wireless sensor nodes (WSNs) used for monitoring of the bridge structures. To target low frequency excitation environment, two cantilever-type BEHs, prototype I and prototype II, are produced having the resonant frequencies of 3.6 Hz, 14.9 Hz, and 17.6 Hz and 7.6 Hz, 33 Hz, and 45 Hz, respectively. The BEHs are characterized in both vibration and wind environments and have showed the capability to produced satisfactory voltage and power levels.

Prototype I is suitable for narrow band vibration environment having the frequency content from 1 to 18 Hz and acceleration levels below $0.4g$ and is capable to generate an open circuit voltage of 810 mV, maximum RMS voltage of 206 mV (at 200 Ω), and an optimum power of 354.51 μW (at 54.5 Ω) in such vibration environment. Moreover, it has also shown the ability to produced adequate voltage and power levels (up to 7.84 μW) from wind surges from 0.5 m/s to 9 m/s.

However, on the other hand, prototype II is more appropriate for vibration surroundings with acceleration levels below $0.6g$ and frequency band from 1 to 45 Hz. It can produce an open circuit voltage of 618 mV, a maximum RMS voltage of 430 mV (at 100 Ω), and an optimum load power of 2214.32 μW (at 28 Ω) from such vibration conditions. Furthermore, it is also able to harvest the power (up to 9.14 μW) from ambient wind with speed from 0.5 m/s to 6 m/s.

In comparison to the other reported BEHs, the energy harvesters developed in this work have showed the ability to produce higher voltage and power than most of the reported BEHs. Moreover, the developed BEHs generated enough power levels to operate the ultralow power sensors and are capable to supplement the power of the batteries in WSNs used for the bridge's health monitoring. Furthermore, the developed BEHs are the only harvests which are capable to extort the energy simultaneously from the bridge's vibration and wind surges (natural or traffic induced).

Conflicts of Interest

The authors declare that there are no conflicts of interest regarding the publication of this paper.

References

[1] P. C. Chang, A. Flatau, and S. C. Liu, "Review paper: health monitoring of civil infrastructure," *Structural Health Monitoring*, vol. 2, no. 3, pp. 257–267, 2003.

[2] C. R. Farrar and N. A. J. Lieven, "Damage prognosis: the future of structural health monitoring," *Philosophical Transactions of the Royal Society A*, vol. 365, no. 1851, pp. 623–632, 2007.

[3] C. B. Williams, A. Pavic, R. S. Crouch, and R. C. Woods, "Feasibility study of vibration-electric generation for bridge vibration sensors," in *Proceedings of the 16th International Modal Analysis Conference (1998 IMAC XVI)*, pp. 1111–1117, Santa Barbara, CA, USA, 1998.

[4] F. Khan, F. Sassani, and B. Stoeber, "Nonlinear behaviour of membrane type electromagnetic energy harvester under harmonic and random vibrations," *Microsystem Technologies*, vol. 20, no. 7, pp. 1323–1335, 2014.

[5] F. U. Khan, "Review of non-resonant vibration based energy harvesters for wireless sensor nodes," *Journal of Renewable and Sustainable Energy*, vol. 8, no. 4, 2016.

[6] F. U. Khan, T. Ali, and K. Jamil, "Development of a low voltage AC to DC converter for meso and micro energy harvesters," *Journal of Engineering and Applied Sciences*, vol. 34, no. 2, pp. 35–46, 2015.

[7] F. U. Khan and A. Khattak, "Development of a power management circuit for micro-energy harvesters," *Journal of Engineering and Applied Sciences*, vol. 34, no. 1, pp. 38–48, 2015.

[8] K. V. Selvan and M. S. M. Ali, "Micro-scale energy harvesting devices: review of methodological performances in the last

decade," *Renewable and Sustainable Energy Reviews*, vol. 54, pp. 1035–1047, 2016.

[9] F. U. Khan and M. U. Khattak, "Contributed review: recent developments in acoustic energy harvesting for autonomous wireless sensor nodes applications," *Review of Scientific Instruments*, vol. 87, no. 2, article 021501, 2016.

[10] F. U. Khan and M. U. Qadir, "State-of-the-art in vibration-based electrostatic energy harvesting," *Journal of Micromechanics and Microengineering*, vol. 26, no. 10, article 103001, 2016.

[11] G. Zhou, L. Huang, W. Li, and Z. Zhu, "Harvesting ambient environmental energy for wireless sensor networks: a survey," *Journal of Sensors*, vol. 2014, Article ID 815467, 20 pages, 2014.

[12] M. Zhang and J. Wang, "Experimental study on piezoelectric energy harvesting from vortex-induced vibrations and wake-induced vibrations," *Journal of Sensors*, vol. 2016, Article ID 2673292, 7 pages, 2016.

[13] F. U. Khan and Izhar, "State of the art in acoustic energy harvesting," *Journal of Micromechanics and Microengineering*, vol. 25, article 023001, p. 13, 2015.

[14] E. Lattanzi, M. Dromedari, V. Freschi, and A. Bogliolo, "A sub-μA ultrasonic wake-up trigger with addressing capability for wireless sensor nodes," *ISRN Sensor Networks*, vol. 2013, Article ID 720817, 10 pages, 2013.

[15] W. Wang, S. Liu, Q. Wang et al., "The impact of traffic-induced bridge vibration on rapid repairing high-performance concrete for bridge deck pavement repairs," *Advances in Materials Science and Engineering*, vol. 2014, Article ID 632051, 9 pages, 2014.

[16] N. G. Elvin, N. Lajnef, and A. A. Elvin, "Feasibility of structural monitoring with vibration powered sensors," *Smart Materials and Structures*, vol. 15, no. 4, pp. 977–986, 2006.

[17] F. U. Khan and I. Ahmad, "Review of energy harvesters utilizing bridge vibrations," *Shock and Vibration*, vol. 2016, Article ID 1340402, 21 pages, 2016.

[18] S. Priya, "Modeling of electric energy harvesting using piezoelectric windmill," *Applied Physics Letters*, vol. 87, no. 18, pp. 184101–184103, 2005.

[19] E. Sazonov, H. Li, D. Curry, and P. Pillay, "Self-powered sensors for monitoring of highway bridges," *IEEE Sensors Journal*, vol. 9, no. 11, pp. 1422–1429, 2009.

[20] D. S. Clair, D. Bibo, V. R. Sennakesavababu, M. F. Daqaq, and G. Li, "A scalable concept for micropower generation using flow-induced self-excited oscillations," *Applied Physics Letters*, vol. 96, no. 14, pp. 144103–144105, 2010.

[21] A. R. Ortiz, C. Ventura, and S. S. Catacoli, "Sensitivity analysis of the lateral damping of bridges for low levels of vibration," in *Conference Proceedings of the Society for Experimental Mechanics Series*, pp. 101–110, San Antonio, Texas, USA, 2013.

[22] T. Galchev, J. McCullagh, R. L. Peterson, and K. Najafi, "A vibration harvesting system for bridge health monitoring applications," in *Proceedings of PowerMEMS*, pp. 179–182, Leuven, Belgium, 2010.

[23] H. Bachmann and W. Ammann, *Vibrations in Structures : Induced by Man and Machines*, International Association for Bridge and Structural Engineering, Zurich, Switzerland, 3rd edition, 1987.

[24] J. Wiberg, *Bridge Monitoring to Allow for Reliable Dynamic FE Modeling: A Case Study of the New Arsta Railway Bridge*, 2006, October 2017, https://www.diva-portal.org/smash/get/diva2:9925/FULLTEXT01.pdf.

[25] F. Neitzel, B. Resnik, S. Weisbrich, and A. Friedrich, "Vibration monitoring of bridges," *Reports on Geodesy*, vol. 1, no. 90, pp. 331–340, 2011.

[26] T. V. Galchev, J. McCullagh, R. L. Peterson, and K. Najafi, "Harvesting traffic-induced vibrations for structural health monitoring of bridges," *Journal of Micromechanics and Microengineering*, vol. 21, no. 10, article 104005, 2011.

[27] M. Peigney and D. Siegert, "Piezoelectric energy harvesting from traffic-induced bridge vibrations," *Smart Materials and Structures*, vol. 22, no. 9, article 095019, 2013.

[28] J. Kala, V. Salajka, and P. Hradil, "Footbridge response on single pedestrian induced vibration analysis," *World Academy of Science, Engineering and Technology*, vol. 3, no. 2, pp. 744–755, 2009.

[29] L. Zhang, X. Yan, and X. yang, "Vehicle-bridge coupled vibration response study of Huanghe cable-stayed bridge due to multiple vehicles parallel with high speed," in *2010 International Conference on Mechanic Automation and Control Engineering (MACE)*, pp. 1134–1137, Wuhan, China, June 2010.

[30] A. M. A. Ghaffar and R. H. Scanlan, "Ambient vibration studies of golden gate bridge: I. Suspended structure," *Journal of Engineering Mechanics*, vol. 111, no. 4, pp. 463–482, 1985.

[31] C. E. Dierks, *Design of an Electromagnetic Vibration Energy Harvester for Structural Health Monitoring of Bridges Employing Wireless Sensor Networks*, MSc dissertation, Department of mechanical engineering, The University of Texas, Austin, TX, USA, 2011.

[32] M. Zhang, D. Brignac, P. Ajmera, and K. Lian, "A low-frequency vibration-to-electrical energy harvester," in *Proceedings Volume 6931, Nanosensors and Microsensors for Bio-Systems 2008*, p. 69310S, San Diego, CA, USA, 2008.

[33] T. McEvoy, E. Dierks, J. Weaver et al., "Developing innovative energy harvesting approaches for infrastructure health monitoring systems," in *ASME 2011 International Design Engineering Technical Conferences and Computers and Information in Engineering Conference*, pp. 325–339, Washington, DC, USA, August 2011.

[34] T. Nagayama, *Dynamic Characteristics Identification for an Arch Bridge using Wireless Sensor Networks before and after Seismic Retrofit; the Application to Model Updating*June 2017, http://www.pwri.go.jp/eng/ujnr/tc/g/pdf/28/28-10-3_Nagayama.pdf.

[35] T. Galchev, J. McCullagh, R. L. Peterson, and K. Najafi, "A vibration harvesting system for bridge health monitoring applications," in *Proceedings of PowerMEMS 2010*, pp. 3–6, Leuven, Belgium, 2010.

[36] B. Jo, Y. Lee, G. Yun, C. Park, and J. Kim, "Vibration-induced energy harvesting for green technology," in *Proceedings of the International Conference on Chemical, Environment and Civil Engineering (ICCECE '2012)*, pp. 167–170, Manila, Philippines, November 2012.

[37] Y. Wang, K. J. Loh, J. P. Lynch, M. Fraser, K. Law, and A. Elgamal, "Vibration monitoring of the Voigt bridge using wired and wireless monitoring systems," in *The Proceeding of 4th China-Japan-US Symposium on Structural Control and Monitoring*, pp. 1–8, China, October 2006.

[38] R. Torah, P. Glynne-Jones, M. Tudor, T. O'Donnell, S. Roy, and S. Beeby, "Self-powered autonomous wireless sensor node using vibration energy harvesting," *Measurement Science and Technology*, vol. 19, no. 12, article 125202, 2008.

[39] J. R. Casas and J. J. Moughty, "Bridge damage detection based on vibration data: past and new developments," *Frontiers in Built Environment*, vol. 3, 2017.

[40] S.-D. Kwon, J. Park, and K. Law, "Electromagnetic energy harvester with repulsively stacked multilayer magnets for low frequency vibrations," *Smart Materials and Structures*, vol. 22, no. 5, article 055007, 2013.

[41] F. Khan, B. Stoeber, and F. Sassani, "Modeling of linear micro electromagnetic energy harvesters with nonuniform magnetic field for sinusoidal vibrations," *Microsystem Technologies*, vol. 21, no. 3, pp. 683–692, 2015.

[42] F. Khan, B. Stoeber, and F. Sassani, "Modeling and simulation of linear and nonlinear MEMS scale electromagnetic energy harvesters for random vibration environments," *The Scientific World Journal*, vol. 2014, Article ID 742580, 15 pages, 2014.

[43] S. Saadon and O. Sidek, "A review of vibration-based MEMS piezoelectric energy harvesters," *Energy Conversion and Management*, vol. 52, no. 1, pp. 500–504, 2011.

[44] S. P. Beeby, R. N. Torah, M. J. Tudor et al., "A micro electromagnetic generator for vibration energy harvesting," *Journal of Micromechanics and Microengineering*, vol. 17, no. 7, pp. 1257–1265, 2007.

[45] F. Khan, F. Sassani, and B. Stoeber, "Copper foil-type vibration-based electromagnetic energy harvester," *Journal of Micromechanics and Microengineering*, vol. 20, no. 12, article 125006, 2010.

[46] C. Cepnik, R. Lausecker, and U. Wallrabe, "Review on electrodynamic energy harvesters—a classification approach," *Micromachines*, vol. 4, no. 2, pp. 168–196, 2013.

[47] W. J. Choi, Y. Jeon, J. –. H. Jeong, R. Sood, and S. G. Kim, "Energy harvesting MEMS device based on thin film piezoelectric cantilevers," *Journal of Electroceramics*, vol. 17, no. 2-4, pp. 543–548, 2006.

[48] P. Basset, D. Galayko, M. Paracha, F. Marty, F. Dudka, and T. Bourouina, "A batch-fabricated and electret-free silicon electrostatic vibration energy harvester," *Journal of Micromechanics and Microengineering*, vol. 19, no. 11, article 115025, 2009.

[49] S. Adhikari, M. I. Friswell, and D. J. Inman, "Piezoelectric energy harvesting from broadband random vibrations," *Smart Materials and Structures*, vol. 18, no. 11, p. 115005, 2009.

[50] J. Wang, S. Wen, X. Zhao, M. Zhang, and J. Ran, "Piezoelectric wind energy harvesting from self-excited vibration of square cylinder," *Journal of Sensors*, vol. 2016, Article ID 2353517, 12 pages, 2016.

[51] F. Khan and Izhar, "Piezoelectric type acoustic energy harvester with a tapered Helmholtz cavity for improved performance," *Journal of Renewable and Sustainable Energy*, vol. 8, no. 5, article 054701, 2016.

[52] F. U. Khan, "Energy harvesting from the stray electromagnetic field around the electrical power cable for smart grid applications," *The Scientific World Journal*, vol. 2016, Article ID 3934289, 20 pages, 2016.

[53] F. U. Khan and Izhar, "Hybrid acoustic energy harvesting using combined electromagnetic and piezoelectric conversion," *Review of Scientific Instruments*, vol. 87, no. 2, article 025003, 2016.

[54] F. Khan and S. Razzaq, "Electrodynamic energy harvester for electrical transformer's temperature monitoring system," *Sadhana*, vol. 40, no. 7, pp. 2001–2019, 2015.

[55] S. Meninger, J. O. Mur-Miranda, R. Amirtharajah, A. P. Chandrakasan, and J. H. Lang, "Vibration-to-electric energy conversion," *IEEE Transactions on Very Large Scale Integration (VLSI) Systems*, vol. 9, no. 1, pp. 64–76, 2001.

[56] F. U. Khan and M. Iqbal, "Development of a testing rig for vibration and wind based energy harvesters," *Journal of Engineering and Applied Science*, vol. 35, no. 2, pp. 101–110, 2016.

[57] T. Galchev, H. Kim, and K. Najafi, "Micro power generator for harvesting low-frequency and nonperiodic vibrations," *Journal of Microelectromechanical Systems*, vol. 20, no. 4, pp. 852–866, 2011.

[58] F. U. Khan and I. Ahmad, "Vibration-based electromagnetic type energy harvester for bridge monitoring sensor application," in *2014 International Conference on Emerging Technologies (ICET)*, pp. 125–129, Islamabad, Pakistan, December 2014.

[59] M. Rhimi and N. Lajnef, "Tunable energy harvesting from ambient vibrations in civil structures," *Journal of Energy Engineering*, vol. 138, no. 4, pp. 185–193, 2012.

[60] J. D. Baldwin, S. Roswurm, J. Nolan, and L. Holliday, *Energy Harvesting on Highway Bridges*, Final Report FHWA-OK-11-01, January 2015, http://www.okladot.state.ok.us/hqdiv/p-r-div/spr-rip/library/reports/rad_spr2-i2224-fy2010-rpt-final-baldwin.pdf.

[61] F. Orfei, H. Vocca, and L. Gammaitoni, "Linear and non linear energy harvesting from bridge vibrations," in *ASME 28th conference on mechanical vibration and noise*, Charlotte, North Carolina, USA, August 2016.

[62] P. Cahill, N. A. N. Nuallain, N. Jackson, A. Mathewson, R. Karoumi, and V. Pakrashi, "Energy harvesting from train-induced response in bridges," *Journal of Bridge Engineering*, vol. 19, no. 9, article 04014034, 2014.

[63] G. Caruso, G. Chirianni, and G. Vairo, "Energy harvesting from wind-induced bridge vibrations via electromagnetic transduction," *Engineering Structures*, vol. 115, pp. 118–128, 2016.

[64] K. Takeya, E. Sasaki, and Y. Kobayashi, "Design and parametric study on energy harvesting from bridge vibration using tuned dual-mass damper systems," *Journal of Sound and Vibration*, vol. 361, pp. 50–65, 2016.

[65] F. U. Khan and M. Iqbal, "Electromagnetic-based bridge energy harvester using traffic- induced bridge's vibrations and ambient wind," in *2016 International Conference on Intelligent Systems Engineering (ICISE)*, pp. 425–430, Islamabad, Pakistan, January 2016.

[66] W. T. Thomson, *Theory of Vibration with Applications, third edition*, Prentice Hall, England, 1998.

[67] B. Yang, C. Lee, W. Xiang et al., "Electromagnetic energy harvesting from vibrations of multiple frequencies," *Journal of Micromechanics and Microengineering*, vol. 19, no. 3, article 035001, 2009.

[68] H. Liu, Y. Qian, and C. Lee, "A multi-frequency vibration-based MEMS electromagnetic energy harvesting device," *Sensors and Actuators A: Physical*, vol. 204, pp. 37–43, 2013.

[69] I. Sari, T. Balkan, and H. Kulah, "An electromagnetic micro power generator for wideband environmental vibrations," *Sensors and Actuators A: Physical*, vol. 145-146, pp. 405–413, 2008.

[70] I. Sari, T. Balkan, and H. Kulah, "A wideband electromagnetic micro power generator for wireless microsystems," in

TRANSDUCERS 2007 - 2007 International Solid-State Sensors, Actuators and Microsystems Conference, pp. 275–278, Lyon, France, June 2007.

[71] H. Liu, T. Chen, L. Sun, and C. Lee, "An electromagnetic MEMS energy harvester array with multiple vibration modes," *Micromachines*, vol. 6, no. 8, pp. 984–992, 2015.

[72] X. Xiong and S. O. Oyadiji, "Design and experimental study of a multi-modal piezoelectric energy harvester," *Journal of Mechanical Science and Technology*, vol. 31, no. 1, pp. 5–15, 2017.

[73] D. Saravanos, H. Wu, L. Tang, Y. Yang, and C. K. Soh, "A novel two-degrees-of-freedom piezoelectric energy harvester," *Journal of Intelligent Material Systems and Structures*, vol. 24, no. 3, pp. 357–368, 2013.

[74] S. Dhote, J. Zu, and Y. Zhu, "A nonlinear multi-mode wideband piezoelectric vibration-based energy harvester using compliant orthoplanar spring," *Applied Physics Letters*, vol. 106, no. 16, article 163903, 2015.

[75] E. E. Aktakka and K. Najafi, "Three-axis piezoelectric vibration energy harvester," in *2015 28th IEEE International Conference on Micro Electro Mechanical Systems (MEMS)*, pp. 1141–1144, Estoril, Portugal, January 2015.

[76] Q. Tang and X. Li, "Two-stage wideband energy harvester driven by multimode coupled vibration," *IEEE/ASME Transactions on Mechatronics*, vol. 20, no. 1, pp. 115–121, 2015.

[77] H. Wang and L. Tang, "Modeling and experiment of bistable two-degree-of-freedom energy harvester with magnetic coupling," *Mechanical Systems and Signal Processing*, vol. 86, pp. 29–39, 2017.

[78] M. Rezaeisaray, M. El Gowini, D. Sameoto, D. Raboud, and W. Moussa, "Low frequency piezoelectric energy harvesting at multi vibration mode shapes," *Sensors and Actuators A: Physical*, vol. 228, pp. 104–111, 2015.

[79] H.-y. Wang, L.-h. Tang, Y. Guo, X.-b. Shan, and T. Xie, "A 2DOF hybrid energy harvester based on combined piezoelectric and electromagnetic conversion mechanisms," *Journal of Zhejiang University SCIENCE A*, vol. 15, no. 9, pp. 711–722, 2014.

[80] Z. Xu, X. Shan, D. Chen, and T. Xie, "A novel tunable multi-frequency hybrid vibration energy harvester using piezoelectric and electromagnetic conversion mechanisms," *Applied Sciences*, vol. 6, no. 1, p. 10, 2016.

14

A Clone Detection Algorithm with Low Resource Expenditure for Wireless Sensor Networks

Zhihua Zhang ⓘ, **Shoushan Luo, Hongliang Zhu** ⓘ, **and Yang Xin**

Beijing University of Posts and Telecommunications, Beijing, China

Correspondence should be addressed to Zhihua Zhang; zhangzhihua@bupt.edu.cn

Academic Editor: Jaime Lloret

Wireless sensor networks (WSNs) are facing the threats of clone attacks which can launch a variety of other attacks to control or damage the networks. In this paper, a novel distributed clone detection protocol with low resource expenditure is proposed for randomly deployed networks. The method consisting of witness chain establishment and clone detection route generation is implemented in the nonhotspot area of the network organized in a ring structure, which balances the resource consumption in the whole network. The witness chains and detection routes are in the centrifugal direction and circumferential direction, respectively, which can ensure the encounter of witnesses and detection routes of nodes with the same ID but different positions to detect clone attacks. Theoretical analysis demonstrates that the detection probability can be up to 1 with reliable witnesses. Moreover, both theoretical analysis and simulation results manifest that the proposed method can achieve better network lifetime and storage requirements with low resource expenditure and outperforms most methods in the literature.

1. Introduction

Wireless sensor networks (WSNs) usually consist of a large number of randomly distributed low-cost sensor nodes with limited resources in the target area. The networks have been widely used in various fields for the purpose of event monitoring and data gathering, including environment monitoring, forest fire monitoring, traffic data collection, and battlefield data gathering [1–3]. However, many WSNs are deployed in harsh or hostile environment which is a challenge for their secure operation. Due to the openness of wireless communication and lack of physical protection, sensor nodes are often compromised or attacked by attackers, making WSNs suffer from various attacks [4]. One of the most challenging attacks is clone attack or node replica attack, which refers to multiple nodes with the same ID. Since the sensor nodes are often unattended and lack of tamper-resistance devices, an attacker could capture a few nodes to obtain all the information materials in them including code and cryptographic mechanism. Hereafter, adversaries could duplicate the captured nodes. The cloned nodes seem to be legal ones for the network because they are exactly the same

as the original ones, thus they could join the network freely without being recognized. It is much easier and cheaper to replicate a compromised node than to capture another normal node. Once the node is captured by an adversary, it could be replicated in large numbers and deployed in different areas of the network to jam or manipulate the network under the control of the adversary. Meanwhile, cloned nodes could initiate other inner attacks [5, 6], including selective forwarding attacks, black-hole attacks, energy exhaustion attacks, and data tampering attacks. With a certain number of cloned nodes that occupy strategic positions, the adversary may take over the whole network [7]. Therefore, it is essential to detect clone attacks effectively to avoid the serious harm to the network. Fortunately, the cloned nodes are often deployed in different positions, because it is not helpful to deploy them in the same location as the original one. It means that the cloned nodes and the original node have the same ID but different positions, which provides favorable conditions for clone detection.

Various methods for clone detection or node replica detection have been proposed up to now. According to different features, we could classify them into different categories: witness-based or not, position dependent or not, centralized or distributed, the witnesses are deterministic or random, and the scheme is for randomly deployed or group deployed, which are detailed in Section 2. The algorithm we proposed is witness-based, location-dependent, distributed, with randomly selected witnesses, and it is for randomly deployed networks. There are some typical methods in the literature similar to our work, such as line-select multicast (LSM) protocol [8], randomized, efficient, and distributed (RED) protocol [9], energy-efficient ring-based clone detection (ERCD) protocol [10], and low-storage clone detection (LSCD) protocol [11]. In these schemes, some witnesses are selected randomly from the network to verify the legitimacy of nodes or detect the cloned nodes according to the private information (ID and location) reported to them. Hence, a clone is identified when at least two nodes possess the same ID but different locations.

However, these solutions have some drawbacks in two aspects. First, the clone detection probability is not high enough. Because all of these methods are distributed, that is, the witness and detection routes (legitimacy verification paths) are distributed, the clone is detected only if the witnesses and the detection routes encounter. Due to the randomness of witness selection and detection routes, the methods LSM, RED and ERCD could not ensure the encounter mentioned above. Whereas the LSCD could ensure the detection probability equal to 1 theoretically, because it adopts ring structure and forms witness arcs with a certain length, meanwhile, the detection routes are perpendicular to the arc (centrifugal) and the distance of each two adjacent detection routes is less than the arc length. Second, the resource consumptions of these methods are relatively high. The resource expenditure of LSM and RED is depending on the number of nodes in the network, and the resource expenditure increases significantly with the increase in the scale of the network. Although the resource expenditure of ERCD and LSCD has been improved to some extent, the resource consumptions are still at a high level which shortens the network lifetime.

In this paper, we propose an effective clone detection algorithm (referred to as CDLR) with a higher detection probability (equal to 1 theoretically) and a lower resource expenditure; meanwhile, the method CDLR avoids consuming the energy of nodes in the hotspot area, and all these measures prolong the network lifetime. Similar to the protocols ERCD and LSCD, a ring structure with the BS as the center is used in our work to ensure the encounter of the witnesses and the detection routes. Different from the two methods, the algorithm CDLR adopts random witness chains in the centrifugal direction (just as the radius of a circle) and detection routes in the circumferential detection, which ensures the encounter of witnesses and detection routes. Comparing with the ERCD with witness rings and circumferential verification paths and LSCD with witness arcs and multiple centrifugal detection routes, the CDLR has a lower resource expenditure and longer lifetime.

The major contributions of this work are as follows:

(1) The CDLR algorithm provides a high detection probability against clone attacks. In the ring structure, the random witness chains run through the entire nonhotspot area in the centrifugal direction, and the random detection routes or verification paths are formed along the circumference. Hence, the witness chains must encounter the detection routes, which ensures the detection probability equal to 1 theoretically.

(2) The resource expenditure of the CDLR algorithm is at a low level. The storage requirements of nodes with CDLR algorithm are not related to the density of nodes, and it almost does not increase with the increment in network scale. Furthermore, the communication load is lower than the similar method ERCD.

(3) The CDLR method makes full use of the energy of nodes in the nonhotspot area and prolongs the network lifetime. The CDLR fully used the residual energy of nodes in outer rings, that is, the nonhotspot area, because all the witness chains and detection routes are formed in outer rings, which effectively prolongs the network lifetime.

The rest of this work is organized as follows. Section 2 reviews the previous related works. Section 3 presents the network model and assumptions. The method CDLR for clone detection is proposed in Section 4. Then in Section 5, the theoretical performance analysis is conducted. Experiments and simulations are given in Section 6. Finally, Section 7 concludes this paper.

2. Related Works

Clone attacks have attracted the attention of researchers, and there has been much effort on clone detection up to now [5–18]. According to different features, we could classify them into different categories: centralized [12–15] or distributed [5–11, 16–18], witness-based [5, 8–11, 16–18] or not, position dependent [5, 8–11, 17, 18] or not, and the scheme is for randomly deployed [5–16] or group-deployed networks [17, 18].

The most common classification in the literature is based on centralized or distributed. For the centralized methods, the BS or sink is responsible for clone detection according to the information reported by nodes [12–15]. The advantages of these methods are that they have low overhead and high detection probability because of the comprehensive information. However, the shortcomings are also explicit: the BS easily suffers from a single point of failure and the nodes around the BS consume much more energy than others due to forwarding packets. In order to solve these problems, distributed schemes are proposed [5–11, 16–18], which assign the detection tasks to different areas and nodes, yet the resource consumptions of nodes are increasing sharply. Most of the works are conducted to balance the detection probability and the resource expenditure.

There are also another categories: according to witness requirements, the schemes are divided into witness-based and no witness-based; based on the location requirements,

TABLE 1: Categories of different clone detection schemes.

Scheme	Detection mechanism		Witness requirements		Location requirements		Network deployment	
	Centralized	Distributed	Witness-based	No witness	Location dependent	Location independent	Randomly deployed	Group deployed
[5]		√	√		√		√	
[7]		√		√		√	√	
[8]		√	√		√		√	
[9]		√	√		√		√	
[10]		√	√		√		√	
[11]		√	√		√		√	
[12]	√			√		√	√	
[13]	√			√		√	√	
[14]	√			√		√	√	
[15]	√			√		√	√	
[16]		√	√			√	√	
[17]		√		√	√			√
[18]		√	√		√			√

the methods are classified into location dependent and location independent; on the basis of network deployment requirements, the protocols are classified into for randomly deployed and for group deployed. All of the categories mentioned above are listed in Table 1.

From Table 1, we could see that most of the distributed methods are witness-based, location-dependent, and for randomly deployed networks. According to the practical application and our work, we focus on the methods with these conditions [5, 8–11].

Randomized multicast (RM) and line-selected multicast (LSM) were proposed in [8]. Both methods are witness-based, and LSM is an improvement of RM. In both methods, the neighbors of each node randomly select a fraction of nodes as its witnesses. Differences are that the clone detection is conducted according to the birthday paradox problem in RM, that is, at least one witness will discover the conflict of nodes with the same ID but different locations; whereas the nodes along the routes from the node to its random witnesses are also selected as witnesses, thus the intersections of different routes could improve the detection probability. However, the detection probability is still in a low level, and the resource consumptions are closely related to the number of nodes, which is not suitable for large-scale networks.

Randomized, efficient, and distributed protocol (RED) was put forward in [9], which is another improvement of RM. It is also witness-based, and some witnesses are randomly selected by the neighbors of the node according to its ID, that is, the witness set of nodes with the same ID will be identical, thus the detection probability is improved. However, the "randomly" selected witnesses of a node based on node ID are always the same ones, which means the witness selection is deterministic in fact, and it is easy to be exploited by adversaries. Besides, the storage overhead is still related to the number of nodes, which is also not suitable for the networks with a large number of nodes.

Similar to LSM, a random walk (RAWL) protocol was proposed in [5], which improves the detection probability by a random walk of all randomly selected witnesses expanding the scope of witnesses. Both the detection probability and the resource consumptions are improved, but the storage overhead is still related to the scale of network.

In order to balance the detection probability and resource consumptions, some other protocols are proposed. To the best of our knowledge, the best two methods are energy-efficient ring-based clone detection (ERCD) protocol and low-storage clone detection (LSCD) protocol described in [10, 11], respectively. Both methods are random witness-based and suitable for large-scale network, which adopt ring structure with the BS (or sink) as the center. The clone detection process in both protocols consists of two phases: witness selection and legitimacy verification (or detection route generation).

In ERCD [10], the witnesses of each node form a witness ring in a randomly selected network ring, and a witness header which is responsible for clone detection is selected randomly from all the witnesses. The legitimacy of each node has to be verified by ERCD before communicating with others. The verification message is transmitted to its corresponding witnesses, and it is broadcasted to witness header in the witness ring and neighbor rings. The clone attack will be detected if the witness header discovers a conflict of nodes with the same ID but different locations in reported verification messages. Theoretical analysis shows that the ERCD has a high detection probability and a constant level of storage overhead. However, the clone can be detected only if the witnesses of two replica nodes are in the same ring or in the neighbor rings. Besides, in the process of witness selection and legitimacy verification, each message is forwarded along the ring, which causes a high communication overhead.

In the process of witness selection of LSCD [11], the witnesses of each node form a witness arc with a certain length in a randomly selected network ring. In detection process, some centrifugal detection routes that are perpendicular to the witness arc are generated from the second ring. The distance between two adjacent routes is less than the witness arc length to ensure the encounter of witnesses and the detection routes. During the establishment of detection routes, it is necessary to check the relationship between the arc length and the adjacent route spacing at any time to increase the detection routes. Thus, the area far from the center has more detection routes. The dynamic mechanism in detection route establishment ensures the high detection probability, and the storage overhead of nodes is relatively low. However, the messages for detection route establishment of all nodes need to be forwarded to the second ring to initiate the detection route generation, which will consume much energy of nodes in the second ring. Hence, the energy consumptions of these nodes in the hotspot area will become the bottleneck of the network lifetime.

Unlike prior works, the algorithm we proposed has a high detection probability (equal to 1 theoretically) and low resource consumptions, which avoids consuming the energy of nodes in the hotspot area and makes full use of the resource of nodes in the nonhotspot area. The random witness chains are formed along the radius of network to run through the whole nonhotspot area. Meanwhile, the circumferential detection routes are generated in randomly selected rings to ensure the encounter of witnesses and detection routes. Note that all these witness chains and detection routes are established in the nonhotspot area, which prolongs the network lifetime effectively. The implementation of CDLR does not introduce a new bottleneck of network lifetime. Besides, different from the most previous works, the storage requirements using CDLR are not relative to the density of nodes in the network.

3. System Model and Assumptions

In this section, the network model, the adversary model, and assumptions are introduced.

3.1. Network Model. Based on ERCD and LSCD, we propose a clone detection algorithm with low resource expenditure (CDLR). Similar to ERCD and LSCD, the ring structure of network with the BS as the center is adopted in our work. In this model, the BS is at the center of the network region, and all nodes are densely and randomly distributed around the BS. The communication radius of each node is r, and the radius of the whole network denoted as $R = hr$. The density of nodes is ρ. Each node has its own relative location to the BS (the hops between a node and the BS) and its 1-hop neighbors; moreover, each node in the network knows its own geographic position by any mature mechanism introduced in [19, 20]. Take the BS as the center, the whole network is virtually divided into concentric circles (or rings), and the width of each ring is r, equal to the communication radius of each node. The identification of rings from inside to outside is from 1 to h. The rings near the BS are considered as hotspot due to heavy traffic load. The nonhotspot area is set as k rings from $h - k + 1$ to h.

Besides, each node in the network is stationary and has a unique ID. Nodes with new ID are not permitted to join the network after the deployment finished. The communication

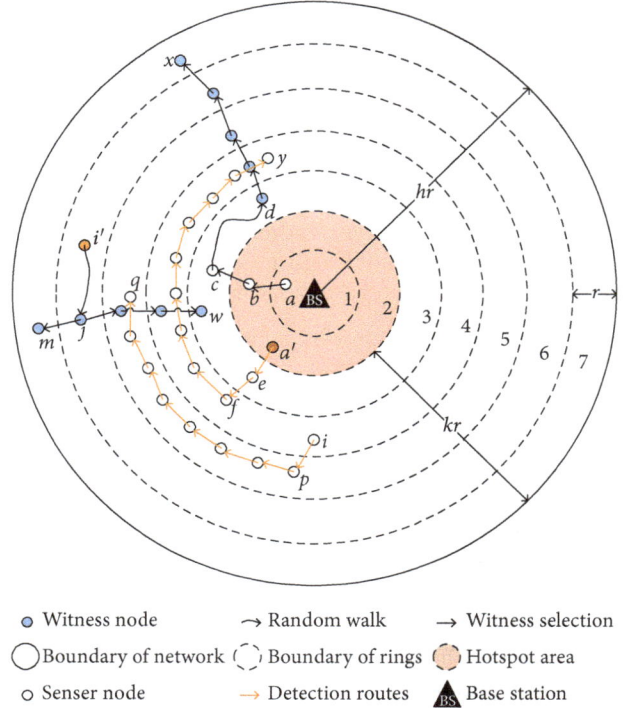

FIGURE 1: The model of a ring structure WSN.

between BS and nodes is through the other intermediate nodes. The nodes report their observing data to the BS periodically. The model of the network is shown in Figure 1.

The information transmitted between nodes are encrypted through a conventional bootstrapping cryptography mechanism. Before deployment, a key pair (ID, secret key) is allocated to each node. All nodes share their ID in the whole network, and the BS is assumed to be secure enough.

3.2. The Adversary Model. The purpose of the adversary is to damage or control the network at a lower price. It is cheaper to duplicate some nodes than to compromise the same amount of nodes. Therefore, the adversary usually compromises only a few sensor nodes, obtains all the information from captured nodes, and then replicates them to a large amount. Henceforth, the adversary could deploy the replica nodes into the network at strategic positions to acquire much more information or damage the network. The replica nodes deployed in different areas of network seem to be normal ones because they have legitimate IDs and encryption authentication mechanism.

We assume that the adversary could capture a limited number of normal sensor nodes at any position and the adversary could not create new IDs to join the network. Meanwhile, the adversary tries to avoid being detected [9].

3.3. Assumptions. We summarize the assumptions used in this work as follows:

(i) For the network

(A1) The nodes in the network are stationary and distributed densely and randomly.

(A2) Each node has a unique ID, and new IDs are forbidden after the network deployed.

(A3) Each node knows its own and neighbors' relative locations and geographic positions.

(A4) The communication between nodes is encrypted through a conventional bootstrapping cryptography mechanism.

(A5) The BS cannot be compromised.

(ii) For the adversary

(A1) The adversary could capture a limited amount of normal nodes.

(A2) The adversary deploys the replica nodes into different positions.

(A3) The adversary could not create new IDs for replica nodes.

The aim of our work is to discover a scheme with higher detection probability and lower resource expenditure; thus, the network lifetime will be maximized.

4. The Proposed Clone Detection Algorithm

In this section, the proposed method, a clone detection algorithm with low resource expenditure (CDLR) is introduced, which has a high detection probability (equal to 1 theoretically) and low resource consumptions and prolongs the network lifetime efficiently.

4.1. Overview of the Proposed CDLR Algorithm. The proposed CDLR approach is a distributed and random witness-based method for WSNs, which consists of two phases: the random witness chain establishment and the clone detection route generation. In order to avoid consuming energy of nodes in the hotspot area, the nonhotspot area should be predefined before deployment. The CDLR is implemented in the nonhotspot area to prolong the network lifetime. Each node in nonhotspot is responsible for both data collection and clone detection. The node should be checked by CDLR procedure before transmitting information to others.

In the first phase, a random witness chain for a node is generated by the node sending its encrypted private information including ID and location to the witness nodes. The witness chain of the node runs through the whole nonhotspot area in the direction of network radius, that is, in the centrifugal direction, and each witness node stores the private information of the source node.

In the second phase, a detection route for the node is established in a ring that is randomly selected from the nonhotspot area, and the route is formulated along the ring, that is, in the circumferential direction. The detection information is broadcast in the ring, and the clone is detected when

the witness node encountering the detection route discovers the conflict of nodes with the same ID but different positions.

The process of CDLR is just shown in Figure 1. It is implemented in the nonhotspot area, that is, the outer k rings. The black arrows demonstrate the witness selection and the witness chain generation, and the blue solid circles represent the witnesses of a node. Similarly, the red arrows indicate the clone detection route establishment, and the red solid circles represent the cloned nodes. The intersection of the witness chain and detection route of the nodes with the same ID demonstrates the clone detection.

In order to facilitate understanding our approach, see Notations for the list of symbols used in this work.

4.2. Witness Chain Establishment. The witness chain establishment is slightly different according to different locations of nodes. Suppose that the whole network is divided into h virtual rings and the nonhotspot area is defined as the outer k rings, that is, the hotspot area is $h-k$ rings near around the BS, just as shown in Figure 1. The witness chain formulation for nodes in hotspot area and nonhotspot area is described in this section.

For nodes in the hotspot area, the witness chain establishment consists of three steps. *Step 1.* The node transmits the encrypted witness selection message (ID, position) to any node in the nearest ring of nonhotspot area, that is, the $(h-k+1)$th ring. In this transmission process, the nodes on the forwarding path do not need to store the forwarded message, because they are not the witnesses of the source node. *Step 2.* In order to confuse the adversary, the node in the $(h-k+1)$th ring does not generate a witness chain directly, instead, it forwards the received message randomly to another node in the same ring, that is, the node randomly walks a few hops denoted as ξ. This node is selected as the first witness and stores the received message. *Step 3.* The first selected witness node initiates the witness chain generation by forwarding the message to the node in its communication range and in the next outer ring. The node that received the message forwards it to the node in the next outer ring until the outermost ring, that is, the hth ring. The nodes on the path of message transmitting formulate the witness chain of the source node. Take node a in Figure 1 as an example, node a locates in the hotspot area of the network, firstly, it should sent its message to node c in the nearest ring of nonhotspot area through node b; then, node c randomly walks ξ hops to node d in the same ring; at last, node d stores and forwards the message to nodes in the outer ring until reaching the node x in the hth ring. The nodes from d to x formulate the witness chain of node a, and the nodes on the chain store the message from node a.

For nodes in the nonhotspot area, the witness chain formulation consists of two steps. They do not need to send messages out of hotspot because they are already in the nonhotspot area. *Step 1.* The node randomly walks ξ hops to another node in the same ring, just like step 2 described in the last paragraph. *Step 2.* The first selected witness node forwards the message in the centrifugal and centripetal directions until reaching the outermost and innermost ring of the nonhotspot area, respectively. Take node i' in Figure 1 as an example, node i' randomly walks ξ hops to node j in the same ring, then node j stores and forwards the message in the centrifugal direction to node m and in the centripetal direction to node w, respectively. The nodes from m to w formulate the witness chain of node i'.

Through the process described above, the random witness chains for nodes are generated. The length of the witness chain is equal to the width of nonhotspot area, which is in a small constant level for the network.

4.3. Detection Route Generation. The detection route generation is also implemented in the nonhotspot area, because the witness chains formulated by the first phase are in the nonhotspot area. Only if the witness chains and the detection routes of nodes with the same ID are encountering, will the clone attacks be detected.

The detection route generation comprises two steps for all nodes in the network. *Step 1.* The node randomly selects a ring in the nonhotspot area to generate detection route and sends the encrypted detection message (ID, position) to any node in the selected ring. *Step 2.* The node received the detection message broadcasts it in the same ring. The witnesses of the source node compare the message they have stored with the received detection message from the same ID, if a conflict of nodes with the same ID but different positions occurs, the clone attack will be detected. Therefore, the revocation procedure for cloned nodes is triggered, and the ID and positions are the evidence.

Take nodes a and a' in Figure 1 as examples. Suppose node a' is a clone of node a, and they are located at different positions. Node a and a' have their own witness chains separately. Node a' first selects a random ring from nonhotspot (the 4th ring in Figure 1) and sends its detection message to node f in the 4th ring. Then, node f broadcasts the received message to the nodes in the same ring. The clone is detected when the detection message of node a' encounters the witness chain of node a, because the witness discovers the conflict that node a has different positions in the network.

From the description above, the random witness chains in the direction of network radius and the detection routes in the circumferential direction must encounter, which ensures the high detection probability. Furthermore, the process of witness selection and clone detection is implemented only in the nonhotspot area, which avoids consuming energy of nodes in hotspot and prolongs the network lifetime.

4.4. The CDLR Algorithm Description. The procedures of CDLR algorithm depicted above are summarized in Algorithm 1.

5. Theoretical Performance Analysis

The performance of CDLR is evaluated and analyzed from the aspects of detection probability, communication load, and storage requirements theoretically.

5.1. Detection Probability Analysis. The clone detection probability refers to whether any witness of a node can discover at least two nodes with the same ID but different positions (if exists) or not. If there exist multiple replica nodes for a source node, the clone attack will be detected successfully when one of the replica nodes is discovered, because all of these replica nodes have the same ID.

Theorem 1. *Given that the randomly selected witnesses of a source node are not compromised, if there exist replica nodes of this source node, the cloned nodes could always be detected successfully.*

Proof. To make the cloned nodes be detected, one of the following conditions should be met:

(1) At least one of the witnesses of the source node encounters one of the detection routes of cloned nodes.

(2) At least one of the witnesses of all cloned nodes encounters the detection route of the source node.

The CDLR algorithm takes some measures to ensure the encounter of the witness chains and the detection routes. First of all, the witness chain of each node (including cloned nodes) runs through the whole nonhotspot area, that is, all the witness chains are in the direction of network radius or in the centrifugal direction and extend to the network boundary. Secondly, the detection route of each node (also including cloned nodes) is generated along one of the virtual rings randomly selected from nonhotspot area by broadcasting the detection message, that is, all the detection routes are in the circumferential detection which are perpendicular to the witness chains. Thus, the detection routes must encounter the witness chains, that is, the witness chain of the source node must encounter the detection routes of cloned nodes and vice versa. Hence, one of the two conditions above must be met during the detection. When the witness chains and detection routes encounter, the conflict of nodes with the same ID but different locations will occur, and the cloned node could always be detected.

5.2. Communication Load and Network Lifetime Analysis. Due to the limited resources of sensor nodes, especially the limited energy supplied by batteries, the network is sensitive to resource consumptions. The energy consumption is related to the communication load of nodes. Moreover, the resource expenditure has a significant impact on the network lifetime. It is necessary to decrease the resource expenditure of nodes to prolong the network lifetime. Here, the network lifetime is defined as the duration from the network deployment to the moment that any node exhausts its energy [10].

In this section, the performance of CDLR is evaluated in terms of communication load of nodes and the network lifetime. The communication load of each node is analyzed at first, and then the network lifetime can be calculated from

```
1 Initialization:
2 Preset encryption mechanism
3 Obtain the relative locations to the BS and the identification of each ring
4 Predefine the nonhotspot area width k
5 Exchange the relative information with neighbors
6 PHASE 1: Witness chain establishment
7 X_a=Encrypt (ID_a, l_a)
8 i = R_a
9 while i < h − k + 1 do
10    Forward X_a to the i + 1 ring
11    i + 1
12 end while
13 if i ≥ h − k + 1 do
14    Forward X_a ξ hops randomly to node b in the same ring
15    node b records (ID_a, l_a) in X_a
16    W_a ← b
17    for j = i; j > h − k + 1; j− do
18       Forward X_a to node x in j-1 ring
19       node x records (ID_a, l_a) in X_a
20       W_a ← x
21    end for
22    for u = i; u < h; u++ do
23       Forward X_a to node x in u + 1 ring
24       node x records (ID_a, l_a) in X_a
25       W_a ← x
26    end for
27 end if
28 PHASE 2: Detection route generation
29 X_a=Encrypt (ID_a, l_a)
30 i = R_a
31 if i < h − k + 1 do
32    Select a ring R_w in nonhotspot randomly
33    while i < R_w do
34       Forward X_a to node x in i + 1 ring
35       i + 1
36    end while
37 end if
38 node x broadcast X_a in the i^{th} ring
39 for each node w in W_a that hears X_a do
40    if (ID_a, l_a) stored in w ≠ (ID_a, l_a) in X_a do
41       Trigger revocation procedure
42    end if
43 end for
```

ALGORITHM 1: A clone detection algorithm with low resource expenditure.

it by the given maximum communication load of each node in its whole lifecycle. In the evaluation, we assume that the communication load of each node in the same ring is the same. The outer (inner) rings of ring i refer to the rings whose identification is larger (smaller) than i.

The communication load of each node consists of three parts: witness selection, clone detection, and observing data collection, which is expressed by Theorems 2, 3, and 4, respectively.

Theorem 2. *Let ε_w, f_w, and ξ represent the size of the request message for witness selection, the frequency of witness selection, and the number of random walk hops, respectively. The*

communication load for witness selection of each node in ring i, $i \in [1, h]$, can be expressed as

$$
C_i^w = \begin{cases} \dfrac{i^2 \varepsilon_w f_w}{2i-1}, & i < h - k + 1, \\[2ex] (\xi + 1)\dfrac{i^2 \varepsilon_w f_w}{2i-1}, & i = h - k + 1, \\[2ex] \left(\xi + 1 + \dfrac{h^2}{2i-1}\right)\varepsilon_w f_w, & i > h - k + 1. \end{cases} \tag{1}
$$

Proof. The communication load of each node for witness selection is different according to its location in the network, that is, the node is located in the hotspot area or nonhotspot area. We assume that the width of nonhotspot area is k in hop counts. If the nodes are located in the hotspot area, they will be responsible for forwarding the request messages from nodes in the inner ring and themselves to nodes in the outer ring. The number of nodes in the ring i and its inner rings is $\pi i^2 r^2 \rho$, and the communication volume is $\pi i^2 r^2 \rho \times \varepsilon_w f_w$. Since the number of nodes in ring i can be expressed as

$$\pi i^2 r^2 \rho - \pi (i-1)^2 r^2 \rho = \pi (2i-1) r^2 \rho. \tag{2}$$

The communication load for witness selection of each node in ring i is

$$C_i^w = \frac{\pi i^2 r^2 \rho \times \varepsilon_w f_w}{\pi (2i-1) r^2 \rho} = \frac{i^2 \varepsilon_w f_w}{2i-1}, \quad i < h - k + 1. \tag{3}$$

If a node is located in the nonhotspot area, there are two cases based on the different locations and responsibilities, that is, the node is located in the innermost ring or other rings of the nonhotspot area.

For the nodes in the innermost ring of the nonhotspot area, they should first forward the request messages from nodes in the inner ring and themselves ξ times in the same ring because of the random walk mechanism and then forward these request messages to nodes in the outer ring. Thus, the communication volume of nodes in ring i is $\pi i^2 r^2 \rho \times \varepsilon_w f_w \xi + \pi i^2 r^2 \rho \times \varepsilon_w f_w$, and the number of nodes in ring i is obtained according to (2); therefore, the communication load for witness selection of each node in ring i is

$$C_i^w = \frac{(\xi+1)\pi i^2 r^2 \rho \times \varepsilon_w f_w}{\pi (2i-1) r^2 \rho} = (\xi+1)\frac{i^2 \varepsilon_w f_w}{2i-1}, \quad i = h - k + 1. \tag{4}$$

For the nodes in other rings except the innermost ring of the nonhotspot area, they have to randomly walk ξ hops and select witnesses in the centrifugal and centripetal directions. Therefore, they have three responsibilities: (1) forward their own messages ξ times in the same ring because of the random walk mechanism; (2) forward the messages from nodes in the inner ring and themselves to nodes in the outer ring; and (3) forward the messages from nodes in the outer ring and themselves to nodes in the inner ring. The communication load of each node for responsibility 1 is

$$\frac{\pi (2i-1) r^2 \rho \varepsilon_w f_w \xi}{\pi (2i-1) r^2 \rho} = \varepsilon_w f_w \xi. \tag{5}$$

The communication load of each node for responsibility 2 is the same as (3), and that for responsibility 3 is

$$\frac{\pi r^2 \left[h^2 - (i-1)^2\right] \rho \varepsilon_w f_w}{\pi (2i-1) r^2 \rho} = \frac{\left[h^2 - (i-1)^2\right] \varepsilon_w f_w}{2i-1}. \tag{6}$$

Thus, the communication load for each node in other rings except the innermost ring of the nonhotspot area is obtained by (3), (5), and (6) as

$$\begin{aligned} C_i^w &= \varepsilon_w f_w \xi + \frac{i^2 \varepsilon_w f_w}{2i-1} + \frac{\left[h^2 - (i-1)^2\right] \varepsilon_w f_w}{2i-1} \\ &= \left(\xi + 1 + \frac{h^2}{2i-1}\right) \varepsilon_w f_w, \quad i > h - k + 1. \end{aligned} \tag{7}$$

Overall, the communication load for witness selection of each node in ring i can be expressed in (1).

The communication load of each node for clone detection is described as follows.

Theorem 3. *Let ε_c and f_c denote the size of the request message for clone detection and the frequency of clone detection, respectively. The communication load for clone detection of each node in ring i, $i \in [1, h]$, can be expressed as*

$$C_i^c = \begin{cases} \dfrac{i^2 \varepsilon_c f_c}{2i-1}, & i < h - k + 1, \\[2ex] \dfrac{\pi i h^2 \varepsilon_c f_c}{k(2i-1)}, & i \geq h - k + 1. \end{cases} \tag{8}$$

Proof. The communication load of each node for clone detection is different according to different locations, that is, the node is located in the hotspot area or nonhotspot area. For nodes in the hotspot area, they only need to forward the detection messages from nodes in the inner ring and themselves to nodes in the outer ring, and this communication load is similar to (3), which can be expressed as

$$C_i^c = \frac{\pi i^2 r^2 \rho \times \varepsilon_c f_c}{\pi (2i-1) r^2 \rho} = \frac{i^2 \varepsilon_c f_c}{2i-1}, \quad i < h - k + 1. \tag{9}$$

For nodes in the nonhotspot area, we consider the following: because the clone detection is implemented in the nonhotspot area, the clone detection messages generated by all nodes in the network should be transmitted to the nonhotspot area, that is, all the k rings, thus each ring i in the nonhotspot area takes $1/k$ of the whole detection messages. Then, the messages are broadcasted in ring i to encounter the witness chains. Thus, the number of detection messages that ring i should take is $\pi h^2 r^2 \rho \varepsilon_c f_c / k$, and these messages are broadcasted in ring i for πi times in an average. Therefore, the communication load of each node in nonhotspot area for clone detection is

$$C_i^c = \frac{\pi h^2 r^2 \rho \varepsilon_c f_c}{k \pi (2i-1) r^2 \rho} \times \pi i = \frac{\pi i h^2 \varepsilon_c f_c}{k(2i-1)}, \quad i \geq h - k + 1. \tag{10}$$

Overall, the communication load of each node for clone detection can be expressed as (8).

The original mission of the network is to collect the observing data periodically, and the communication load of each node in ring i for collecting data is calculated as follows.

Theorem 4. *Let ε_d and f_d denote the size of the message for observing data and the frequency of observing data collection, respectively. The communication load for observing data collection of each node in ring i, $i \in [1, h]$, can be expressed as*

$$C_i^d = \frac{\left[h^2 - (i-1)^2\right]\varepsilon_d f_d}{2i - 1}. \qquad (11)$$

Proof. In the process of normal data collection, the observing data is transmitted to the BS through the intermediate nodes; thus, the nodes in ring i are responsible for transmitting the messages of their own and from the outer rings to the nodes in the inner ring. The number of nodes in ring i and its outer rings is $\pi r^2 [h^2 - (i-1)^2]\rho$, and the number of nodes in ring i can be obtained by (2). Therefore, the communication load of each node for observing data collection is

$$C_i^d = \frac{\pi r^2 \left[h^2 - (i-1)^2\right]\rho\varepsilon_d f_d}{\pi(2i-1)r^2\rho} = \frac{\left[h^2 - (i-1)^2\right]\varepsilon_d f_d}{2i - 1}. \qquad (12)$$

According to Theorem 2, 3, and 4, the overall communication load for each node in ring i can be obtained by

$$C_i^o = C_i^w + C_i^c + C_i^d. \qquad (13)$$

If the parameters of ε_w, f_w, ε_c, f_c, ε_d, f_d, and ξ are known for the network, the optimal k will be acquired based on (13) to maximize the network lifetime. Figure 2 demonstrates the communication load of each node in ring i with different width of nonhotspot area k, from which we can see that the parameter k has a significant impact on the communication load of nodes; thus, it can also affect the energy consumption and network lifetime. Here in Figure 2, under the condition that $h = 20$, $f_c = 10$, and other parameters are set to 1, when k is 19, 18, and 17, the communication load of nodes in rings 2, 4, and 5 is the highest, that is, consumes the most energy, respectively. From the evaluation of the communication load, we can obtain the optimal k.

Under the same conditions, a comparison of the communication load of nodes in ring i with different k using CDLR and ERCD is conducted according to (13) and the evaluation described in [10]. As shown in Figure 3, no matter what value is assigned to k, the corresponding communication load of each node in ring i using CDLR is lower than that using ERCD. The most straightforward reason is that the communication load of nodes in witness section using CDLR is much lower than that using ERCD, because the witness selection of ERCD is along the circumferential direction whereas that of CDLR is along the radius of the network; thus, the number of witnesses of ERCD is much more than that of CDLR.

Based on the communication load evaluation above, the network lifetime using CDLR is analyzed and compared with the lifetime using LSM [8] and ERCD [10] under the same conditions. According to the definition of network lifetime, the energy exhaustion of any node means the end of network lifetime in the evaluation. To simplify the evaluation, we assume that the communication capability of each node in network is the same and the total size of messages a node can transmit in its life cycle denoted as T_z is 1 million. In the evaluation, the

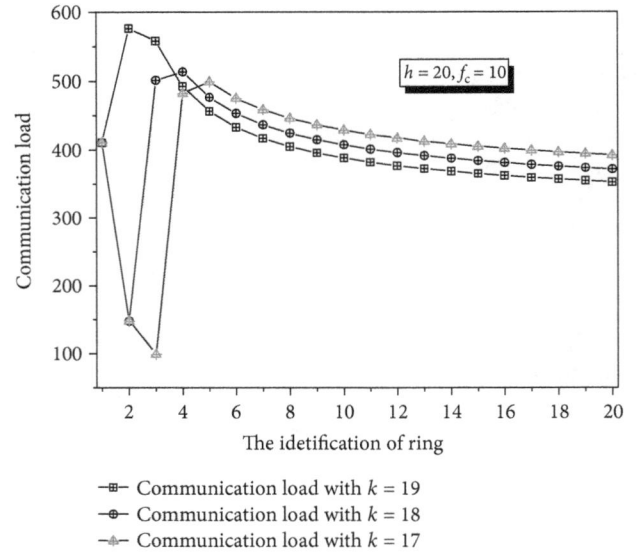

FIGURE 2: The communication load of each node in ring i with different k.

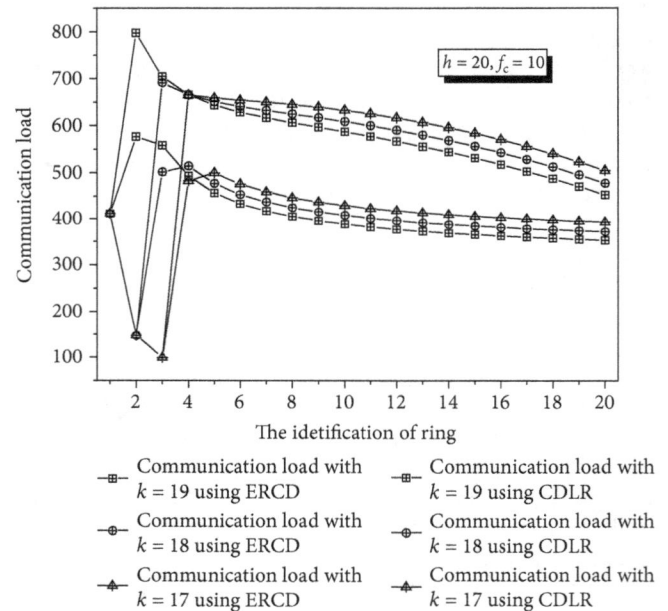

FIGURE 3: The comparison of communication load of each node using ERCD and CDLR.

parameters used are assumed as $\varepsilon_w = \varepsilon_c = \varepsilon_d = 1$, $f_w = f_d = 1$, and $\xi = 1$. Meanwhile, in LSM, the variable g denotes the number of witnesses selected by each neighbor of a source node, p represents the probability that the neighbor transmits the message from the source node, and δ is the average degree of each node or the average number of neighbors of each node.

Theorem 5. *If $\varepsilon_w = \varepsilon_c = \varepsilon_d = 1$, $f_w = f_d = 1$, and $\xi = 1$, the ratio of network lifetime using CDLR algorithm and LSM algorithm is*

$$\frac{h^2 + gp\delta h f_c \sqrt{\delta + 1}}{\max\left((1 + f_c + h^2), \max\left(\left((\pi i h^2 f_c/3k) + (h^2 + i^2/2i - 1) + 1\right), \max\left((\pi i h^2 f_c/3k) + (2h^2 - i^2/2i - 1) + 3\right)\right)\right)}. \tag{14}$$

Proof. In LSM, the communication load consists of observing data collection and clone detection, because the witness selection and clone detection are the same process. In observing data collection, the communication load of each node in ring i can be obtained by (11), from which we can see that the communication load of nodes in ring 1 is the maximum, that is, h^2. Thus, the communication load of nodes in ring 1 determines the network lifetime. The communication load for clone detection in ring 1 is $gp\delta\sqrt{n}$, where n is the number of nodes in the whole network and $n = \pi h^2 r^2 \rho$. All nodes in ring 1 are neighbors to each other; thus, the number of nodes in ring 1 is equal to the number of neighbors of a node plus itself, that is, $\pi r^2 \rho = \delta + 1$. Therefore, the overall communication load of each node in ring 1 is $h^2 + gp\delta h f_c \sqrt{\delta + 1}$, and the network lifetime using LSM is expressed as

$$\text{LT}_{\text{LSM}} = \frac{T_z}{h^2 + gp\delta h f_c \sqrt{\delta + 1}}. \tag{15}$$

In CDLR algorithm, suppose the node in ring i has the maximum overall communication load. Due to the difference of nodes in the hotspot area and nonhotspot area, the maximum communication load is different according to the locations of nodes. If the node is located in the hotspot area, that

is, $i < h-k+1$, the maximum communication load is the node in ring 1 calculated as $1 + f_c + h^2$, where 1, f_c, and h^2 are the communication load for witness selection, clone detection, and observing data collection, respectively. If the node is in the nonhotspot area, there are two cases for communication load calculation according to whether the node is in the innermost ring of nonhotspot area or not. For the node in the innermost ring of nonhotspot area, that is, the node in ring $i = h-k+1$, the overall communication load is calculated according to (13) as

$$C_i^o = \frac{2i^2}{2i - 1} + \frac{\pi i h^2 f_c}{3k} + \frac{h^2 - (i - 1)^2}{2i - 1}, \quad i = h - k + 1. \tag{16}$$

For the node in other rings except the innermost ring of nonhotspot area, that is, $i > h-k+1$, the overall communication load is calculated according to (13) as

$$C_i^o = 2 + \frac{h^2}{2i - 1} + \frac{\pi i h^2 f_c}{3k} + \frac{h^2 - (i - 1)^2}{2i - 1}, \quad i > h - k + 1. \tag{17}$$

The network lifetime is determined by the lifetime of the node in ring i which has the maximum communication load; therefore, the network lifetime using CDLR is expressed as

$$\text{LT}_{\text{CDLR}} = \frac{T_z}{\max\left((1 + f_c + h^2), \left((2i^2/2i - 1) + (\pi i h^2 f_c/3k) + (h^2 - (i - 1)^2/2i - 1)\right), \max\left(2 + (h^2/2i - 1) + (\pi i h^2 f_c/3k) + (h^2 - (i - 1)^2/2i - 1)\right)\right)}. \tag{18}$$

Based on (18) and (15), the ratio of network lifetime using CDLR algorithm and LSM algorithm is obtained as shown in (14).

Theorem 6. *If $\varepsilon_w = \varepsilon_c = \varepsilon_d = 1$, $f_w = f_d = 1$, and $\xi = 1$, the ratio of network lifetime using CDLR algorithm and ERCD algorithm is*

$$\frac{\max\left((1 + f_c + h^2), \max\left((h^2 - (i - 1)^2/2i - 1) + h f_c + \left(2\pi i h^2/k(2i - 1)\right)\right)\right)}{\max\left((1 + f_c + h^2), \max\left(\left((\pi i h^2 f_c/3k) + (h^2 + i^2/2i - 1) + 1\right), \max\left((\pi i h^2 f_c/3k) + (2h^2 - i^2/2i - 1) + 3\right)\right)\right)}. \tag{19}$$

Proof. The network lifetime using CDLR is proved in Theorem 5. For ERCD protocol, the communication load for nodes is also different based on the locations. For nodes in the hotspot area, the communication load of nodes is the same as CDLR, that is, when $i < h-k+1$, the maximum communication load is the node in ring 1, and the load is

$1 + f_c + h^2$. For nodes in the nonhotspot area, that is, when $i \geq h-k+1$, the communication load of each node for witness selection, legitimacy verification, and data collection in ring i is $2\pi i h^2/(k(2i - 1))$, $h f_c$, and $(h^2 - (i - 1)^2)/(2i - 1)$, respectively [10]. Thus the network lifetime using ERCD is expressed as

$$\text{LT}_{\text{ERCD}} = \frac{T_z}{\max\left(\left(1 + f_c + h^2\right), \max\left(\left(h^2 - (i-1)^2/2i - 1\right) + hf_c + \left(2\pi i h^2/k(2i-1)\right)\right)\right)}. \qquad (20)$$

Therefore, based on (18) and (20), the ratio of network lifetime using CDLR algorithm and ERCD algorithm is obtained as shown in (19).

The comparison of network lifetime using CDLR, LSM, and ERCD and their ratio under the same conditions with different parameters is shown in Figures 4–6 according to (18), (15), (20), (14), and (19), respectively.

Figures 4–6 manifest that the network lifetime using CDLR obviously outperforms other methods including LSM and ERCD under the same conditions. Figures 4 and 5 demonstrate that the network lifetime using CDLR is about average 2.4 times and 1.4 times that of using LSM and ERCD, respectively. The main reasons are as follows: first of all, the CDLR implemented in the nonhotspot area makes full use of the energy of nodes in the nonhotspot area and avoids introducing new bottleneck by clone detection. Secondly, the communication load of nodes using CDLR is lower than other methods due to the random witness chains in the centrifugal direction and clone detection routes in the circumferential direction. Furthermore, the network lifetime is not related to the density of nodes in the network, which can be obtained by Figure 6(a). The network lifetime using CDLR and ERCD is constant with the change of node degree, whereas that using LSM is sensitive to the density of nodes. Therefore, the theoretical analysis declares that the performance of CDLR is better than other approaches LSM and ERCD under the same conditions.

5.3. Storage Requirement Analysis. Due to the constraints of the storage of sensor nodes, it is necessary to decrease the storage requirements during the designing of a protocol. Here, the storage requirements of sensor nodes using CDLR are evaluated.

Theorem 7. *The storage requirements of each node using CDLR algorithm are O(h).*

Proof. In CDLR, the number of witnesses of each source node in its witness chain is k, and the witness selection message from the source node is stored by these k witnesses. There are $n = \pi h^2 r^2 \rho$ nodes in the whole network, and all the witness selection messages generated by all nodes should be stored by nodes in the nonhotspot area, the number of which is $\pi h^2 r^2 \rho - \pi (h-k)^2 r^2 \rho$. Therefore, the storage requirements of each node implementing the clone detection are calculated as

$$\frac{\pi h^2 r^2 \rho k}{\pi h^2 r^2 \rho - \pi (h-k)^2 r^2 \rho} = \frac{h^2 k}{h^2 - (h-k)^2} = \frac{h^2}{2h-k} = O(h). \qquad (21)$$

From (21), we can see that the storage requirements of each sensor node are not related to the number of nodes in the network, it is only related to the radius in hops of the network. Therefore, the storage requirements are not related to the density of node in network, which is similar to ERCD, but different from many previous works.

6. Experiments and Simulations

In this section, the performance of CDLR is evaluated by experiments, and the performance comparison of different methods including LSM [8], ERCD [10], and LSCD [11] is also conducted on OMNET++ [21].

6.1. Experiments Description. In experiments, the network consisting of 2000 nodes is deployed in a circular shape whose radius is 600 m, and the BS is at the position near the center. The communication range of each node is 40 m, and ring 1 is set as the hotspot area in our experiments. In the simulation, the frequency of witness selection and observing data collection is set the same, that is, $f_w = f_d = 1$, and the frequency of clone detection is set as $f_c = 10$. To simplify the experiments, the different messages transmitted among nodes are in the same size, that is, $\varepsilon_w = \varepsilon_c = \varepsilon_d = 100$ bytes. The widely accepted energy consumption model in WSN detailed in [2] is adopted in our experiments. The simulation parameters and the range of values used in the experiments are listed in Table 2.

6.2. Experiment Results and Analysis. The experiment results in terms of clone detection probability and the network lifetime are displayed and analyzed in this section.

The detection probability of CDLR is equal to 1 theoretically under the condition that the witnesses selected are normal and trusted. However, a compromised node or a cloned node may be selected in practice situation, the behavior of which cannot be trusted. Therefore, the witness nodes controlled by adversary will cause the failure of the clone detection. Suppose that there are 10% cloned nodes in the network, the actual detection probability of CDLR is shown in Figure 7, from which we can see that the detection probability decreases with the increment in malicious nodes, but it is still in a high detection rate reaching more than 98%. Because in the set of random witnesses, the probability of a malicious node being selected as witness is very low, which is also confirmed by experiment results.

The clone detection probability of four methods including LSM, ERCD, LSCD, and CDLR is compared under the same conditions with different sensor node density (the node degree or the number of neighbors) and network scale (the radius of network). Figures 8 and 9 display the experiment results, from which we can draw a conclusion that the clone detection probability using CDLR is higher than that using other three methods under the same conditions.

Figure 8 demonstrates that the clone detection probability using CDLR, LSCD, and ERCD increases from 89% to

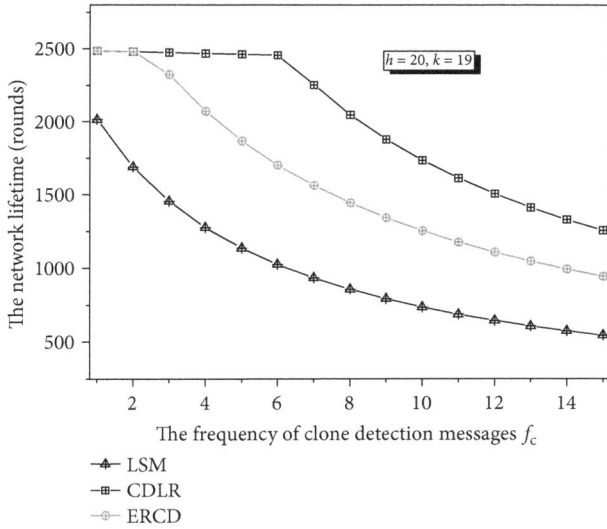

(a) The network lifetime

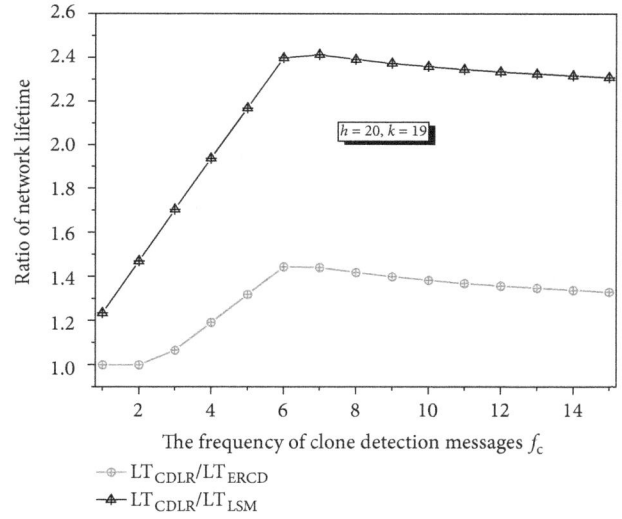

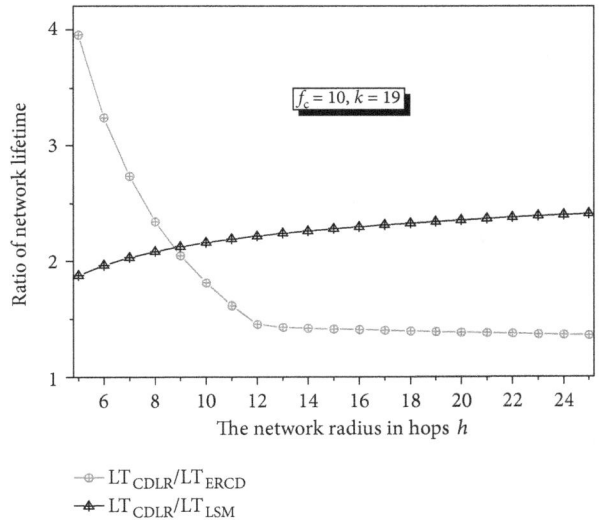

(b) Ratio of network lifetime

FIGURE 4: The comparison of network lifetime of three methods with different f_c.

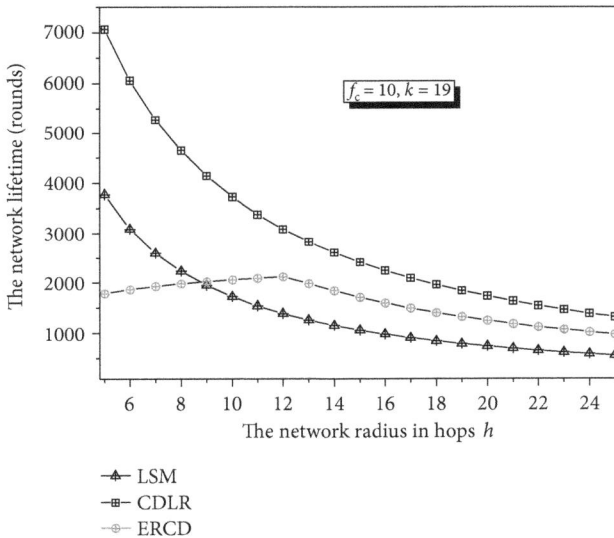

(a) The network lifetime

(b) Ratio of network lifetime

FIGURE 5: The comparison of network lifetime of three methods with different h.

98%, from 86% to 96%, and from 85% to 95% with the increment in node degree, respectively. Whereas the detection probability using LSM decreases from 70% to 60% with the increment in node degree. The reason for better performance of CDLR is that the probability of success in witness selection and clone detection message broadcasting is higher when node density is larger. Moreover, the encounter probability of the witnesses and detection routes of nodes with the same ID under CDLR is higher than that under ERCD and LSCD because of the mechanism of witness selection. As for LSM, the reason for detection probability decreasing is that the LSM uses at least two-path intersection at the same node to detect clone attacks. If the node density is large, there will be more options for the next hop

and the probability of two-path intersection at the same node will decrease.

Figure 9 manifests that the detection probability of four methods with the increment in network scale, that is, the network radius in hops, from which we can see that the detection probability using CDLR, LSCD, and ERCD is around 97%, 96%, and 95%, respectively, and there is not much fluctuation. It indicates that the detection probability of CDLR and LSCD is not depending on the network scale. The detection probability under ERCD has declined a little with the increase in the network radius, because both the witnesses and verification paths in ERCD are in ring structure, the probability of selecting the same ring or neighbor rings is lower with the increase in the number of rings. Overall, the

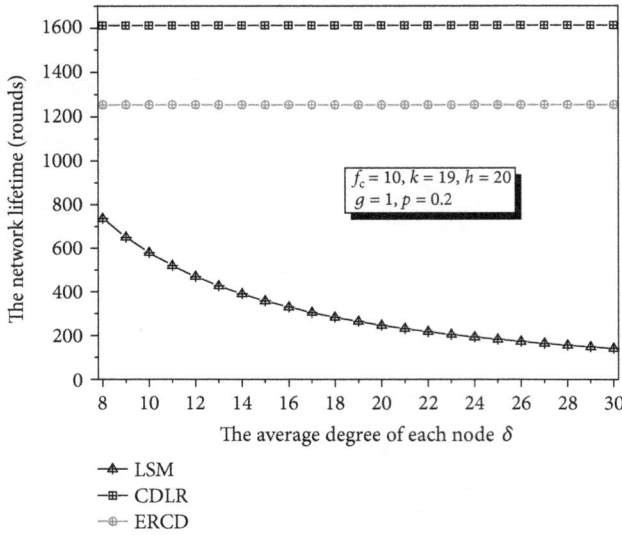

(a) The network lifetime

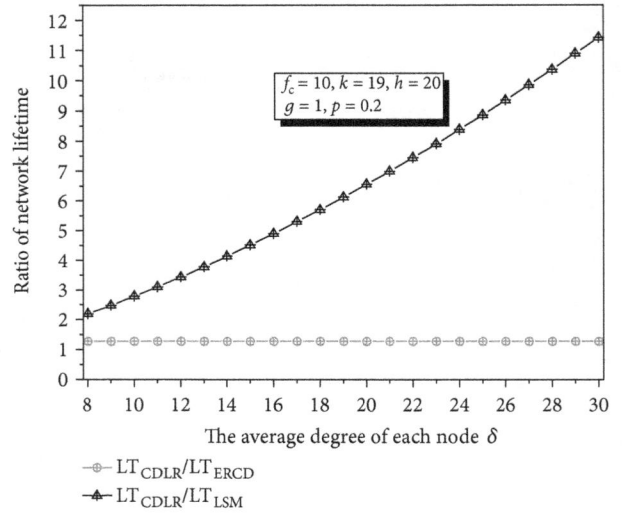

(b) Ratio of network lifetime

FIGURE 6: The comparison of network lifetime of three methods with different δ.

TABLE 2: The simulation parameters and the range of values in experiments.

Items	Values or ranges
The radius of WSNs deployed (m)	600
The number of members in WSNs	2000
Transmission range of nodes (m)	40
The frequency of witness selection	1
The frequency of observing data collection	1
The frequency of clone detection	10
The size of the messages transmitted in WSNs (bytes)	100
The number of random hops in witness selection	1
The proportion of replica nodes	1%–10%
Initial energy of a node (J)	0.5

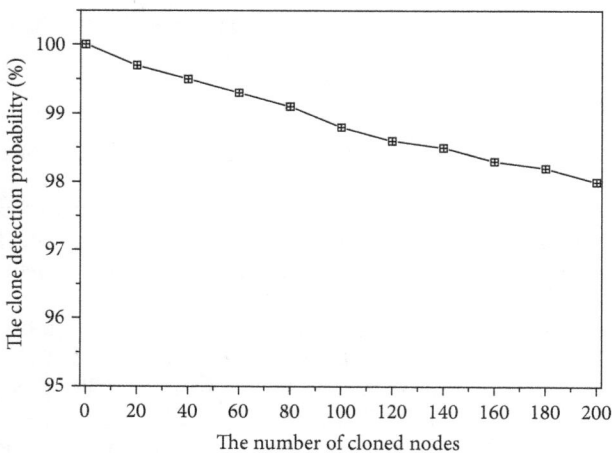

FIGURE 8: Comparison of detection probability of four methods with different node density.

performance of CDLR in detection probability is better than that of ERCD, LSCD, and LSM.

The network lifetime is compared under four different protocols, and the results are displayed in Figures 10–12. As the theoretical analysis results, the network lifetime using different methods is affected by clone detection frequency, network scale (network radius in hops), and node density. The experiments are conducted in these three aspects to compare the performance of four methods.

During the comparison, the differences between theoretical and experimental data are also considered. The differences are mainly from the evaluation of energy consumption. In theoretical analysis, we only consider the energy consumption caused by message sending for simplicity, because message sending consumes the most energy of nodes. In fact,

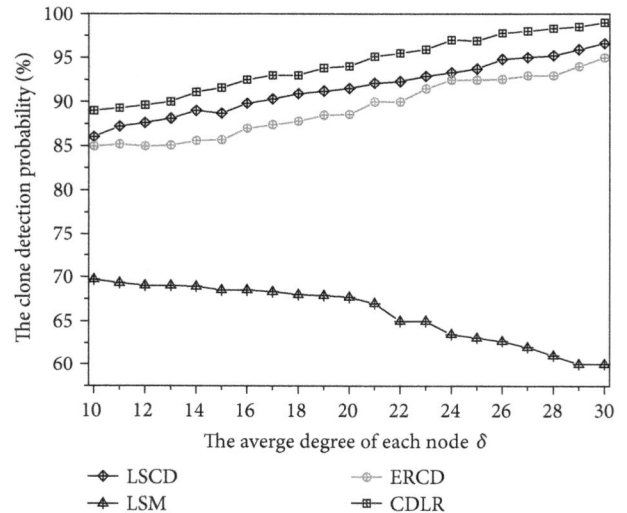

FIGURE 7: The clone detection probability with the number of cloned nodes.

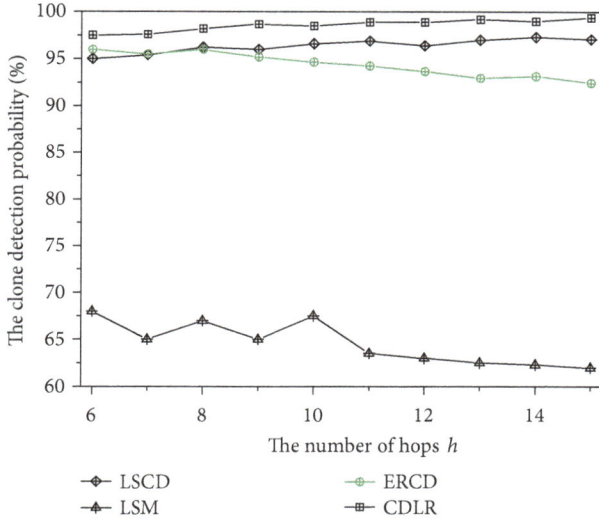

FIGURE 9: Comparison of detection probability of four methods with different network radius in hops.

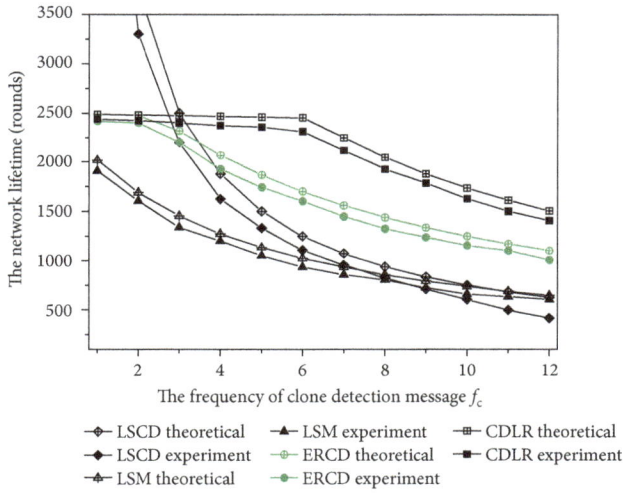

FIGURE 10: Comparison of network lifetime using four methods with different frequency of detection.

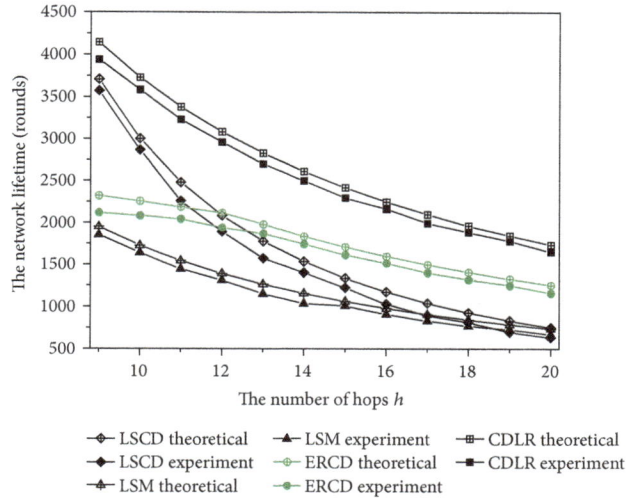

FIGURE 11: Comparison of network lifetime using four methods with different network radius in hops.

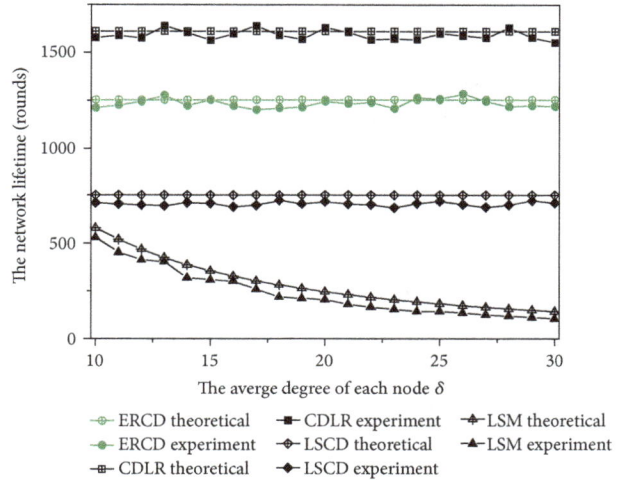

FIGURE 12: Comparison of network lifetime using four methods with different node density.

the energy of nodes is depleted by many factors, such as message receiving, sending, and data computing.

Figure 10 depicts the network lifetime with the increase in the frequency of detection messages using four methods, from which we can see that the network lifetime using all these methods decreases with the increase in the frequency of clone detection. However, the network lifetime using CDLR begins decreasing significantly when the frequency of clone detection is more than 6, because most of the energy consumption occurs in ring 1 mainly caused by observing data collection when the frequency of clone detection messages is less than or equal to 6. That means, in this situation, the clone detection does not affect the network lifetime. The same phenomenon occurs for ERCD when the frequency of clone detection messages is less than or equal to 2. However, the network lifetime using LSM and LSCD decreases sharply with the increase in the frequency of clone detection

messages, especially for LSCD, because each clone detection message of all the nodes should start from the nodes in the second ring, whose lifetime is the bottleneck of the whole network. Thus, the network lifetime using CDLR is an average of 1.4 times, 1.7 times, and 2 times of that using ERCD, LSCD, and LSM, respectively.

Figure 11 demonstrates the comparison of network lifetime under four methods with different network scales, that is, different network radius in hops. We can get a conclusion that the network lifetime decreases with the increase in the network scale, because both the number of witnesses and the length of clone detection routes are increasing as the expansion of the network scale; thus, the communication load is also increasing. However, the network lifetime using CDLR is longer than that using other three methods, because under the same conditions, the communication load of network using CDLR is the minimal in these methods. In general, the network lifetime using CDLR is improved by an average of 50%, 75%, and 120% compared to that using ERCD, LSCD, and LSM, respectively.

The comparison of network lifetime under four methods with different node densities is shown in Figure 12, from which we can see that the network lifetime using CDLR, LSCD, and ERCD is not related to the node density and the lifetime fluctuates in a small range, whereas the network lifetime using LSM decreases significantly with the increase in the node density. Because there is no obvious change for both the number of witnesses and the length of clone detection routes, thus the communication load is also stable. As for LSM, the number of witnesses and routes increases with the increase in node density; hence, the communication load is also increasing. The number of witnesses and routes using CDLR is the minimal in these methods, and the communication load of CDLR is the lowest of all; therefore, the network lifetime using CDLR is the longest in these methods.

7. Conclusion

In this paper, we have proposed a distributed clone detection algorithm with low resource expenditure for randomly deployed WSNs with ring structure, which consists of two phases: random witness chain establishment and detection route generation. In the proposed method, the witness chains are in the direction of network radius or in the centrifugal direction, and the detection routes are in the circumferential direction in each ring, which could ensure the encounter of the witness chains and the detection routes of nodes with the same ID but different positions. The detection probability is equal to 1 according to the theoretical analysis under the conditions that the witness nodes are not compromised. Furthermore, the performance of proposed method in terms of network lifetime and storage requirements is better than most other existing methods, such as LSM, LSCD, and ERCD. The communication load of proposed method is low, and the detection process is implemented in the nonhotspot area, which makes full use of the resource of nodes far from the BS and avoids consuming the energy of nodes in the hotspot area. Experiments and simulations have demonstrated that the proposed method outperforms most other methods in detection probability, network lifetime, and storage requirements, at the same time, the method is suitable for large-scale or densely deployed networks.

Notations

r:　The communication range of sensor nodes
h:　The radius of the network in hop counts
ρ:　The density of deployed sensor nodes
k:　The width of the nonhotspot area in hop counts
ξ:　The number of random walk hops
ID_a:　The ID of node a
l_a:　The location/position of node a
X_a:　The encrypted message of node a
R_a:　The identification of the ring that node a locates in
W_a:　The witness set of node a
ε_d:　The size of the message for observing data

f_d:　The frequency of observing data collection
ε_w:　The size of the request message for witness selection
f_w:　The frequency of witness selection
ε_c:　The size of the request message for clone detection
f_c:　The frequency of clone detection
C_i^d:　The communication load of each node in ring i during data collection
C_i^w:　The communication load of each node in ring i during witness selection
C_i^c:　The communication load of each node in ring i during clone detection
C_i^o:　The overall communication load of each node in ring i
T_z:　The total size of messages a node can transmit in its life cycle
LT_A:　The network lifetime using algorithm A.

Conflicts of Interest

Zhihua Zhang, Shoushan Luo, Hongliang Zhu, and Yang Xin are with the Information Security Center in School of Cyberspace Security, Beijing University of Posts and Telecommunications, and National Engineering Laboratory for Disaster Backup and Recovery, Beijing 100876, China.

Acknowledgments

This research is supported in part by the National Key R&D Program of China under Grant 2017YFB0802300, in part by the National High Technology Research and Development Program of China (863 Program) under Grant 2015AA017201, in part by the National Natural Science Foundation of China under Grant U1536119, and in part by Applied Sci-Tech R&D Special Fund Program of Guangdong Province of China under Grant 2015B010131007.

References

[1] Y. Zhang, S. He, and J. Chen, "Data gathering optimization by dynamic sensing and routing in rechargeable sensor networks," in *2013 IEEE International Conference on Sensing, Communications and Networking (SECON)*, vol. 24no. 3, pp. 273–281, New Orleans, LA, USA, June 2013.

[2] Y. Hu and A. Liu, "An efficient heuristic subtraction deployment strategy to guarantee quality of event detection for WSNs," *The Computer Journal*, vol. 58, no. 8, pp. 1747–1762, 2015.

[3] L. Jiang, A. Liu, Y. Hu, and Z. Chen, "Lifetime maximization through dynamic ring-based routing scheme for correlated data collecting in WSNs," *Computers and Electrical Engineering*, vol. 41, pp. 191–215, 2015.

[4] I. Butun, S. Morgera, and R. Sankar, "A survey of intrusion detection systems in wireless sensor networks," in *2015 6th International Conference on Modeling, Simulation, and Applied Optimization (ICMSAO)*, vol. 16no. 1, pp. 266–282, Istanbul, Turkey, May 2014.

[5] Y. Zeng, J. Cao, S. Zhang, S. Guo, and L. Xie, "Random-walk based approach to detect clone attacks in wireless sensor networks," *IEEE Journal on Selected Areas in Communications*, vol. 28, no. 5, pp. 677–691, 2010.

[6] J. Ho, M. Wright, and S. K. Das, "Fast detection of mobile replica node attacks in wireless sensor networks using sequential hypothesis testing," *IEEE Transactions on Mobile Computing*, vol. 10, no. 6, pp. 767–782, 2011.

[7] Z. Li and G. Gong, "On the node clone detection in wireless sensor networks," *IEEE/ACM Transactions on Networking*, vol. 21, no. 6, pp. 1799–1811, 2013.

[8] B. Parno, A. Perrig, and V. Gligor, "Distributed detection of node replication attacks in sensor networks," in *2005 IEEE Symposium on Security and Privacy (S&P'05)*, pp. 49–63, Oakland, CA, USA, May 2005.

[9] M. Conti, R. D. Pietro, L. Mancini, and A. Mei, "Distributed detection of clone attacks in wireless sensor networks," *IEEE Transactions on Dependable and Secure Computing*, vol. 8, no. 5, pp. 685–698, 2011.

[10] Z. Zheng, A. Liu, L. Cai, Z. Chen, and X. Shen, "Energy and memory efficient clone detection in wireless sensor networks," *IEEE Transactions on Mobile Computing*, vol. 15, no. 5, pp. 1130–1143, 2016.

[11] M. Dong, K. OTA, L. Yang, A. Liu, and M. Guo, "LSCD: a low-storage clone detection protocol for cyber-physical systems," *IEEE Transactions on Computer-Aided Design of Integrated Circuits and Systems*, vol. 35, no. 5, pp. 712–723, 2016.

[12] H. Choi, S. Zhu, and P. TFL, "SET: detecting node clones in sensor networks," in *2007 Third International Conference on Security and Privacy in Communications Networks and the Workshops - SecureComm 2007*, pp. 341–350, Nice, France, France, September 2007.

[13] R. Brooks, P. Y. Govindaraju, M. Pirretti, N. Vijaykrishnan, and M. T. Kandemir, "On the detection of clones in sensor networks using random key predistribution," *IEEE Transactions on Systems, Man and Cybernetics, Part C (Applications and Reviews)*, vol. 37, no. 6, pp. 1246–1258, 2007.

[14] W. Naruephiphat, Y. Ji, and C. Charnsripinyo, "An area-based approach for node replica detection in wireless sensor networks," in *2012 IEEE 11th International Conference on Trust, Security and Privacy in Computing and Communications*, pp. 745–750, Liverpool, UK, June 2012.

[15] C. Yu, C. Lu, and S. Kuo, "Compressed sensing-based clone identification in sensor networks," *IEEE Transactions on Wireless Communications*, vol. 15, no. 4, pp. 3071–3084, 2016.

[16] A. KumarMishra and A. Turuk, "A zone-based node replica detection scheme for wireless sensor networks," *Wireless Personal Communications*, vol. 69, no. 2, pp. 601–621, 2012.

[17] N. Shashidhar, C. Kari, and R. Verma, "The efficacy of epidemic algorithms on detecting node replicas in wireless sensor networks," *Journal of Sensor and Actuator Networks*, vol. 4, no. 4, pp. 378–409, 2015.

[18] C. Ding, L. Yang, and M. Wu, "Localization-free detection of replica node attacks in wireless sensor networks using similarity estimation with group deployment knowledge," *Sensors*, vol. 17, no. 1, 2017.

[19] N. Bulusu, J. Heidemann, and D. Estrin, "GPS-less low-cost outdoor localization for very small devices," *IEEE Personal Communications*, vol. 7, no. 5, pp. 28–34, 2000.

[20] J. Newsome and D. Song, "GEM: graph embedding for routing and data-centric storage in sensor networks with-out geographic information," in *ACM Conference on Embedded Networked Sensor Systems (SenSys)*, November 2003.

[21] "OMNet++ network simulation framework," http://www.omnetpp.org.

15

An Improved Niche Chaotic Genetic Algorithm for Low-Energy Clustering Problem in Large-Scale Wireless Sensor Networks

Min Tian [ID],[1] Jie Zhou,[1] and Xin Lv [ID][2]

[1]*College of Information Science and Technology, Shihezi University, Shihezi, Xinjiang 832003, China*
[2]*The Key Laboratory of Oasis Ecological Agriculture of Xinjiang Production and Construction Group, Shihezi University, Shihezi, Xinjiang 832003, China*

Correspondence should be addressed to Xin Lv; lxshz@126.com

Academic Editor: Jaime Lloret

Large-scale wireless sensor networks consist of a large number of tiny sensors that have sensing, computation, wireless communication, and free-infrastructure abilities. The low-energy clustering scheme is usually designed for large-scale wireless sensor networks to improve the communication energy efficiency. However, the low-energy clustering problem can be formulated as a nonlinear mixed integer combinatorial optimization problem. In this paper, we propose a low-energy clustering approach based on improved niche chaotic genetic algorithm (INCGA) for minimizing the communication energy consumption. We formulate our objective function to minimize the communication energy consumption under multiple constraints. Although suboptimal for LSWSN systems, simulation results show that the proposed INCGA algorithm allows to reduce the communication energy consumption with lower complexity compared to the QEA (quantum evolutionary algorithm) and PSO (particle swarm optimization) approaches.

1. Introduction

The advancement of microsensors, microelectromechanical systems, and communication make economically and technically feasible a network composed of low computational complexity, numerous, vulnerable, and fast-response wireless sensors [1]. Large-scale wireless sensor networks are composed of a large number of sensing units that have limited computing, communication, sensing, and free-infrastructure capabilities [2–4]. Each sensor node consists of four units, including the transmitter unit, the computation unit, the signals sensing unit, the data storage unit, and the battery unit [5]. Large-scale wireless sensor networks (LSWSNs) have been widely studied and usefully employed in many areas such as target detection and tracking, advanced health care delivery, intelligent family, military affairs, multimedia surveillance, and environmental monitoring [6].

2. Problem Statement and Our Contributions

Generally, sensor nodes are small devices with limited monitoring capabilities [7]. Because sensor nodes are equipped with restricted power device, it is very important and necessary to explore new low-energy clustering algorithms [8]. Such a problem is important in enhancing network lifetime [9]. Generating an optimal low-energy clustering for large-scale wireless sensor networks with restricted power device and communication energy consumption constraints is very useful but is an NP-hard problem [10]. While exhaustive search is recognized as one possible solution to low-energy clustering problem [11], its computational complexity is too high to be implemented for practical real-time applications [12].

Due to its computational complexity, many heuristics have been proposed to obtain near-optimal solutions in reasonable time [13–18]. For the low-energy clustering problem, the authors proposed in [19] a PSO-based low-energy clustering method, which is found to be more efficient than GA methods. It provides a wider search space by randomly selecting low-energy clustering solutions when each solution in population gets updated. However, PSO suffer from either low convergence rate or high computational complexity. In [20], a low-energy clustering technique that enables trade-offs between the computation cost and communication energy consumption is investigated using fuzzy logic. Their work focused on a low-energy clustering optimization with fuzzy logic. The fuzzy logic approach is simple and fast, but it usually yields high computational cost with large-scale WSNs. There are several approaches based on the intelligent clustering protocol, which attempt to solve this issue by using the concept of heuristic [21]. The proposed method uses a distance-aware clustering protocol instead of a traditional clustering method. However, the protocol is complex, and the computation complexity is still high.

Based on a combination of quantum theory and evolution theory, such as rotation gates, superposition of states, and quantum bits, the authors applied the QEA to the low-energy clustering problem [22]. Compared with traditional genetic algorithm, QEA is robust and global in operation. The QEA with Q-bit representation can explore the search space with quicker convergence speed. This enables the solution of low-energy clustering problem and reduces the computational burden.

Recently, the concepts of niche computing and chaos theory have motivated the new development of evolutionary theory. Evolutionary algorithms have been shown to be effective in optimizing multidimensional problems. Hence, an improved niche chaotic genetic algorithm- (INCGA-) based low-energy clustering approach is proposed in this article. To evaluate the algorithm's performance, we first model the low-energy clustering problem as an integer programming that is proved to be NP complete. Also, in order to enhance the search ability of the proposed genetic algorithm on the low-energy clustering problem, we add the niche operator and chaotic operators to speed up the convergence rate. It combines the merits of niche selection algorithm that considers various features and a chaotic generator that enhances the convergence rate. This can maintain the diversity of the algorithm and improve the algorithm's global search ability. INCGA could adjust the parameters automatically as well as appropriate partitions of the populations and avoid local optima. INCGA also uses chaotic generator strategy, which aims to avoid local optima. Simulations are conducted for the LSWSN with low-energy clustering methods based on INCGA, QEA, and PSO. Simulation results show that the proposed INCGA method significantly outperforms the QEA and PSO method.

3. System Model

This section describes a model of low-energy clustering in LSWSN with respect to the constraints of cluster head percentage and communication energy consumption. A typical low-energy clustering model in LSWSN is represented in [23]. The proposed model is of max–min type with nonlinear constraints. Later in [24], the authors showed a similar model by a fewer number of sensors and cluster head nodes. We design an optimizing model minimizing the communication energy consumption for LSWSN.

In LSWSN, the transmission radio energy could be represented as

$$\mathrm{Cost}_{\mathrm{send}}(x, y) = E_{\mathrm{elec}} \cdot x + \varepsilon_{\mathrm{amp}} \cdot x \cdot y^n, \tag{1}$$

where y is the distance between two nodes, $\mathrm{cost}_{\mathrm{send}}(x, y)$ is the communication radio energy, x is the size of transmission bits, n is the path-loss exponent, $\varepsilon_{\mathrm{amp}}$ is the power amplification parameter, and E_{elec} is the electronics energy parameter. The received communication energy of x bits data can be represented as

$$\mathrm{Cost}_{\mathrm{rev}}(x) = E_{\mathrm{elec}} \cdot x, \tag{2}$$

where $\mathrm{cost}_{\mathrm{rev}}(x)$ is the receiver dissipated energy for receiving x bits. In the actual system, E_{elec} for sending and receiving are not exactly the same. This thesis uses a simplified model and we assume that the E_{elec} for sending and receiving are the same. In this paper, we set $k = 1$ Mbit, $\varepsilon_{\mathrm{amp}} = 100\,\mathrm{pJ/bit/m}^2$, $E_{\mathrm{elec}} = 50\,\mathrm{nJ/bit}$, and $n = 3$.

Consider a LSWSN system with M sensors in the monitoring area. The gateway node is located in the center of the field. The distance y represents the distance between two random sensors. Each sensor node sends x bit data to its cluster head node and then to the gateway node. We need to choose the cluster head according to the predetermined ratio in advance. We number each sensor node and indicate the set of cluster head node. When the cluster head node is determined, the energy consumption of the communication is determined according to (1) and (2). So the input of the system is the set of cluster head nodes, and the output is the total communication energy consumption of the system. The total communication energy consumption can be represented as

$$E = \sum_{n=1}^{N} \mathrm{cost}_{\mathrm{send}} + \mathrm{cost}_{\mathrm{rev}}, \tag{3}$$

where N is the total number of sensor nodes in LSWSN and n is the order of the sensor.

4. INCGA for Low-Energy Clustering Problem in Large-Scale Wireless Sensor Networks

The genetic algorithm (GA) is a stochastic optimization algorithm, which is inspired by the biological principles of evolution theory, which include natural selection and mutation. GA was originally proposed by Holland in 1975 as a computational strategy to artificially model biological evolution. GA is usually utilized by lots of function optimization issues and is confirmed to be wonderful in finding suitable and near-optimal answers. Typically, GA has comprehensive explore skills, and the technique is routinely not successful for large-size complex problems. In most cases, GAs are iterative, and typically, they converge to a local optimum.

Currently, chaos theory- and evolutionary theory-inspired strategies are actually being formed. They turned out to be very effective in researching for effective solutions, aiming at an enhanced convergence and resolving complex optimization issues. Here, an improved niche chaotic genetic algorithm (INCGA) is proposed, that is, an alteration of GA, for minimizing the communication energy consumption in large-scale WSNs. In INCGA, according to the conception and strategies of chaos theory and evolutionary theory, novel niche approaches are being designed, which could have a much better stability between searching and get more attractive results in contrast to GAs. In this mixed strategy, the scan abilities of niche selection algorithm and chaotic generator are incorporated to rapidly get convergence. It undergoes parallel searching in complex search places.

In the time of every iteration, INCGA creates a new population using the last population by utilizing different operations, including mutation, crossover, and selection. In addition to that, the presented niche selection operator is utilizing niche theory, and the provided chaos strategy is using the chaotic generator to help improve the global searching ability of INCGA.

4.1. Encoding.
In INCGA, an answer is shown by an individual. We use a vector as a chromosome to represent an answer to the low-energy clustering problem. Every solution might depict a number of arbitrarily chosen cluster head nodes. The low-energy clustering problem is transformed into finding out a cluster head node set in the prospect solution space using the INCGA. In this way, a potential solution will be shown as a string. Therefore, binary encoding is used on every individual. Every single a sensor node within the system corresponds to a gene. The chromosome contains Boolean factors denoting whether the corresponding sensor node is selected as the cluster head node or not. The gene quantity is equal to the quantity of sensor nodes in LSWSN, for instance, assuming a chromosome is C1 = (0, 1, 0, 0, 0, 0, 0, 1). There are 8 sensors in the region and No. 2 and No. 8 nodes are selected as cluster head nodes.

Such encoding is adequate and powerful since it contains the whole searching region. Moreover, the length of the string is the number of design variables. As an example, for LSWSN with 100 sensor nodes, the entire string length is 100. Here the lookup region is the searching area of all solutions, which meet the cluster head proportion constraint. That is, the ith gene matches the ith sensor node. As stated, the number of all combinations is 2^{100}.

4.2. Generation of Initial Population.
INCGA keeps a population of individuals, and the size of the population is called the population size. The INCGA gets started by producing a set of random applicant solutions called the initial population. In every individual, a randomly selected amount bit within the individual is set to random binary numbers with logistic map in (4). According to the proportion and the number of cluster heads, sensors corresponding to largest random number will be selected as the cluster head. For implementing this, the initial set is constructed of randomly designed individuals. Every individual in the set would be assigned

randomly. In this way, the initial population is constructed. This procedure will then be employed over and over again until every individual is ready. The initialization operation is kept random to ensure population diversity.

$$z_{v+1} = 4z_v(1 - z_v). \tag{4}$$

4.3. Selection.
For every single generation, there will be a number of children produced by merging the features of mother or father chromosomes. The algorithm selects an individual inside the latest population with chances relative to its valuation result.

Solutions in the present population are considered dependent on their advantage to survive over the next population. This implies that each solution in a population is connected with a amount of advantage as well as a fitness value. When a couple of parent solutions are to be selected through the current population, foremost, the fitness is assessed by (3).

The selection of parents would be determined by arbitrary and specific strategies. In each generation, mothers and fathers are chosen by function value. High-value chromosomes are often employed to produce children compared to other chromosomes, which likely enhances the population quality. Considering that the purpose is to minimize the communication energy consumption, a binary vector with fairly small fitness must have basically superior valuation, which means the individuals that are better would have a greater probability to be selected. It imitates natural selection process.

The fitness-based selection is put into practice for selecting the individuals. Fitter individuals will get a fairly greater chance to be picked for processing. The idea of fitness-based selection is easy. The chance of every individual is based on the fitness by using the scaling. Making use of the scaling, the chance of picking out the ith chromosome for processing is proportionate to the fitness.

In every stage, mothers and fathers are chosen foremost in accordance with the fitness-based selection. The procedure is going to be employed continuously 1/2*population size times until each and every solution becomes achievable. With this, in the crossover operation, vectors were separated into 1/2*population size couples, with each and every two producing a couple of new individuals. After that, the components could be merged to develop a new individual (child). The previously mentioned selection operation enables reliable constraint, boosting the quality of solutions by selecting the child solutions from high fitness locations.

4.4. Crossover.
While selection decides which reproduce, the crossover operation produces novel patterns to search the pattern area. To produce novel binary chromosomes, as well as layout features, the crossover operation picks chromosomes within the mating pool to generate children. The reproduction procedure permits the genes belonging to the mating pool being forwarded to the next iteration while producing novel structures, which might turn out to be effective. The children as a consequence of crossover obtain characteristics coming from the mating pool and

therefore maintain an enhanced possibility of including an alternative pattern having a improved fitness compared to the past iteration. If the children pattern is bad, alternatively, it is going to be removed over the following selection operation. By doing this, crossover tries to take advantage of the information which was included in the last generation with the development of the parents, as well as discovering unknown parts of the searching area.

Joined together, crossover and selection offer INCGA to be able to effectively identify good regions of the searching area. Generally, there are a number of various types of the crossover operation. Nevertheless, scientific research has revealed that uniform crossover is quite efficient for the detection of novel individual styles by generating various children. Uniform crossover, similar to several crossover strategies, starts with an arbitrary choice of two individuals from the mating pool produced by selection. The couple of the individuals will be mated to generate two offspring, which would consist of a portion of the following iteration. When individuals get mated, they will be removed, and a couple of novel individuals are chosen randomly from the mating pool to create children. The operation is repetitive till a novel children population is full.

The main distinction between different types of the crossover operation is the procedure during which a couple of the individuals generate children. In uniform crossover, individuals are mixed so that every bit spot within the individuals can be a crossover position. Each position within the two offspring's genes will then be packed with a random binary mask to find out which offspring ought to obtain its bit through the left individual and which offspring ought to obtain its bit through the other individual. For instance, every bit spot in the individual of the left offspring has a 0.5 possibility of obtaining the bit from the left individual. In case the left offspring doesn't get its bit from the left individual, this will make it acquire from the right individual, while the right offspring gets from the left individual. Nevertheless, in case a couple of individuals offer the equal bit at the same position, both offspring would receive that matching bit. This procedure enables styles present with both mom and dad to be sustained to the following iteration.

In uniform crossover, chromosomes 1 and 2 are chosen randomly from the mating pool to build a couple of individuals. The individual masks of the offspring are dependent on a random binary number generator: "0" suggests that the left offspring will get its bit from the left individual and the right offspring will get its bit from the right individual, and the opposite way round if the mask is "1."

When both individuals share an identical bit, no switch is necessary, and both offspring get the equal bit. Other types of the crossover operation will not provide so much exploration capability, like uniform crossover, but tend to be not so collapsing to effective binary styles previously in the mating pool. For instance, in a single-point crossover, a single crossover location is chosen arbitrarily since the position for a exchange of entire parts of the individual's genes. In a similar way, with double-point crossover, a pair of locations is chosen for swapping. Both double- and single-point crossover

maintain many patterns of individuals of the last iteration. Because of this, just uniform crossover is utilized in simulation, although the utilization of other crossover strategies needs additional assessment.

4.5. Mutation Operation. The next operation in INCGA is mutation. The mutation operation promotes diversity by marginally transforming the individuals in the new generation. This operation offers novel changes into the genes for a more substantial searching capability. Mutation usually happens at a reasonable possibility, arbitrarily changing a small proportion of the genes in the individuals. For instance, take into account the individual of 1110000. In case the third bit of the individual is mutated, the new individual turns into 1100000.

Mutation could be carried out in a number of strategies, the most basic of which is a bit flipping with a certain possibility. If the choice from the bit flipping would be to mutate, the present bit is going to be transformed to its opposite number of binary. Nevertheless, this technique requires creating an arbitrary binary for each bit in each individual. A reduced complexity strategy is the random mutation operator, which employs a uniform distribution to get the position of the upcoming bit to get mutated. Other sorts of mutation strategies that take advantage of structure of problem may also be applied.

4.6. Recombination. After applying the crossover operator and the mutation operator, there had been 2 * population size chromosomes in the population that were comprised of population size parent strings and population size offspring strings. Distinct from the application in GA, the selection and recombination operations in INCGA are blended thoroughly to develop a more suitable population. Recombination techniques are a type of uniform crossover which makes one offspring for each two strings. Right after recombination, the new offspring fitness will be assessed based on (5).

4.7. Fitness Computation. In the evaluate stage, the fitness values of every solution are measured. Every solution within the population is evaluated according to the fitness of the problem. The solutions with the most desirable fitness are chosen. In low-energy clustering problem, the target is to decrease the communication energy consumption. So the solution fitness will rely on the cluster head set selection. The fitness associated with the cluster head set is then evaluated.

$$\text{Fit}(S) = \sum_{n=1}^{N} \text{cost}_{\text{send}} + \text{cost}_{\text{rev}}, \qquad (5)$$

where N is the number of sensor nodes in LSWSN.

4.8. Niche and Elite Operation. The objective of niche and elitism is to ensure that the optimal solutions from past iterations are forwarded to upcoming iterations. Therefore, the optimal individual within the population will be better from iteration to iteration. Keeping the formerly found best solution makes sure that the INCGA will never be necessary to

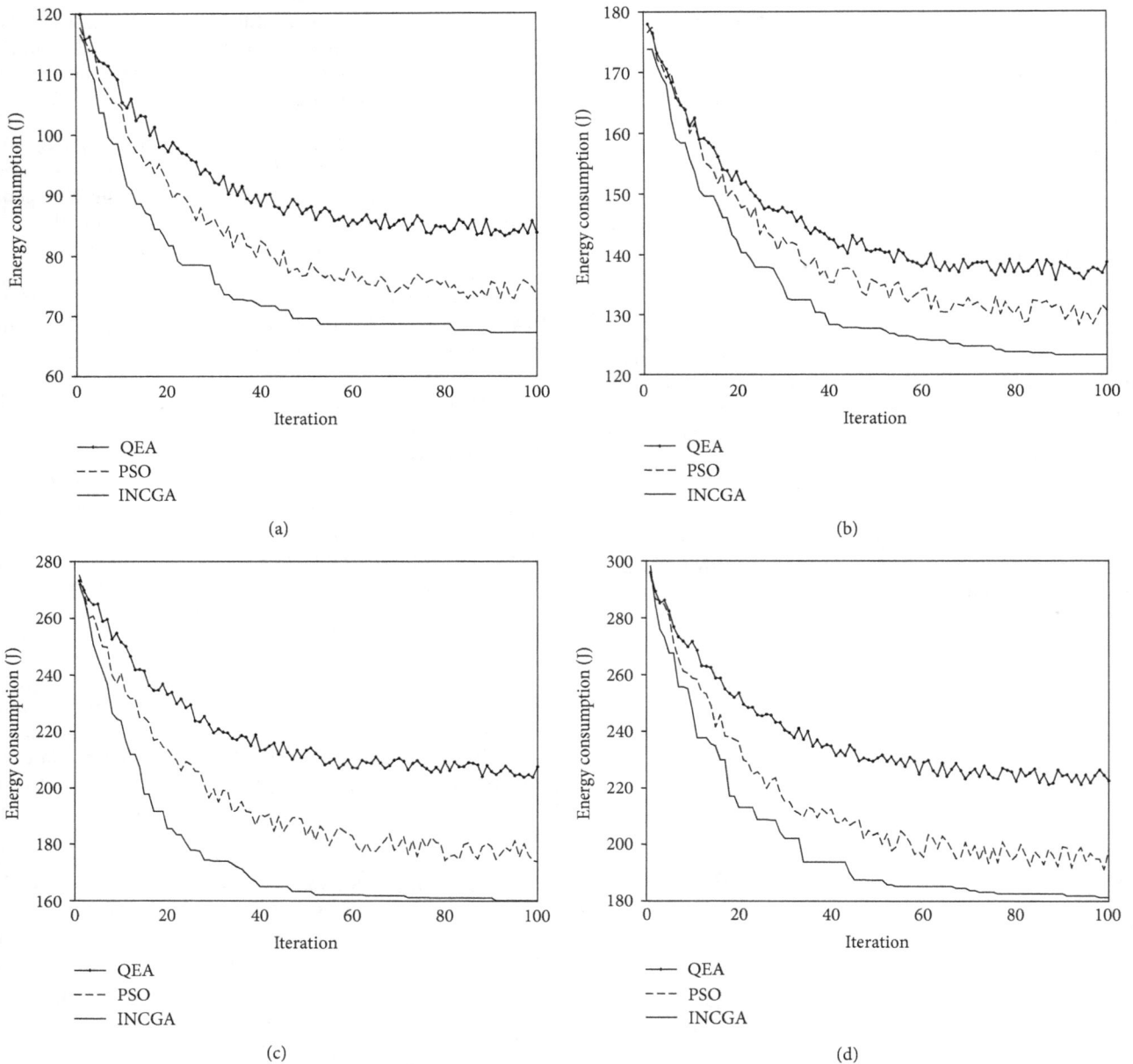

FIGURE 1: Communication energy consumption with 100 (a), 200 (b), 300 (c), and 400 (d) sensors and 10% of cluster heads.

search again for good regions of the space, and in addition, it provides the elite patterns additional mating chances.

Similar to the other operations, various editions of elite maintenance strategy have already been designed, and a variety of the most suitable edition is problem specified. For this application, a particular proportion of the top performing patterns from the past iteration switches the equal proportion of the worst patterns of the present population. The installation of the elite patterns takes place following the fitness assessment of the present population, just prior to selection. This strategy boosts selection stress for the elites, which may increase overall performance but could additionally negatively influence diversity. Nevertheless, uniform crossover may counterbalance the decrease of diversity, which might be caused by this ingredients of elitism.

4.9. Clone Operation. The INCGA gets underway using the creation of an initial population, commonly by distributing arbitrary binary throughout the research area, which means that this fitness is graded in reducing sequence. The best 10% points are then specified for clone. The solutions are then copied. The number of copies is proportional to the fitness. After that, the algorithms carry out the mutation strategy. To save the data, the mutation operation only works on the cloned strings. For implementing this, the effective use of the clonal mutation will permit a neighborhood survey around the original string, while the utilization of the chaotic mutation permits a global lookup around the chromosome. Afterwards, the cloned individuals are examined with the objective valuation, and basically, the top of each clone is allowed to transfer to the next generation, holding precisely

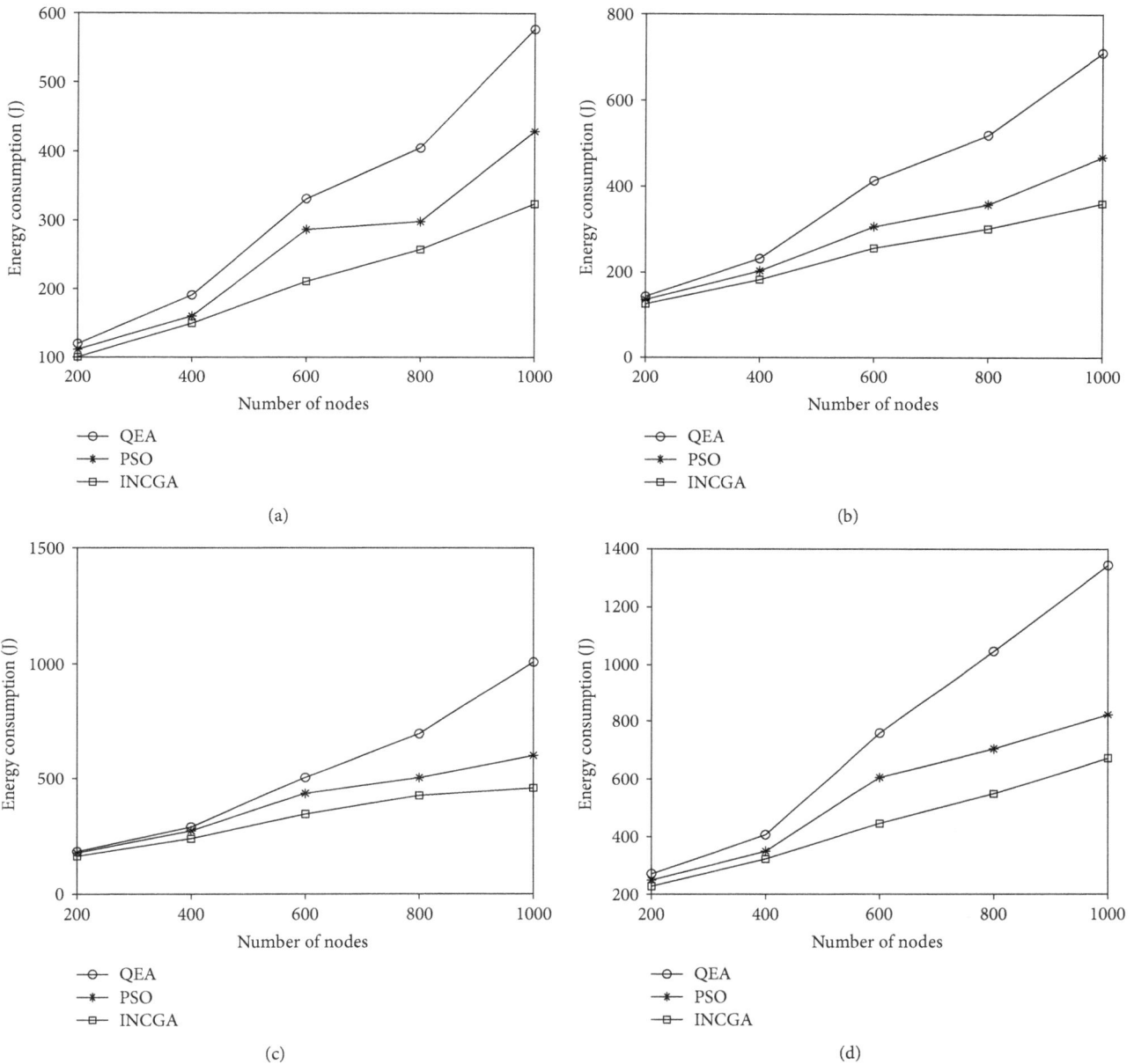

FIGURE 2: Communication energy consumption with different numbers of sensors and 5% (a), 10% (b), 15% (c), and 20% (d) of cluster heads.

the same dimensions of the population. Using this type of replacement, the variety is preserved, and new areas of searching space can be probably researched.

5. Simulation Results and Discussion

In this section, we test the performances of INCGA method with the QEA and PSO method for low-energy clustering in LSWSN. With different numbers of sensors and cluster head nodes, we demonstrate the capabilities of the INCGA via simulations. We implemented the simulations in MATLAB on a personal computer with 2.5 GHz dual-core Intel Core i5 processor and 4 GB RAM. We used a fitness function (4) in order to compare the efficiency and convergence speed of the INCGA with respect to other two

heuristics. The sensors are deployed in a 100×100 rectangle area, and the coordinates of sensors are randomly generated within the area.

In the simulation experiments, we implement INCGA with QEA and PSO. The maximum generation of INCGA, QEA, and PSO is set to 100. In INCGA, the population size is 50 (the same as that in QEA and PSO). In INCGA, the crossover rate is 0.8 and the mutation rate is 0.08. The PSO parameters values are $c1 = c2 = 2$, the dynamic range of the particle has been set to 0.4, and the maximum velocity is 4.

Figure 1 illustrates the convergence curve for INCGA, QEA, and PSO with 100 iterations. It compares the communication energy consumption computed by INCGA, QEA, and PSO on the low-energy clustering problem with 100,

TABLE 1: Communication energy consumption with 10% of cluster head nodes.

Number of nodes	INCGA	PSO	QEA
200	125.53	135.66	143.28
400	182.98	203.08	232.16
600	256.27	306.32	413.48
800	301.54	357.68	518.93
1000	360.21	467.59	710.20

200, 300, and 400 sensors, respectively. The ratio of cluster heads is 10%. Only the global best solution at each iteration is recorded. In order to show the convergence, the result is the value of just 1 run. The comparison between the given methods reveals that the performance of INCGA method is significantly superior to that of niche computing, evolutionary theory, and chaos theory. From Figure 1(a), at the initial iterations, the initial communication energy consumption found by INCGA is lower than that by QEA and PSO. It can be seen that the proposed INCGA converges after around 50 iterations, which is so fast that it can meet real-time requirements of low-energy clustering. QEA and PSO show stagnation. Actually, it is proved that INCGA can jump out of the local optimal solutions. From 50 to 100 iterations, INCGA has approached close to 67 J, while QEA and PSO are still higher than that. Over all 100 iterations, INCGA provides a lower energy consumption than QEA and PSO, which means that it converges with a faster rate. Similar conclusions can be obtained from Figures 1(b)–1(d).

Figure 2 compares the communication energy consumption computed by INCGA, QEA, and PSO on the low-energy clustering problem with the cluster head proportion of 5%, 10%, 15%, and 20%, respectively. For each experiment, the methods have been executed 100 times, and all the results are averaged over 100 Monte Carlo runs. Table 1 shows the best results of INCGA, QEA, and PSO after 100 iterations (corresponding to Figure 2(b)). In all simulations, the performance of INCGA is better than that of the other two heuristic algorithms. For example, as shown in Table 1, the proposed INCGA achieves a communication energy consumption at 360.21 J with 1000 sensor nodes. Meanwhile, the communication energy consumption obtained by the PSO and QEA are 467.59 and 710.20, respectively. Similar conclusions can be obtained from Figures 2(a), 2(c), and 2(d).

According to the simulation results, when the number of sensors in wireless sensor networks is close to or less than 200, the performance of PSO and QEA is close to INCGA. The scale of LSWSN is generally larger than 400 nodes. When the number of sensors is greater than 400, the INCGA has obvious advantages. Therefore, the algorithm should be selected according to the network size. The rules are as follows: when the number of sensors is close to or less than 200, the performance of all algorithms is close, and all three algorithms can be selected. When the number of sensors is greater than 400, INCGA should be selected.

6. Conclusion

In this paper we propose an improved niche chaotic genetic algorithm (INCGA) for low-energy clustering problem applied in LSWSN. To evaluate the algorithm's performance, we first model the low-energy clustering problem as an integer programming that is proved to be NP complete. Simulations are conducted to show the performance of our proposed INCGA scheme against QEA and PSO. Simulation results show that the proposed INCGA scheme outperforms the conventional QEA and PSO schemes with less communication energy consumption.

Conflicts of Interest

The authors declare that they have no conflicts of interest.

References

[1] H. Pei, X. Li, S. Soltani, M. W. Mutka, and X. Ning, "The evolution of MAC protocols in wireless sensor networks: a survey," *IEEE Communications Surveys & Tutorials*, vol. 15, no. 1, pp. 101–120, 2013.

[2] Z. Xue-jian, Z. Yi, and W. Jin, "Local adaptive transmit power assignment strategy for wireless sensor networks," *Journal of Central South University*, vol. 19, pp. 1909–1920, 2012.

[3] J. Lloret, M. Garcia, J. Tomás, and F. Boronat, "GBP-WAHSN: a group-based protocol for large wireless ad hoc and sensor networks," *Journal of Computer Science and Technology*, vol. 23, no. 3, pp. 461–480, 2008.

[4] J. Lloret, M. Garcia, F. Boronat, and J. Tomas, "A group-based protocol for large wireless AD-HOC and sensor networks," in *NOMS Workshops 2008 - IEEE Network Operations and Management Symposium Workshops*, Salvador Da Bahia, Brazil, 2008.

[5] D. Feng, C. Jiang, G. Lim, L. J. Cimini, G. Feng, and G. Y. Li, "A survey of energy-efficient wireless communications," *IEEE Communications Surveys & Tutorials*, vol. 15, no. 1, pp. 167–178, 2013.

[6] D.-z. Dong, X. Liao, K. Liu, Y. Liu, and W. Xu, "Distributed coverage in wireless ad hoc and sensor networks by topological graph approaches," *IEEE Transactions on Computers*, vol. 61, no. 10, pp. 1417–1428, 2012.

[7] N. Sun, Y.-s. Jeong, and S.-h. Lee, "Energy efficient mechanism using flexible medium access control protocol for hybrid wireless sensor networks," *Journal of Central South University*, vol. 20, no. 8, pp. 2165–2174, 2013.

[8] A. Ajith Kumar S, K. Ovsthus, and L. M. Kristensen, "An industrial perspective on wireless sensor networks – a survey of requirements, protocols, and challenges," *IEEE Communications Surveys & Tutorials*, vol. 16, no. 3, pp. 1391–1412, 2014.

[9] I. F. Akyildiz, T. Melodia, and K. R. Chowdury, "Wireless multimedia sensor networks: a survey," *IEEE Wireless Communications*, vol. 14, no. 6, pp. 32–39, 2007.

[10] J. Zhang, F. Ren, S. Gao, H. Yang, and C. Lin, "Dynamic routing for data integrity and delay differentiated services in wireless sensor networks," *IEEE Transactions on Mobile Computing*, vol. 14, no. 2, pp. 328–343, 2015.

[11] V. Akbarzadeh, C. Gagne, M. Parizeau, M. Argany, and M. A. Mostafavi, "Probabilistic sensing model for sensor placement

optimization based on line-of-sight coverage," *IEEE Transactions on Instrumentation and Measurement*, vol. 62, no. 2, pp. 293–303, 2013.

[12] T. Back, U. Hammel, and H. P. Schwefel, "Evolutionary computation: comments on the history and current state," *IEEE Transactions on Evolutionary Computation*, vol. 1, no. 1, pp. 3–17, 1997.

[13] K. Rajeswari and S. Neduncheliyan, "Genetic algorithm based fault tolerant clustering in wireless sensor network," *IET Communications*, vol. 11, no. 12, pp. 1927–1932, 2017.

[14] A. Shokrollahi and B. Mazloom-Nezhad Maybodi, "An energy-efficient clustering algorithm using fuzzy C-means and genetic fuzzy system for wireless sensor network," *Journal of Circuits Systems and Computers*, vol. 26, no. 1, p. 1750004, 2017.

[15] X.-Y. Zhang, J. Zhang, Y.-J. Gong, Z.-H. Zhan, W.-N. Chen, and Y. Li, "Kuhn-Munkres parallel genetic algorithm for the set cover problem and Its application to large-scale wireless sensor networks," *IEEE Transactions on Evolutionary Computation*, vol. 20, no. 5, pp. 695–710, 2016.

[16] M. Elhoseny, X. Yuan, Z. Yu, C. Mao, H. K. El-Minir, and A. M. Riad, "Balancing energy consumption in heterogeneous wireless sensor networks using genetic algorithm," *IEEE Communications Letters*, vol. 19, no. 12, pp. 2194–2197, 2015.

[17] D. He, G. Mujica, J. Portilla, and T. Riesgo, "Modelling and planning reliable wireless sensor networks based on multi-objective optimization genetic algorithm with changeable length," *Journal of Heuristics*, vol. 21, no. 2, pp. 257–300, 2015.

[18] M. F. Abdulhalim and B.'a. A. Attea, "Multi-layer genetic algorithm for maximum disjoint reliable set covers problem in wireless sensor networks," *Wireless Personal Communications*, vol. 80, no. 1, pp. 203–227, 2015.

[19] X. Wang, S. Wang, and J.-J. Ma, "An improved co-evolutionary particle swarm optimization for wireless sensor networks with dynamic deployment," *Sensors*, vol. 7, no. 12, pp. 354–370, 2007.

[20] P. Nayak and A. Devulapalli, "A fuzzy logic-based clustering algorithm for WSN to extend the network lifetime," *IEEE Sensors Journal*, vol. 16, no. 1, pp. 137–144, 2016.

[21] N. Gautam and J.-Y. Pyun, "Distance aware intelligent clustering protocol for wireless sensor networks," *Journal of Communications and Networks*, vol. 12, no. 2, pp. 122–129, 2010.

[22] L.-l. Wang and C. Wang, "A self-organizing wireless sensor networks based on quantum ant Colony evolutionary algorithm," *International Journal of Online Engineering*, vol. 13, no. 7, pp. 69–80, 2017.

[23] B.-C. Cheng, H.-H. Yeh, and P.-H. Hsu, "Schedulability analysis for hard network lifetime wireless sensor networks with high energy first clustering," *IEEE Transactions on Reliability*, vol. 60, no. 3, pp. 675–688, 2011.

[24] L. Kong, M. Zhao, X.-Y. Liu et al., "Surface coverage in sensor networks," *IEEE Transactions on Parallel and Distributed Systems*, vol. 25, no. 1, pp. 234–243, 2014.

16

Distributed Particle Flow Filter for Target Tracking in Wireless Sensor Networks

Junjie Wang ⓘ **, Lingling Zhao** ⓘ **, and Xiaohong Su** ⓘ

School of Computer, Science and Technology, Harbin Institute of Technology, Harbin, China

Correspondence should be addressed to Lingling Zhao; zhaoll@hit.edu.cn

Academic Editor: Hana Vaisocherova - Lisalova

We propose, in this paper, a fully distributed tracking algorithm based on particle flow filter over sensor networks based on the max-consensus. The presented distributed particle flow filter is particularly suitable for the sensor network with limited sensing range and consists of two phases: the estimation phase and consensus phase. The local estimation results are obtained via particle flow filter in the estimation phase; then the sensor nodes agree on the best estimation based on max-consensus protocol in the consensus phase. Numerical simulations and comparisons with other distributed target tracking algorithms are carried out to show the effectiveness and feasibility of our approach.

1. Introduction

Distributed target tracking focuses on using a group of sensors to collect and process information about environment status. Compared with the central target tracking, the distributed target tracking has the following characteristics: scalability, flexibility, robustness, and fault tolerance. Due to these characteristics, the distributed target tracking has played an import role in many applications such as pedestrian tracking [1], biology [2], and environmental monitoring [3].

The distributed target tracking algorithms can be classified into three types: fusion center (FC) based, leader agent (LA) based, and consensus based [4]. In the FC-based approaches, each sensor node uses the local measurement to estimate the local states by filtering algorithms and then transmits the local estimation to a single FC, where a global estimate is calculated based on all the local estimates. In a LA-based filter, only a subset of sensors are activated in a special manner and the information about the target is accumulated along a path formed by selected sensors. While in a consensus-based filter, all sensor nodes are simultaneously active and process the local data to get local posterior; then each one communicates with its neighboring agents [4] to agree with global posterior using consensus algorithms.

The Kalman filter is an optimal target tracking algorithm in the linear Gaussian situation. Recently, the distributed Kalman filter (DKF) [5] for the track-to-track fusion has been proposed as an optimal solution. However, the track-to-track DKF needs to handle the multiple information paths. Another well-known distributed Kalman filter is the Kalman consensus filter (KCF) [6] based on the consensus fusion algorithm which can avoid addressing the multiple information paths. The KCF performs well when all the nodes can get the measurement of a target. In the realistic scenario with limited sensing capability of sensors, some nodes become naive about the target state at some time instants [7]; the performance of the KCF will deteriorate as each node weighs its neighbors' estimates in an equal manner. To overcome this issue, the generalized Kalman consensus filter (GKCF) [7] was proposed utilizing the weighted averaging consensus. A node selection [8, 9] strategy was proposed to select the best estimate and propagate it to all the nodes rather than fusing all the sensor-estimated results. The distributed Kalman filter with node selection [9] selects the most accurate estimate to propagate through the network rather than fusing all the sensor-estimated results. The distributed Kalman filters have the desirable property of computational simplicity in linear systems, but it is still needed to develop distributed target tracking algorithms for nonlinear non-Gaussian systems.

The distributed particle filters (DPFs) [4, 10–12] have been proposed to track a target in a wireless sensor network for the nonlinear non-Gaussian system. In [11], a distributed particle filter computes the product of likelihood function over the network using iterative average consensus. A kind of information-weighted consensus-based distributed particle filter [12] can avoid the divergence of the consensus error introduced by the naive nodes, but it performs in a low convergence rate. In the distributed particle filter proposed in [13, 14], the local posterior probability density function (PDF) is approximated by a Gaussian distribution. Then the local PDF parameters are fused into the global posterior PDF's parameters by average consensus. The work in [15] approximates the local likelihood functions by a Gaussian function and builds the global likelihood through exchange of information with neighboring nodes.

The DPFs based on averaging consensus mentioned above have two defects. The first drawback is that they require a large number of particles or samples for a given level of accuracy. To address this drawback of DPFs, we adapt another nonlinear filter, namely, particle flow filter proposed by Daum and Huang [16–19] to estimate the target state. The particle flow filter can achieve a good performance with fewer particles compared with the particle filter, especially in the high-dimensional state space. The principle behind the particle flow filter is to sample a set of particles from the prior distribution and use a stochastic method to move them such that they are then distributed according to the posterior. In other words, particles are migrated smoothly using a particle flow derived from a log homotopy relating the prior and the posterior [20]. Compared with the particle filter, the particle flow filter can yield a significant reduction of the number of particles especially in the high-dimensional case. Another issue of DPFs based on averaged consensus is that they are not suitable for the scenario which there exist some naive nodes in the wireless sensor network (WSN). In this scenario, it may happen that only a minority of sensors have measurements. Therefore, the WSN involves many unreliable sensors in the fusion step which may cause a divergent error. This is because of the fact that the average consensus algorithm gives all the nodes equal weights; even the naive nodes get less information about the target.

In this paper, inspired by the particle flow filter framework in [17] and the fusion rule in [21], we propose a distributed particle flow filter (DPFF) algorithm for WSN. We approximate local posterior as a Gaussian distribution and fuse the local posterior via a max-consensus protocol. To address the challenge in limited energy and sensor range of sensor nodes, the particle flow filter [17] is utilized to approximate local posterior. Also, the proposed DPFF seeks consensus on the best local posterior, rather than on the average of local posteriori. To the best of our knowledge, particle flow filter has not been yet investigated in WSNs.

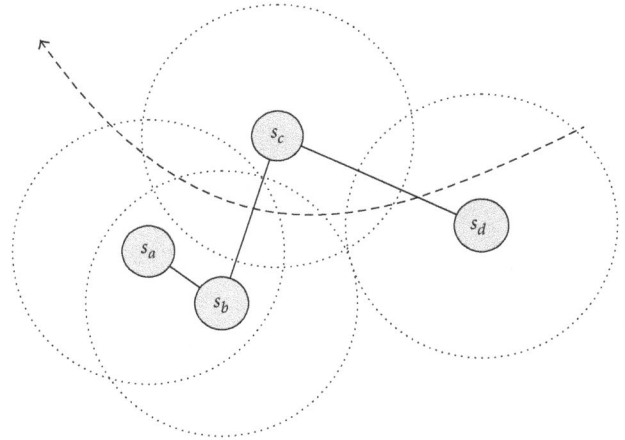

FIGURE 1: Abstract representation of the distributed tracking scenario. Four nodes (s_a, s_b, s_c, and s_d) are randomly placed in the tracking environment. Each node has a local sensing area (dotted circles around the node). The target trajectory is represented by the dashed line. The communication topology is denoted by the solid black line between nodes.

The rest of the paper is organized as follows. Section 2 introduces the background of the sensor network model and consensus theory. The details of DPFF are presented in Section 3. Section 4 evaluates the DPFF performance via two numerical examples. Finally, Section 5 gives the conclusion of this paper and the possible future work.

2. Background

2.1. Network Model. We consider a single-target tracking in the WSN which consists of N_c sensors with restricted monitoring area and communication ranges. Assume that the WSN can be modeled as an undirected connected graph $\mathcal{G} = (\mathcal{C}, \mathcal{E})$, in which each vertex $\mathbf{c} \in \mathcal{C}$ represents a sensor node and each edge $\{i, j\} \in \mathcal{E}$ denotes the link between different nodes. If an edge $\{i, j\} \in \mathcal{E}$ denotes that the node j can get information from the node i and vice versa, the set of neighbors of node i is denoted as $\mathcal{N}_i = \{j \in \mathcal{C} : \{j, i\} \in \mathcal{E}\}$.

Figure 1 shows a network with four nodes. The distributed WSN has no central unit, and, thus, the sensors locally process their measurements. We restrict the monitoring region of each sensor node defined as an area within a dotted circle of radius ρ, in the sense that a sensor may not detect the target over the tracking period. Also, we assume that each sensor can only directly communicate with its neighbors in a certain communication range.

2.2. Target Model. The state of a target is represented by a vector $\mathbf{x}$. For a maneuvering target, the state vector $\mathbf{x}$ contains the information about the position and velocity. The dynamic transition of a target is modeled as

$$\mathbf{x}_k = f(\mathbf{x}_{k-1}) + w_k, \tag{1}$$

where f is the state transition function and w_k is the processing noise.

At time k, each sensor can only obtain measurement when a target appears in its sensing area. The measurement function of sensor i is

$$\mathbf{z}_k^i = h_k^i(\mathbf{x}_k) + v_k^i, \tag{2}$$

where h_k^i denotes the observation function for the ith sensor node and v_k^i stands for the measurement noise. The measurements of nodes are assumed independent over the network.

2.3. Max-Consensus. The max-consensus [21] is a well-known distributed algorithm which makes all the sensor nodes agree on the maximum of the value of their initial state through finite iterations. In the max-consensus algorithm, each sensor node initializes its state value as $u_i(0) = u_i$ and iteratively communicates with its neighbors $\mathcal{N}_i$ based on the update rule as follows to update its state:

$$u_i(t+1) = \max_{j \in \mathcal{N}_i \cup i}\left(u_j(t)\right), \tag{3}$$

where u_i is the state value of node i and t is the iterative step index.

According to [21], we define the max-consensus as follows:

Definition 1 (max-consensus). Consider a WSN with $\mathscr{C}$ nodes, connected over an undirected graph $\mathscr{G} = (\mathscr{C}, \mathscr{E})$. Each node has a state variable $u_i(0)$, $i \in \mathscr{C}$. The discrete time max-consensus protocol is defined as

$$u_i(\mathbf{x}) = u_j(k) = \max\{u_1(0), \dots, u_n(0)\} \quad \forall k \geq K, \ \forall i, j \in \mathscr{C}. \tag{4}$$

If (4) holds for all $u_0 \in \mathbb{R}^n$, strong max-consensus is achieved. If (4) only holds for a subset of initial states, weak max-consensus is achieved [22].

2.4. Average Consensus-Based Distributed Particle Filter in WSNs. The average consensus-based distributed particle filter consists two steps: local particle filter and average consensus filter. The local particle filter uses the local observation to get the local estimation. The output of the local particle filter is the local posteriori approximated as a Gaussian distribution $\mathcal{N}(\cdot|m_k^{i,\text{local}}, \mathbf{P}_k^{i,\text{local}})$. For the sake of using outputs of local filter among networks more effectively, each node maintains an average consensus filter. The aim of the average consensus filter is to fuse these local posteriori between neighbor nodes. According to [13], node i at time k, based on the local measurement $\mathbf{z}_k^i$, runs a local particle filter to obtain the parameters of local posterior $m_k^{i,\text{local}}, \mathbf{P}_k^{i,\text{local}}$ as follows:

$$m_k^{i,\text{local}} = \sum_{j=1}^{M} w_k^{ij} \mathbf{x}_k^{ij},$$

$$\mathbf{P}_k^{i,\text{local}} = \sum_{j=1}^{M} w_k^{ij}\left(\mathbf{x}_k^{ij} - m_k^{i,\text{local}}\right)\left(\mathbf{x}_k^{ij} - m_k^{i,\text{local}}\right)^T, \tag{5}$$

where w_k^{ij} and $\mathbf{x}_k^{ij}$ denote the weights and particles of the local particle filter, respectively. Then the average consensus algorithm is run between linked neighbor nodes as the following equation to obtain the global estimation results:

$$\mathbf{y}_{i,l+1} = \mathbf{y}_{i,l} + \epsilon\left[\sum_{j \in N_i}\left(\mathbf{y}_{j,l} - \mathbf{j}_{i,l}\right) + \left(\mathbf{u}_{i,l} - \mathbf{y}_{i,l}\right)\right], \tag{6}$$

where ϵ is the updating rate. $y_{i,l}$ can be represented as a parameter of local posterior such as $m_k^{i,\text{local}}$ or $\mathbf{P}_k^{i,\text{local}}$.

3. Distributed Particle Flow Filter

The average consensus-based distributed particle filter is robust to time-varying network topologies [4]. In the average-based distributed particle filter, each node runs a local particle filter to estimate the target state. This method inherits the drawbacks of the particle filter, namely, the curse of dimensionality, and particle degeneracy in the highly informative scenario. The other drawback of the particle filter is its requirement to maintain a large number of particles to attain good performance, leading to the waste of the sensors' energy.

Motivated by some desirable properties of the particle flow filter (see for example, [23, 24]) such as sufficient accuracy and low computational complexity, uniqueness of the solutions, we propose a consensus-based distributed particle flow filter (DPFF) algorithm, which consists of two main phases: *estimation* and *consensus*. According to the sensing range of sensors, the estimation phase can be divided into two branches. If a target is measured by the node i at time k, its estimation is carried out via a particle flow filter. On the contrary, the node can not detect the target at time k; only the prediction part of the particle flow filter is run. Then each sensor computes its perception confidence value $\gamma_i(k)$ (refer to (15)) based on the estimated posterior covariance matrix or prior covariance matrix. In the consensus phase, a max-consensus algorithm is utilized to make all the sensors agree on the best estimated sensor.

Note that our algorithm requires the synchronization of clocks over the sensor networks. We now give the details of the two phases of DPFF at time k and assumed that each node obtains the best estimation results $(\bar{\mu}(k-1), \bar{\mathbf{P}}(k-1))$ over the WSN at time $k-1$.

3.1. Estimation Phase. In the DPFF, the posterior distribution is approximated by a particle set $\{\mathbf{x}_{k-1}^i, w_{k-1}^i\}_{i=1}^{N_p}$ which is sampled from Gaussian distribution $\mathcal{N}(\cdot, \bar{\mu}(k-1), \bar{\mathbf{P}}(k-1))$ with same weights $w_{k-1}^i = 1/N_p$. Then all particles are transmitted to $\{\hat{\mathbf{x}}_k^i\}_{i=1}^{N_p}$ at time step k by the dynamic model. Therefore, the prior distribution can be represented by particles $\{\hat{\mathbf{x}}_k^i\}_{i=1}^{N_p}$.

The overall process of the estimation phase for each node is outlined in Figure 2. As a target may move in or out of the sensing area of the node i, if the sensor i has detected the target in its sensing range, then its estimation is carried out by the particle flow filter (Algorithm 1). The

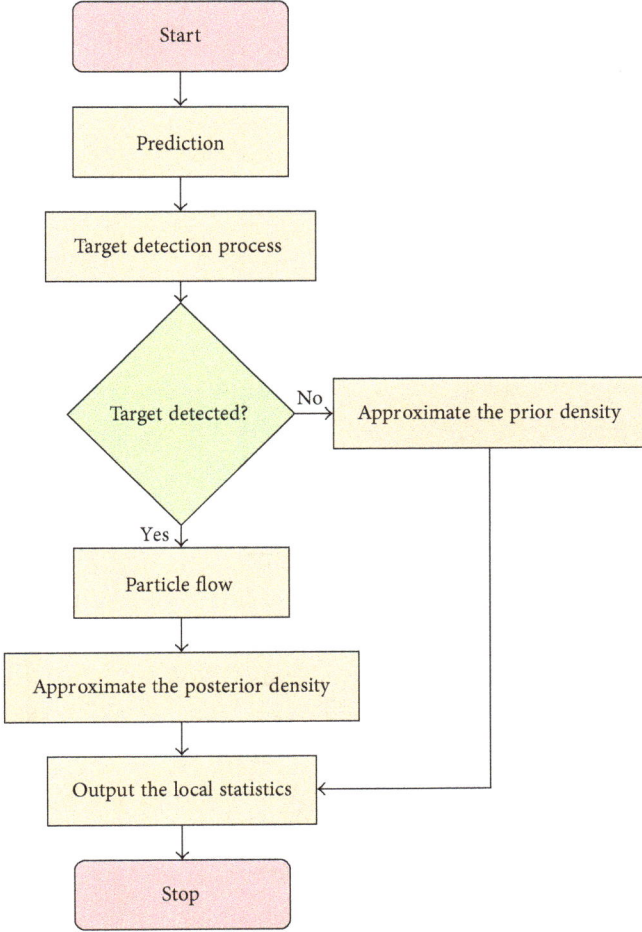

FIGURE 2: The block diagram of the estimation phase for each node.

$$\frac{\partial p(\mathbf{x}, \lambda)}{\partial \lambda} = -\nabla \cdot [f(\mathbf{x}, \lambda), p(\mathbf{x}, \lambda)] + \frac{1}{2} \mathrm{tr} \left[Q \frac{\partial^2 p(\mathbf{x}, \lambda)}{\partial \mathbf{x}^2} \right], \quad (9)$$

where Q is the covariance of the process noise. For simplicity, it is assumed that $Q = 0$. According to (7) and (9), the following equation can be derived:

$$\log h(\mathbf{x}) + [\nabla \log p(\mathbf{x}, \lambda)]^T \cdot f(\mathbf{x}, \lambda) = -\nabla f(\mathbf{x}, \lambda). \quad (10)$$

Assuming that $g(\mathbf{x}), h(\mathbf{x})$ are Gaussian PDFs, then a closed-form solution termed as exact flow filter [25] can be derived as

$$\frac{d\mathbf{x}}{d\lambda} = A(\lambda)\mathbf{x} + b(\lambda), \quad (11)$$

where

$$A = -\frac{1}{2}\mathbf{P}\mathbf{H}^T \left(\lambda \mathbf{H}\mathbf{P}\mathbf{H}^T + R \right)^{-1} \mathbf{H}, \quad (12)$$

$$b = (I + 2\lambda A)\left[(I + \lambda A)\mathbf{P}\mathbf{H}^T R^{-1} z + A\bar{\mathbf{x}} \right]. \quad (13)$$

Let $\bar{\mathbf{x}}$ and $\mathbf{P}$ represent the predicted mean vector and the prior covariance matrix, respectively. $\mathbf{H}$ denotes the measurement function matrix and R is the covariance of the measurement noise. For nonlinear models, the measurement function matrix $\mathbf{H}$ can be obtained by linearization of the measurement model. For more details on the implementation and analysis of the exact flow filter, please refer to [20, 25]. We summarize the exact flow filter in Algorithm 1.

It is important to note that there exist several different realizations for the particle flow filter, such as nonzero diffusion particle flow filter [26], incompressible flow filter [16], and exact Daum and Huang (EDH) filter [18]. In some cases, the Gromov's method was explained in [27–29], which would improve the particle flow filter performance. The performance of the particle flow filter also is influenced by the discretization of pseudotime λ. As the particle flow filter is described by an ordinary differential equation, a suitable discretization is essential to capture the flow dynamics [30]. In this paper, we set a sequence of discrete steps with uniform step size $\Delta\lambda = 0.1$.

In summary, when the sensor node received the measurement, the node can obtain the estimation result $\{\hat{\mu}_i(k), \hat{\mathbf{P}}_i(k)\}$ according to the particle flow filter (Algorithm 1). On the other hand, if the target is not detected by the ith sensor node, the target state is estimated according to the particles $\{\hat{\mathbf{x}}_k^{i,j}\}_{j=1}^{N_p}$ from prediction. Then the estimated state and error covariance matrix can be computed as

$$\hat{\mu}_i(k) = \frac{1}{N_p} \sum_{j=1}^{N_p} \hat{\mathbf{x}}_k^{i,j},$$

$$\hat{\mathbf{P}}_i(k) = \frac{1}{N_p} \sum_{j=1}^{N_p} \left(\hat{\mathbf{x}}_k^{i,j} - \hat{\mu}_i(k) \right) \left(\hat{\mathbf{x}}_k^{i,j} - \hat{\mu}_i(k) \right)^T. \quad (14)$$

In order to measure the quality of the estimated target state of each node, the perception confidence value $\gamma_i(k)$

local output of each node is the local posterior approximated as a Gaussian distribution. On the contrary, in the case that the measurement is not available to the node i, then only the prediction of the particle flow filter will be executed (lines 1–3 of Algorithm 1).

In the following, the particle flow filter is presented in details. To avoid confusion, we will omit particle state indices. Particle flow filter is used to guide particles by the current measurement so that they can more accurately approximate the posterior distribution. Let $h(\mathbf{x}_k) = p(y_k \mid \mathbf{x}_k)$ and $g(\mathbf{x}_k) = p(\mathbf{x}_k \mid y_{1:k-1})$ denote the likelihood and prior functions, respectively. The log homotopy is defined as

$$\log(p(\mathbf{x}, \lambda)) = \log g(\mathbf{x}_k) + \lambda \log(h(\mathbf{x})) - \log(K(\lambda)), \quad (7)$$

where λ is a real number that varies from 0 to 1. $K(\lambda)$ is the normalization constant. When $\lambda = 0$, we obtain the predicted density function $g(\mathbf{x}_k)$, and when $\lambda = 1$, we get the posterior distribution. Suppose that the flow of particles is guided by the Bayes' rule according to

$$d\mathbf{x} = f(\mathbf{x}, \lambda) + dw. \quad (8)$$

Combined by (7), $f(\mathbf{x}, \lambda)$ can be computed by the Fokker-Planck equation

---

Input: $\bar{\mu}_i(k-1), \bar{\mathbf{P}}_i(k-1)$ **Output:** $\hat{\mu}_i(k), \hat{\mathbf{P}}_i(k)$

1: redraw particles $\{\hat{\mathbf{x}}_i^j(k)\}_{j=1}^{N_p}$ from the Gaussian
    distribution which $\bar{\mu}_i(k-1), \bar{\mathbf{P}}_i(k-1)$ are the mean and
    covariance, respectively.
2: **for** $j = 1, \ldots, N_p$ **do**
3:    propagate particles $\mathbf{x}_i^j(k) = f_k(\hat{\mathbf{x}}_i^j(k)) + \nu_k$
4: compute $\mathbf{P}_i^-(k) = \mathrm{cov}(\mathbf{x}(k))$
5: **for** $j = 1, \ldots, N_\lambda$ **do**
6:    Calculate the mean particles value $\bar{\mathbf{x}}_k$
7:    set $\lambda = j\Delta\lambda$
8:    linearize the measurement function at $\bar{\mathbf{x}}_k$ to get
    $\mathbf{H}_{\mathbf{x}_k}$
9:    Calculate A and b from (12) and (13), respectively, using
    $\mathbf{P}_i^-(k), \bar{\mathbf{x}}_k, \mathbf{H}_{\mathbf{x}_k}$;
10:   **for** $j = 1, \ldots, N_p$ **do**
11:      migrate particles $\mathbf{x}_i^j(k) = \mathbf{x}_i^j(k) + \Delta\lambda(A\mathbf{x}_i^j(k) + b)$
12: Compute the mean value and error matrix $\hat{\mu}_i(k), \hat{\mathbf{P}}_i(k)$ of the particles by (14)

ALGORITHM 1: Particle flow filter for node i at time k.

needs to be calculated at the end of the estimation phase. If the target is detected by the sensor i, $\gamma_i(k)$ is calculated based on the posterior error covariance matrix $\hat{\mathbf{P}}_i(k)$; otherwise, $\gamma_i(k)$ is calculated based on the prior error covariance matrix $\bar{\mathbf{P}}_i(k)$. The specific calculation formula of $\gamma_i(k)$ is as follows

$$\gamma_i(k) = \frac{1}{\mathrm{trace}(\hat{\mathbf{P}}_i(k))}, \qquad (15)$$

where $\mathrm{trace}(\cdot)$ is the matrix trace operator. It is clear that $\gamma_i(k)$ grows with the reliability of the estimation performed by sensor node i at time k [22].

At the end of the estimation phase, each sensor node will obtain the value of perception confidence value $\gamma_i(k)$, the local estimate $\hat{\mu}_i(k)$, and the local error covariance matrix $\hat{\mathbf{P}}_i(k)$. These values can be employed to reach consensus in the consensus phase.

3.2. Consensus Phase. The aim of the consensus phase is to select the best estimation over the sensor networks and propagate the selected estimation with correlation information (error covariance matrix). It is noted that, in order to reduce the communication cost, we exchange the error covariance matrix and state estimate rather than the whole particle set. Therefore, at the next time step, each sensor node needs to redraw particles based on the best state estimate and error covariance matrix. The max-consensus for node i at time k is reported in Algorithm 2.

Algorithm 2 works as follows: node i obtains the values from the estimation phase, $\gamma_i(k), \hat{\mu}_i(k), \hat{\mathbf{P}}_i(k)$. And then sensor node i initializes its variables $\Gamma_i(0), \mathbf{U}_i(0), \prod_i(0)$ with $\gamma_i(k), \hat{\mu}_i(k), \hat{\mathbf{P}}_i(k)$, respectively. After initializing all variables, node i exchanges the variables $\Gamma_i(t-1), \mathbf{U}_i(t-1), \prod_i(t-1)$ with its neighbors (lines 5-6 of Algorithm 2). Then node i selects the max perception confidence value which corresponds to the best estimate from its neighbors (lines 7-8 of

---

Input: $\gamma_i(k), \hat{\mu}_i(k), \hat{\mathbf{P}}_i(k)$ **Output:** $\bar{\mu}_i(k), \bar{\mathbf{P}}_i(k)$

1: $\Gamma_i(0) = \gamma_i(k)$
2: $\mathbf{U}_i(0) = \hat{\mu}_i(k)$
3: $\prod_i(0) = \hat{\mathbf{P}}_i(k)$
4: **for** $t = 1, \ldots, D$ **do**
5:    send $\Gamma_i(t-1), \mathbf{U}_i(t-1), \prod_i(t-1)$
6:    receive the information with connected sensors
    sets $\mathcal{N}_i$ to obtain $\Gamma_j(t-1), \mathbf{U}_j(t-1), \prod_j(t-1), \forall j \in$
    $\mathcal{N}_i \cup i$
7:    $\Gamma_i(t) = \max_{j \in \mathcal{N}_i \cup i}\{\Gamma_j(t-1)\}$
8:    $\alpha = \arg\max_{j \in \mathcal{N}_i \cup i}\{\Gamma_j(t-1)\}$
9:    $\mathbf{U}_i(t) = \mathbf{U}_\alpha(t-1)$
10:   $\prod_i(t) = \prod_\alpha(t-1)$
11: $\bar{\mu}_i(k) = \mathbf{U}_i(D)$
12: $\bar{\mathbf{P}}_i(k) = \prod_i(D)$

ALGORITHM 2: Max-consensus for node i at time k.

Algorithm 2). The node i will replace its estimation results to the corresponding estimate and covariance estimate matrices $\mathbf{U}_i(t)$ and $\prod_i(t)$ (lines 9 and 10, resp.). At the end of the phase, each sensor will agree with the best state estimate of the target $\bar{\mu}_i(k)$ with related error covariance matrix $\bar{\mathbf{P}}_i(k)$(lines 12-13 of Algorithm 2). These two variables will be used in the next time $k+1$ of the estimation phase, in order to let the particle flow filter of each node start from the best estimation results and therefore to improve the algorithm's performance.

It can be proved [21] that the node converges during D steps, where D is the diameter of the sensor graph.

3.3. Convergence Analysis. In the DPFF algorithm, each node will converge within finite discrete steps through the max-consensus algorithm. The convergence of the DPFF is proved as follows:

Assuming that each node runs the DPFF algorithm, after consensus phases at time k, node i and node j hold the equation

$$\bar{\mu}_i(k) = \bar{\mu}_j(k) = \bar{\mu}_{max}(k), \quad \forall i, j \in \mathscr{C}, \tag{16}$$

where $\mathscr{C}$ is the set of nodes and $\bar{\mu}_{max}(k)$ is estimated mean value corresponding to the max perception confidence value $\gamma(k)$.

In each iteration of Algorithm 2, line 7 is to select the max $\gamma(k)$ from the neighbor nodes which is the update rule for the max-consensus algorithm. The max-consensus algorithm is guaranteed to converge in a finite number of iterations. Therefore, in D steps, Algorithm 2 will be guaranteed that

$$\Gamma_i(n) = \Gamma_j(n) = \max_{l \in \mathscr{J}} \gamma_l(k), \quad \forall i, j \in \mathscr{C}. \tag{17}$$

In each iteration, the variables $\mathbf{U}_i$ and $\prod_i$ store the corresponding mean value and error covariance matrix, respectively. Thus, at the end of iterations, (16) is workable. The convergence issues are discussed in [9] in detail.

Remark 1. In this paper, we approximate the posterior distribution with a Gaussian distribution. In the non-Gaussian system, the posterior distribution can be represented by the mixture of multiple Gaussian distributions. So, the DPFF can not be applied directly. A new particle flow filter called Gaussian mixture particle flow [31] can be employed to cater for the non-Gaussian situation. When the posterior distribution is represented by the Gaussian mixture model (GMM) $\sum_{j=1}^{N} (\mathscr{N} \cdot |m_k^j, \mathbf{P}_k^j)$, the corresponding $\gamma(k)$ is calculated by the following:

$$\gamma(k) = \frac{1}{\sum_{j=1}^{N} \text{trace}\left(\mathbf{P}_k^j\right)}. \tag{18}$$

4. Experiments

In this section, we evaluate the performance of the proposed DPFF algorithm in the simulated environment and compare it with other approaches: the centralized particle filter (CPF) where we use the performance of the CPF as the base performance, the distributed particle filter based on average consensus [15] (DPF-AV), and the information weight average consensus-based distributed particle filter (DPF-WAV) [12].

4.1. Example 1: Grid Network. Consider such a simulation that a target moves in a $300\,\text{m} \times 500\,\text{m}$ area with 15 sensors. The sensors have overlapped monitored regions. The monitored region of each sensor node is assumed to be a circle region of $75\,\text{m}$ radius whose center is at the sensor's location. A sensor has a measurement of the target only if the target appears in the sensor's sensing area. Figure 3 illustrates the sensors and network connectivity. Connections between the sensors are shown as grey dashed lines. Each sensor can only communicate directly with its neighboring sensors whose distance to it is less than the communication range. The state vector of the target is represented as $\mathbf{x}_k = [\mathbf{x}_k, y_k, v_{x,k}, v_{y,k}, w_k]$,

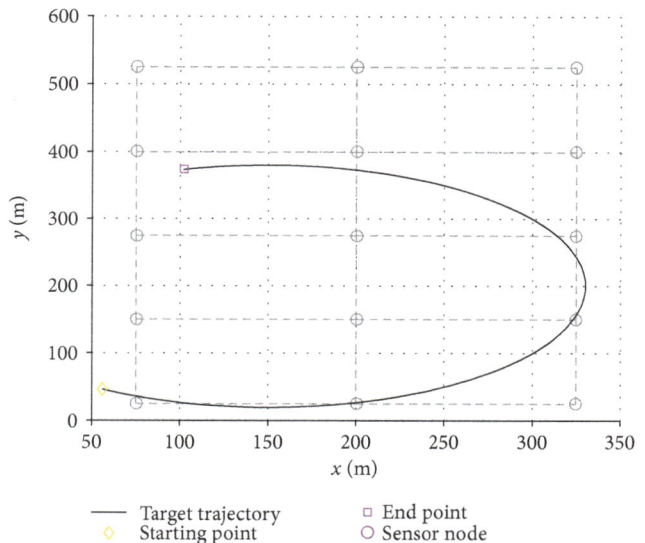

FIGURE 3: Object trajectory, sensor node, and connectivity considered in the experiments. The connectivity among the nodes is shown using the grey dashed lines between nodes.

where $\mathbf{x}_k, y_k$ and $v_{x,k}, v_{y,k}$ represent the target position and velocity, respectively. w_k denotes the turn rate of the target. In this case, a nearly coordinated turn model with the known constant turn rate w_k and the unknown velocity v_k is considered. This model is able to account for the motion of complicated maneuverable targets. The nonlinear scenario is used in [32], and the motion of target is modeled according to

$$\begin{aligned} \mathbf{x}_{k+1} &= \mathbf{F}(\omega_k)\mathbf{x}_k + \mathbf{G}w_k, \\ \omega_{k+1} &= \omega_k + \Delta u_k, \end{aligned} \tag{19}$$

where

$$\mathbf{F}(\omega) = \begin{bmatrix} 1 & \dfrac{\sin \omega\Delta}{\omega} & 0 & -\dfrac{1 - \cos \omega\Delta}{\omega} \\ 0 & \cos \omega\Delta & 0 & -\sin \omega\Delta \\ 0 & \dfrac{1 - \cos \omega\Delta}{\omega} & 1 & \dfrac{\sin \omega\Delta}{\omega} \\ 0 & \sin \omega\Delta & 0 & \cos \omega\Delta \end{bmatrix},$$

$$\mathbf{G}(\omega) = \begin{bmatrix} \dfrac{\Delta^2}{2} & 0 \\ \Delta & 0 \\ 0 & \dfrac{\Delta^2}{2} \\ 0 & \Delta \end{bmatrix}. \tag{20}$$

$w_k \sim \mathscr{N}(\cdot; 0, \sigma_w^2 I)$, $u_k \sim \mathscr{N}(\cdot; 0, \sigma_u^2 I)$, $\Delta = 1$ s, $\sigma_w = 15$ m/s^2, and $\sigma_u = \pi/180$ rad/s. The observation is a noisy bearing and range vector given by

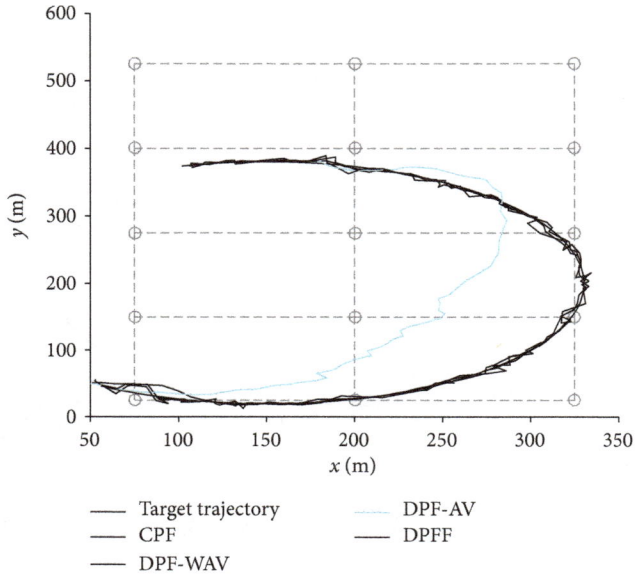

- —— Target trajectory   ——— DPF-AV
- —— CPF   —— DPFF
- —— DPF-WAV

FIGURE 4: One example of estimated trajectory using DPFF (black curve), CPF (green curve), DPF-AV (cyan curve), and DPF-WAV (blue curve). The red curve indicates the ground truth.

$$\mathbf{z}_k = \begin{bmatrix} \arctan\left(\dfrac{p_{x,k}}{p_{y,k}}\right) \\ \sqrt{p_{x,k}^2 + p_{y,k}^2} \end{bmatrix} + \varepsilon_k, \qquad (21)$$

where $\varepsilon_k \sim \mathcal{N}(\cdot; 0, R_k)$, with $R_k = \text{diag}([\sigma_\theta^2, \sigma_r^2]^T)$ and $\sigma_\theta = 2\pi/180, \sigma_r = 10$. p_x, p_y are the distance from the target to the sensor.

The number of particles for each node is to set 500 in the DPF-WAV and DPF-AV algorithms, while the DPFF algorithm only uses 50 particles in each node, while the CPF algorithm collects the entire available measurements from all sensor nodes and uses 500 particles to estimate.

The error metric that we have computed is the root mean square error (RMSE) between the true and estimated target positions at each time instant k from all sensors. Let (p_k^x, p_k^y) and $(\hat{p}_k^x, \hat{p}_k^y)$ denote the true and estimated target positions, respectively, at time k. The RMSE value at time k is calculated over a number of Monte Carlo (N_{MC}) runs according to

$$\text{RMSE} = \sqrt{\frac{1}{N_{MC}} \sum_{i=1}^{N_{MC}} (p_k^x - \hat{p}_k^x)^2 + (p_k^y - \hat{p}_k^y)^2}, \qquad (22)$$

where $N_{MC} = 500$ is the number of Monte Carlo runs.

A sample of the estimated target track of DPFF is shown in Figure 4, which also shows the estimated track from the CPF, DPF-WAV, and DPF-AV algorithms. It is obvious that the DPF-AV lost the track while other algorithms work well when tracking the target.

Figure 5 shows the temporal evolution of RMSE. It is observed that the DPF-AV algorithm performs worst in these methods, while results of CPF, DPFF, and DPF-WAV are

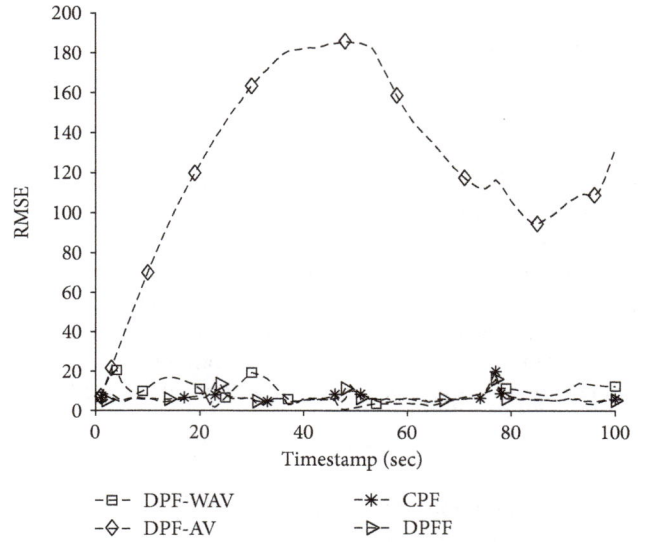

- -□- DPF-WAV   -*- CPF
- -◇- DPF-AV   -▷- DPFF

FIGURE 5: Root mean square error (RMSE) versus time.

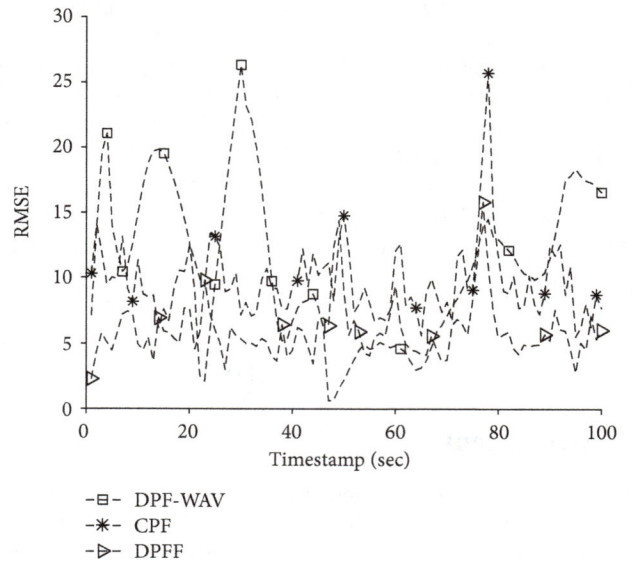

- -□- DPF-WAV
- -*- CPF
- -▷- DPFF

FIGURE 6: Root mean square error (RMSE) versus time.

fairly close to each other. The performance of DPFF and DPF-WAV is almost as good as the CPF and better than DPF-AV. Particularly, there is serious divergence with DPF-AV. For clear comparison of the proposed algorithm with DPF-WAV, we plot only DPF-WAV, CPF, and DPFF in Figure 6. Figure 6 illustrates the relationship of RMSE with timestamp of the DPF-WAV, CPF, and DPFF algorithms. As the DPF-WAV has a sharp fluctuation, it is obvious that the DPFF is better than the DPF-WAV and closer to the CPF. Especially, the DPF-WAV needs more particles than the DPFF and usually requires a significant amount of computing resources.

Moreover, we also computed the averaged RMSE (ARMSE) for all methods. The ARMSE is given by

TABLE 1: Average results over 500 runs with different methods.

Number	Methods	Average RMSE	Std RMSE	Average runtime (sec)
500	CPF	6.33	4.43	0.3919
	DPF-WAV	9.16	4.35	0.2851
	DPF-AV	126.62	27.51	0.2705
1000	CPF	5.98	4.03	0.7815
	DPF-WAV	8.97	4.12	0.3334
	DPF-AV	124.12	21.48	0.3102
50	DPFF	6.32	4.76	0.0672

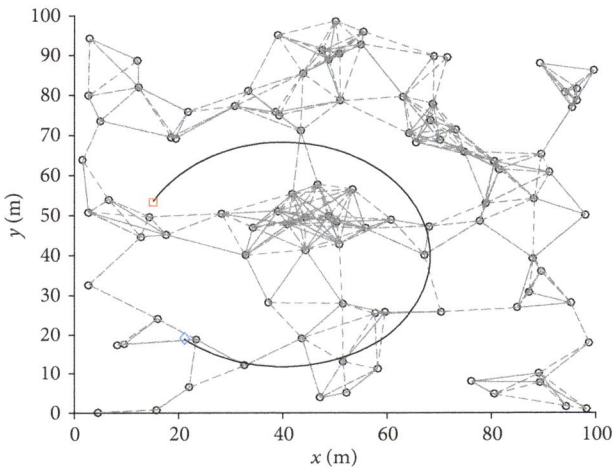

FIGURE 7: Large-scale sparse WSN.

$$\text{ARMSE} = \sqrt{\frac{1}{T N_{\text{MC}}} \sum_{i=1}^{N_{\text{MC}}} \sum_{j=1}^{T} \left(p_k^x - \widehat{p}_k^x\right)^2 + \left(p_k^y - \widehat{p}_k^y\right)^2}. \quad (23)$$

Table 1 lists the ARMSE, standard deviation of RMSE values, and the average execution time per node over 500 simulation trials. The first column in Table 1 is the number of particles for different algorithms. It can be seen that the proposed DPFF only with 50 particles gets better estimate results than the DPF-WAV and DPF-AV, almost equal to the CPF with one-tenth the number of particles. For the average execution time per node, the DPFF is less than a quarter of the DPF-WAV and DPF-AV. In summary, the DPFF has higher accuracy and less execution time.

4.2. Example 2: Random Network. In the second scenario, we consider a large-scale sparse WSN as shown in Figure 7. That is a good benchmark since most of the WSN is spread over a large scale in the real world. Figure 7 shows the large scale sparse WSN in our simulation. There exists 100 sensor nodes and positions of sensors are randomly placed. The communication and sensing ranges of each sensor node are 20 m and 10 m, respectively. In Figure 7, the red line denotes true trajectory of the target. The state dynamic function and measurement function are similar to (19) and (21) whereas

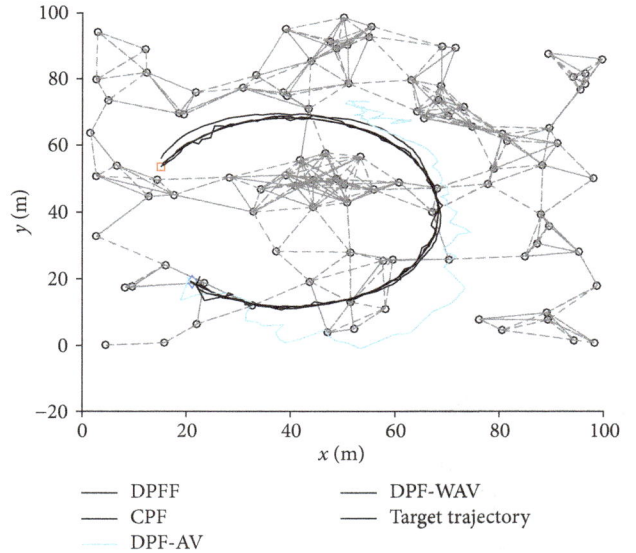

FIGURE 8: One example of estimated trajectory using DPFF (black curve), CPF (green curve), DPF-AV (cyan curve), and DPF-WAV (blue curve). The red curve indicates the ground truth in the large-scale sparse WSN.

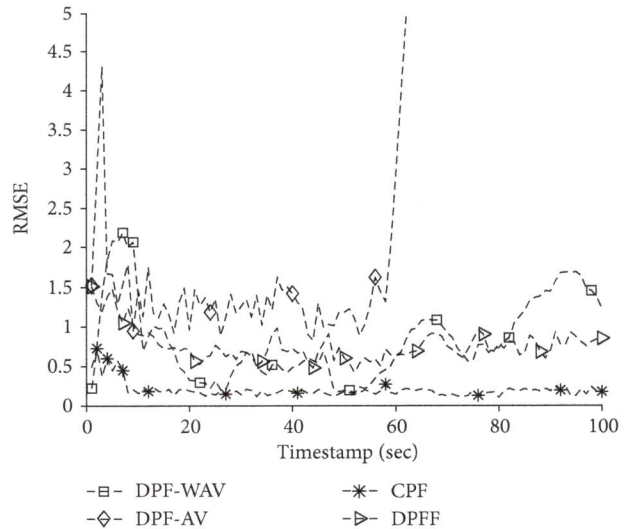

FIGURE 9: RMSE versus time.

the parameter of process noise $\sigma_w = 1$ m/s^2 and the parameters of measurement noise $\sigma_\theta = pi/180$, $\sigma_r = 2$. Then we compared the CPF, DPF-WAV, and DPF-AV with the DPFF as well. In this simulation, the DPFF maintains 50 particles per node for each time step while CPF, DPF-WAV, and DPF-AV use 1000 particles.

Figure 8 illustrates the estimated tracks of all algorithms. It can be seen that the estimated trajectory of DPF-AV (cyan curve) is far away from the target trajectory (red curve). Compared with the DPF-WAV (blue curve), the estimated result (black curve) of the DPFF is more close to the target trajectory. The RMSE values of four algorithms versus time

TABLE 2: Average results over 500 runs with different methods for random network.

Number	Methods	Average RMSE	Std RMSE	Average runtime (sec)
	CPF	0.19	0.13	5.7027
1000	DPF-WAV	1.29	1.73	0.9928
	DPF-AV	14.84	9.29	2.6869
50	DPFF	0.97	0.43	0.1295

are shown in Figure 9. A remarkable fact is that the DPF-AV clearly diverge after 60 time steps. We note from Figure 9 that the RMSE of DPF-WAV is fluctuating with time depending on the number of nodes which detected the target. In addition, to provide an overall indication of the comparative performance of the different methods, Table 2 gives the average and standard deviation RMSE and the average runtime over four methods. In Table 2, it can be seen that the average runtime per node of DPFF is 0.1295 seconds, which accounts for approximately one-tenth of the DPF-WAV (0.9928 s). The most time-consuming algorithm is the CPF, which runs a particle filter in one central node based on all the measurements. In the end, the proposed algorithm significantly outperforms the alternative algorithms in this experiment.

As a whole, the DPFF algorithm achieves a steady tracking with high accuracy with a few particles.

5. Conclusions

We presented a distributed particle flow filter algorithm for wireless sensor networks. At each sensor, a local particle flow filter computes a local state estimate that only depends on the local measurement. Then a perception confidence value is calculated from the particle flow filter. A max-consensus is used to make all the nodes agree on the best estimate of the target position. In the proposed distributed particle flow filter, each node just communicates with its neighboring sensor nodes and does not require any routing protocols. We applied the proposed distributed particle flow filter in two target tracking scenarios and demonstrated experimentally that its performance is better than the distributed averaged-based particle filter; moreover, it needs less computation time and samples. An extension of the distributed particle flow filter to multiple target tracking in the WSN remains a potential topic for the future.

Conflicts of Interest

The authors declare that they have no conflicts of interest.

Acknowledgments

This work is supported by the National Natural Science Foundation of China (NSFC; Grant no. 61305013).

References

[1] H. Wang, H. Lenz, A. Szabo, J. Bamberger, and U. D. Hanebeck, "Wlan-based pedestrian tracking using particle filters and low-cost MEMS sensors," in *2007 4th Workshop on Positioning, Navigation and Communication*, pp. 1–7, Hannover, Germany, 2007, IEEE.

[2] A. Mainwaring, D. Culler, J. Polastre, R. Szewczyk, and J. Anderson, "Wireless sensor networks for habitat monitoring," in *Proceedings of the 1st ACM International Workshop on Wireless Sensor Networks and Applications - WSNA '02 ACM*, pp. 88–97, Atlanta, GA, USA, 2002, ACM.

[3] S. Santini, B. Ostermaier, and A. Vitaletti, "First experiences using wireless sensor networks for noise pollution monitoring," in *Proceedings of the Workshop on Real-World Wireless Sensor Networks - REALWSN '08*, pp. 61–65, Glasgow, Scotland, 2008, ACM.

[4] O. Hlinka, F. Hlawatsch, and P. M. Djuric, "Distributed particle filtering in agent networks: a survey, classification, and comparison," *IEEE Signal Processing Magazine*, vol. 30, no. 1, pp. 61–81, 2013.

[5] F. Govaers and W. Koch, "An exact solution to track-to-track-fusion at arbitrary communication rates," *IEEE Transactions on Aerospace and Electronic Systems*, vol. 48, no. 3, pp. 2718–2729, 2012.

[6] R. Olfati-Saber, "Kalman-consensus filter : optimality, stability, and performance," in *Proceedings of the 48h IEEE Conference on Decision and Control (CDC) held jointly with 2009 28th Chinese Control Conference*, pp. 7036–7042, Shanghai, China, 2009, IEEE.

[7] A. T. Kamal, C. Ding, B. Song, J. A. Farrell, and A. Roy-Chowdhury, "A generalized Kalman consensus filter for wide-area video networks," in *IEEE Conference on Decision and Control and European Control Conference*, pp. 7863–7869, Orlando, FL, USA, 2011, IEEE.

[8] W. Yang and H. Shi, "Sensor selection schemes for consensus based distributed estimation over energy constrained wireless sensor networks," *Neurocomputing*, vol. 87, pp. 132–137, 2012.

[9] D. Di Paola, A. Petitti, and A. Rizzo, "Distributed Kalman filtering via node selection in heterogeneous sensor networks," *International Journal of Systems Science*, vol. 46, no. 14, pp. 2572–2583, 2015.

[10] J. Read, K. Achutegui, and J. Míguez, "A distributed particle filter for nonlinear tracking in wireless sensor networks," *Signal Processing*, vol. 98, pp. 121–134, 2014.

[11] O. Hlinka, O. Sluciak, F. Hlawatsch, P. M. Djuric, and M. Rupp, "Likelihood consensus and its application to distributed particle filtering," *IEEE Transactions on Signal Processing*, vol. 60, no. 8, pp. 4334–4349, 2012.

[12] W. Tang, G. Zhang, J. Zeng, and Y. Yue, "Information weighted consensus-based distributed particle filter for large-scale sparse wireless sensor networks," *IET Communications*, vol. 8, no. 17, pp. 3113–3121, 2014.

[13] D. Gu, J. Sun, Z. Hu, and H. Li, "Consensus based distributed particle filter in sensor networks," in *2008 International Conference on Information and Automation*, pp. 302–307, Changsha, China, 2008, IEEE.

[14] A. Mohammadi and A. Asif, "Consensus-based distributed unscented particle filter," in *2011 IEEE Statistical Signal Processing Workshop (SSP)*, pp. 237–240, Nice, France, 2011, IEEE.

[15] T. Ghirmai, "Distributed particle filter for target tracking: with reduced sensor communications," *Sensors*, vol. 16, no. 9, p. 1454, 2016.

[16] F. Daum and J. Huang, "Nonlinear filters with log-homotopy," in *Optical Engineering+ Applications*, p. 669918, San Diego, CA, USA, 2007, International Society for Optics and Photonics.

[17] F. Daum and J. Huang, "Particle flow for nonlinear filters with log-homotopy," in *SPIE Defense and Security Symposium*, p. 696918, Orlando, FL, USA, 2008, International Society for Optics and Photonics.

[18] F. Daum and J. Huang, "Nonlinear filters with particle flow induced by log-homotopy," in *SPIE Defense, Security, and Sensing*, p. 733603, Orlando, FL, USA, 2009, International Society for Optics and Photonics.

[19] F. Daum and J. Huang, "Generalized particle flow for nonlinear filters," in *SPIE Defense, Security, and Sensing*, p. 76980I, Orlando, FL, USA, 2010, International Society for Optics and Photonics.

[20] T. Ding and M. Coates, "Implementation of the Daum-Huang exact-flow particle filter," in *2012 IEEE Statistical Signal Processing Workshop (SSP)*, pp. 257–260, Ann Arbor, MI, USA, 2012.

[21] B. M. Nejad, S. A. Attia, and J. Raisch, "Max-consensus in a max-plus algebraic setting: the case of fixed communication topologies," in *2009 XXII International Symposium on Information, Communication and Automation Technologies*, pp. 1–7, Bosnia, Serbia, 2009, IEEE.

[22] A. Petitti, D. Di Paola, A. Rizzo, and G. Cicirelli, "Consensus-based distributed estimation for target tracking in heterogeneous sensor networks," in *IEEE Conference on Decision and Control and European Control Conference*, pp. 6648–6653, Orlando, FL, USA, 2011, IEEE.

[23] Y. Li and M. Coates, "Particle filtering with invertible particle flow," 2016, http://arxiv.org/abs/1607.08799.

[24] Y. Li, L. Zhao, and M. Coates, "Particle flow for particle filtering," in *2016 IEEE International Conference on Acoustics, Speech and Signal Processing (ICASSP)*, pp. 3979–3983, Shanghai, China, 2016, IEEE.

[25] F. Daum, J. Huang, and A. Noushin, "Exact particle flow for nonlinear filters," in *SPIE Defense, Security, and Sensing*, p. 769704, Orlando, FL, USA, 2010, International Society for Optics and Photonics.

[26] F. Daum and J. Huang, "Particle flow with non-zero diffusion for nonlinear filters," in *Proceedings of SPIE: Signal Processing, Sensor Fusion and Target Recognition XXII*, p. 87450P, Baltimore, MD, USA, 2013.

[27] F. Daum, J. Huang, and A. Noushin, "Gromov's method for Bayesian stochastic particle flow: a simple exact formula for Q," in *2016 IEEE International Conference on Multisensor Fusion and Integration for Intelligent Systems (MFI)*, pp. 540–545, Baden-Baden, Germany, 2016, IEEE.

[28] F. Daum, A. Noushin, and J. Huang, "Numerical experiments for Gromov's stochastic particle flow filters," in *SPIE Defense+ Security*, p. 102000J, Anaheim, CA, USA, 2017, International Society for Optics and Photonics.

[29] F. Daum, J. Huang, and A. Noushin, "Generalized Gromov method for stochastic particle flow filters," in *SPIE Defense+ Security*, p. 102000I, Anaheim, CA, USA, 2017, International Society for Optics and Photonics.

[30] M. A. Khan and M. Ulmke, "Improvements in the implementation of log-homotopy based particle flow filters," in *2015 18th International Conference on Information Fusion (Fusion)*, pp. 74–81, Washington, DC, USA, July 2015.

[31] M. A. Khan, M. Ulmke, and W. Koch, "A log homotopy based particle flow solution for mixture of Gaussian prior densities," in *2016 IEEE International Conference on Multisensor Fusion and Integration for Intelligent Systems (MFI)*, pp. 546–551, Baden-Baden, Germany, Sept 2016.

[32] B.-T. Vo, B.-N. Vo, and A. Cantoni, "The cardinality balanced multi-target multi-Bernoulli filter and its implementations," *IEEE Transactions on Signal Processing*, vol. 57, no. 2, pp. 409–423, 2009.

17

Intrusion Detection System based on Evolving Rules for Wireless Sensor Networks

Nannan Lu,[1] **Yanjing Sun,**[1] **Hui Liu,**[2] **and Song Li**[1]

[1]*School of Information and Electrical Engineering, China University of Mining and Technology, Xuzhou, Jiangsu, China*
[2]*Institute of Information Photonics and Optical Communications, Beijing University of Posts and Telecommunications, Beijing, China*

Correspondence should be addressed to Nannan Lu; lnn_921@126.com

Academic Editor: Mucheol Kim

Human care services, as one of the classical Internet of things applications, enable various kinds of things to connect with each other through wireless sensor networks (WSNs). Owing to the lack of physical defense devices, data exchanged through WSNs such as personal information is exposed to malicious attacks. Therefore, intrusion detection is urgently needed to actively defend against such attacks. Intrusion detection as a data mining procedure cannot control the size of rule sets and distinguish the similarity between normal and intrusion network behaviors. Therefore, in this paper, an evolving mechanism is introduced to extract the rules for intrusion detection. To extract diversified rules as well as control the quantity of rulesets, the extracted rules are examined according to the distance between the rules in the rule set of the same class and the rules in the rule set of different classes. Thereby, it alleviates the problem that the quantity of rules expands unexpectedly with the evolving genetic network programming. The simulations are conducted on a benchmark intrusion dataset, and the results show that the proposed method provides an effective solution to evolve the class association rules and improves the intrusion detection performance.

1. Introduction

The Internet of things (IoT) enables a large number of physical things or objects to connect, communicate, and exchange data with each other. IoT techniques span from health care to tactical military, in which human care is a type of classical application. The objects of human care services could include various kinds of medical equipment, even body parts. Wireless sensor networks (WSNs) are crucial for connecting, communicating, and exchanging data among such a large number of things. Although WSNs have the advantages of low installation cost, unattended network operation, and flexible deployment, their deficiency in physical defense devices renders both network and information vulnerable for malicious attacks [1]. To protect human care services from the internal or external attacks, prevention and detection are two main components involved in WSN security. However, as a passive network security mechanism, prevention is aimed at preventing any attack before it occurs and is therefore not sufficient. Thus, an active technique is urgently required to perceive malicious intrusions. Naturally, the focus shifts on the intrusion detection that can detect the actions attempting to compromise the confidentiality, integrity, or availability of one resource.

In general, the intrusion detection has two main techniques: misuse detection and anomaly detection. Misuse detection essentially identifies the previously known attacks from the normal network behaviors, while anomaly detection establishes the normal profiles to detect the new attacks. The combination of these two intrusion detection techniques is the hybrid intrusion detection. All the three techniques have been widely used in IoT. For example, Faisal et al. implemented anomaly detection to detect the external and internal attacks on smart meters [2]. Wang et al. and Pan et al. utilized the hybrid intrusion detection framework to protect the heterogeneous WSN, which was applied to power systems [3, 4]. Whereas, the specific methods of intrusion detection must be reviewed from the classical applications in the wired networks. Early studies on intrusion detection were conducted

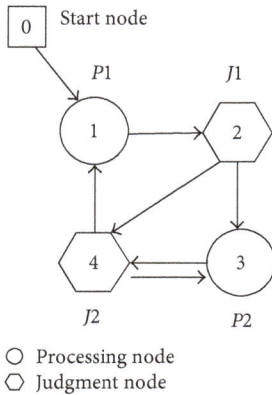

FIGURE 1: Phenotype of GNP.

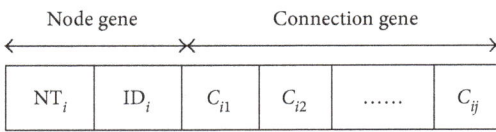

FIGURE 2: Genotype expression of GNP.

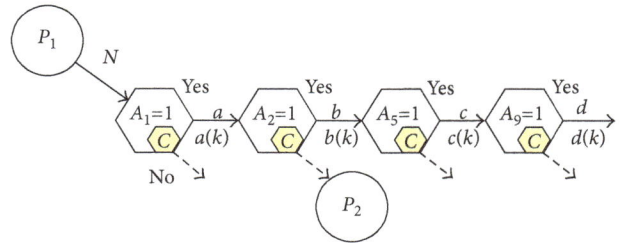

FIGURE 3: Class association rule mining based on GNP.

by Denning and Anderson [5, 6]. They aimed to build the monitoring systems for computer security, so that utilized statistics and rules to recognize attacks or viruses from the audited data. Since then, machine learning, data mining, statistic modeling, and pattern matching have been used to construct intrusion detection systems [7]. SVM was used to classify and select the audited data for the intrusion detection [8, 9]. The k-nearest neighbors (KNN) method has also been used to identify the intrusions through measuring distance [10]. Neural networks were also used to realize the intrusion detection systems, such as multilayer perceptron (MLP) [11]. Moreover, MLP was considered the basic unit to form the ensemble classifiers [12] such as AdaBoost. In [13], decision stumps were utilized as the weak classifiers to form a strong classifier.

Data mining is a successful solution to actively detect intrusive attacks based on the rules hidden in the network behavior data. Association rule mining is used to discover the correlations among the attribute sets in the data set for intrusion detection. The rules usually form as "$X \rightarrow Y$," which means that the -tuples in the dataset satisfy X is likely to satisfy Y. A RIPPER approach is proposed to generate frequent episodes firstly and then form the rules by associating the frequent episodes [14]. To extract the diversified rules, the fuzzy set theory was widely used to extract compact association rules. Tajbakhsh et al. [15] proposed a fuzzy association rule induction algorithm with two steps. The first step involved finding the significant itemsets with a higher significance factor than the user-specified threshold, and the second step involved generating rules by using the large itemsets induced in the first step. From the other perspective, intrusion detection generally distinguishes normal behavior, known intrusions and unknown intrusions, respectively, which can be taken as a classification procedure. Thus, the classes are considered with association rules

to form the class association rules. Different from the association rule, class association rule has the specified class label as its consequent part. Ozyer et al. [16] proposed to use GA boosting to find fuzzy class association rules. They encoded the rules as strings and used GA to evolve them. To extract as many rules as possible for identifying various kinds of intrusions, many algorithms were designed and implemented. Tsang et al. [17] employed a hierarchical GA structure. Each chromosome comprises control and parameter genes, and the parameters of fuzzy member functions were encoded as the parameter genes; the activations of which are managed by the control genes. Thus, the method was also used as a genetic feature selection wrapper to search for an optimal feature subset for dimensionality reduction. Feature selection can be used not only to alleviate the disadvantage of dimensionality and minimize the classification errors but also to improve the interpretability of the rule-based classifiers. Genetic programming (GP) has also been applied to intrusion detection. Some researches utilized GP to extract rules for intrusion detection based on its linear genomes and homologous crossover operators [18, 19]. In conclusion, most of current researches generally pursue the extraction of a large number of rules and overlook the discrimination of the rules [20]. Therefore, it is difficult to identify various types of intrusions with a high detection rate and a low false alarm rate. This could be due to the following two reasons. First, the network behavior data generated rapidly prompts the increase of the rules. Second, the similarity between the normal behavior and the new intrusion behavior limits the discrimination of the rules. Furthermore, this also brings about a considerable amount of redundant, irrelevant, and obvious information into the rule sets. In this case, the important rules are overwhelmed with the useless information. Therefore, the balance between the quality and quantity of rules is crucial for improving the intrusion detection performance.

To improve the quality of the rule sets as well as reserve the diversity of the rules, a new class association rule selection method is proposed based on genetic network programming (GNP) to solve the intrusion detection problem in smart human care services. Specifically, the similarities between rules and between rule sets are checked based on the distances during GNP evolution. The distance between the rules in the rule set of the same class is minimized, and the distance between the rules in the rule sets of the different classes is maximized by adding the newly extracted rules into the rule sets. If the above minimization and maximization criteria are

TABLE 1: Measurements of class association rules.

Class association rules	Support	Confidence
$(A_1 = 1) \Rightarrow (C = k)$	$a(k)/N$	$a(k)/a$
$(A_1 = 1) \wedge (A_2 = 1) \Rightarrow (C = k)$	$b(k)/N$	$b(k)/b$
$(A_1 = 1) \wedge (A_2 = 1) \wedge (A_5 = 1) \Rightarrow (C = k)$	$c(k)/N$	$c(k)/c$
$(A_1 = 1) \wedge (A_2 = 1) \wedge (A_5 = 1) \wedge (A_9 = 1) \Rightarrow (C = k)$	$d(k)/N$	$d(k)/d$

satisfied, the extracted rules are added to the rule sets; otherwise, they are discarded. In this way, redundant information can be avoided during the rule evolution, and the discrimination of the rule sets would be enhanced. In addition, this method also reserves the diversity of the rule sets according to the evolving mechanism. Thus, the GNP evolving method has the ability to discover discriminative class association rules for intrusion detection, which can further improve the intrusion detection performance.

The remainder of this paper is organized as follows. Section 2 describes the GNP structure and GNP-based class association rule mining in detail, and Section 3 introduces how to evolve class association rules based on distance. The simulation results are shown in Section 4, and Section 5 concludes this paper.

2. Genetic Network Programming

2.1. Basic Structure of GNP. GA has a string structure, and GP has a tree structure. With the complexity of problems increasing, it is difficult to express the problem using GA, and the GP structure starts bloating. As an extension of GA and GP, GNP has a quite different structure from GA and GP, which is the directed graph structure [21]. Figure 1 shows the phenotype of GNP, and there are three kinds of nodes in each individual. The start node is used to determine the first node to be executed. Judgment nodes work as the decision-making functions and are represented as $J_1, J_2, \ldots, J_m$. Processing nodes represent the functions of actions or processes and are expressed as $P_1, P_2, \ldots, P_n$. Node transition starts from the start node, and then, the next node to be executed is determined by the node transition. In addition, the number of nodes and their functions depend on the specific problem, which are determined by designers. In addition, judgment nodes have conditional branches, whereas processing nodes do not.

Figure 2 illustrates the genotype of the GNP structure. N T_i indicates the node type, the values of which are 0, 1, or 2. 0 is the start node, 1 is the processing node, and 2 is the judgment node. ID_i serves as an identification number. And C_{ij} denotes the node connection between node i and j.

2.2. Class Association Rule Mining Based on GNP. When GNP is used to extract class association rules, the function of the judgment node corresponds to the attribute of each tuple in the dataset, and the processing nodes are used to calculate the measurements of the class association rules. The specific procedure of class association rule mining using GNP is shown in Figure 3. GNP examines the attribute values of tuples in the dataset using judgment nodes and

calculates the measurements using processing nodes. The judgment node determines the next node by the judgment result of yes or no, corresponding to the yes side or no side.

The yes side of the judgment node is connected to another judgment node. Judgment nodes can be reused and shared with other class association rules because of GNP's reusability. The no side of the judgment node is connected to another processing node, which represents the end of the current rule and the start of another new rule. The start node is connected to the first processing node. The connections of judgment nodes in Figure 3 are extracted as the candidate class association rules, which are shown below. There are four class association rules that correspond to four connections.

$$
\begin{aligned}
(A_1 = 1) &\Rightarrow (C = k), \\
(A_1 = 1) \wedge (A_2 = 1) &\Rightarrow (C = k), \\
(A_1 = 1) \wedge (A_2 = 1) \wedge (A_5 = 1) &\Rightarrow (C = k), \\
(A_1 = 1) \wedge (A_2 = 1) \wedge (A_5 = 1) \wedge (A_9 = 1) &\Rightarrow (C = k).
\end{aligned}
\tag{1}
$$

To evaluate the above candidates of class association rules, we can calculate the corresponding support and confidence, which are shown in Table 1.

Let A_i be the item in the data set, and let its value be 1 or 0. Let C be the class label. So, the class association rule can be represented as the following unified form.

$$
(A_p = 1) \wedge \cdots \wedge (A_q = 1) \Rightarrow (C = k), \quad k = 0, 1, 2, \ldots, K, \tag{2}
$$

where $A_m = 1$ means that attribute A_m equals to 1 and C is the set of suffixes of classes.

3. Evolving Class Association Rules

GNP can extract a great number of class association rules for intrusion detection. However, with an increase in the amount of rules, the detection performance is not always enhanced by the extracted rules. Lots of rules bring redundant and irrelevant information into rule sets. This section describes how to implement the new evolving mechanism on the class association rule mining by GNP.

3.1. Jaccard Distance. Evolving class association rules are aimed at pruning the redundant and irrelevant rules for intrusion detection and at reserving the discriminative rules. In fact, a class association rule is composed of a set of attributes. Thus, the difference between two rules can be regarded as the distance between two sets of attributes, which is computed by the definition of Jaccard distance [22] shown as Definition 1.

```
Input: Target generation of GNP, N;
          Training data base, TrainDB;
Output: Accumulated ruleset, R;
1: for   i = 0; i < N; i + +   do
2:            while   ith GNP generation   do
3:                  Extract next rule r based on TrainDB
4:                  Distance_N ← CalculateDistance(r, R_N)
5:                  // Calculate distance using the latest normal ruleset
6:                  if   Distance_N > CalculateDistance(r, R_N)   then
7:                        if   the class of rule r equals normal   then
8:                              AddOneRule(r, R_N)
9:                        else
10:                              AddOneRule(r, R_I)
11:                       end if
12:                       break
13:                 end if
14:                 Distance_I ← CalculateDistance(r, R_I)
15:                 // Calculate distance using the latest intrusion ruleset.
16:                 if   Distance_I > CalculateDistance(r, R_I)   then
17:                       if   the class of rule r equals normal   then
18:                             AddOneRule(r, R_N)
19:                       else
20:                             AddOneRule(r, R_I)
21:                       end if
22:                 end if
23:           end while
24:           GNP population comets to (i + 1)th generation
25: end for
26: R ← MergeRulePool(R_N, R_I)
27: return R
```

ALGORITHM 1: (GNP with rule evolving).

Definition 1. Given two sets A and B, the Jaccard distance is defined as

$$D_J(A, B) = \frac{|A \cup B| - |A \cap B|}{|A \cup B|}, \quad (3)$$

where $|A \cup B|$ states the union of set A and set B, and $|A \cap B|$ indicates the intersection between set A and set B. The Jaccard distance can measure the degree of overlap between the two sets.

3.2. Rule Selection Based on Distance. Pruning the redundant and irrelevant rules is achieved by minimizing the distance of rules in the same class rule set as well as maximizing the distance between the different class rule sets. Therefore, the generated rules are checked according to the similarity of the rules and that of the rule sets. Specifically, when a newly extracted rule is added into the rule set, the distance between the rules in the rule set of the same class is minimized and the distance between the rules in the rule sets of different classes are simultaneously maximized. In this case, the rule is regarded as a distinguishable class association rule.

As the description of a class association rule, it comprises a group of attributes, which can be regarded as the mathematic theory "set." Thus, the distance either between the rules or between the rule sets can be described by the difference of two sets. Based on this principle, the distance between rule r and r' is defined as (4). And the distance between the rule sets can be calculated based on (4). The detailed definition is shown as (5).

$$d(r, r') = \frac{|A_r \cup A_{r'}| - |A_r \cap A_{r'}| + \sum_{a \in A_r \cap A_{r'}} d(v(r, a), v(r', a))}{|A_r \cup A_{r'}|}, \quad (4)$$

$$d(R, R') = \frac{\sum_{r \in R} \sum_{r' \in R'} d(r, r')}{|R||R'|}, \quad (5)$$

where R and R' denote the rule set with different classes. r and r' represent the two rules. A_r and $A_{r'}$ are the corresponding attribute sets of rule r and rule r', respectively. a denotes the attribute in the rule. $v(r, a)$ is the value of attribute a of rule r. $d(R, R')$ stands for the distance between rule set R and rule set R'. $d(r, r')$ represents the distance between rule r and rule r', whose range is $[0,1]$. $d(v(r, a), v(r', a))$ is defined as

$$d(v(r, a), v(r', a)) = \begin{cases} 1, & v(r, a) \neq v(r', a), \\ 0, & v(r, a) = v(r', a). \end{cases} \quad (6)$$

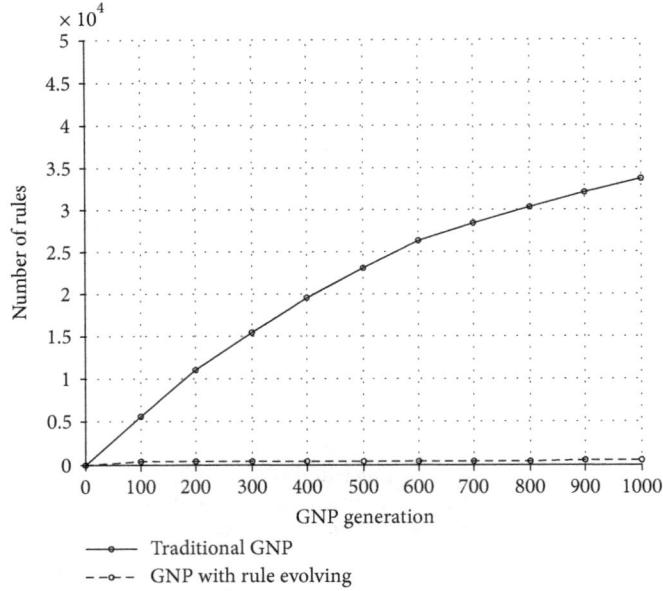

FIGURE 4: The comparison of rule quantity between the traditional GNP and GNP with rule evolving.

From (4) and (5), the modified distance considers the actual value of each attribute by adding $d(v(r, a), v(r', a))$ to the traditional Jaccard distance. $d(v(r, a), v(r', a)) = 0$ indicates that the attributes of rule r are completely the same with those of rule r', whereas $d(v(r, a), v(r', a)) = 1$ means that the attributes of rule r are completely different from those of rule r'. Therefore, the larger is the number of the same pairs (attribute, value), the shorter is the distance between r and r'. In this paper, the thresholds of intradistance between the rules in the same rule sets and interdistance between the rules in the different rule sets are all set as 0.98.

3.3. Evolving Class Association Rules. Except for support and confidence, χ^2 is also used to measure the significance of a rule. The class association rule is abbreviated as the form $X \rightarrow Y$, where $X, Y \subseteq I$ and $X \cap Y = \varnothing$, with I being the set of attributes. X and Y are the antecedent and consequent-of the association rule, respectively. Unlike the association rule, the class association rule has a class label as the consequent part. In this way, the support is defined as support$(X) = x$. x is the fraction of tuples containing X in the database. Confidence is defined by support$(X \cup Y)$/support(X). Therefore, χ^2 of the rule is given by

$$\chi^2 = \frac{N(z - xy)^2}{xy(1 - x)(1 - y)}, \qquad (7)$$

where N is the total number of tuples in the database, z is the value of support$(X \cup Y)$, and x and y are supports of X and Y, respectively.

Then, the minimum support, minimum confidence, and minimum χ^2 are used to select the candidate rules. After calculating the support, confidence, and χ^2 values of the above candidate class association rules, and if they satisfy the following conditions, support $\geq$ support$_{\min}$, confidence $\geq$

TABLE 2: Confusion matrix of classification results.

	Normal (C)	Misuse (C)	Anomaly (C)	Total
Normal (A)	8202	31	1478	9711
Misuse (A)	0	4776	546	5322
Anomaly (A)	1292	17	6202	7511
Total	9494	4824	8226	22,544

confidence$_{\min}$, and $\chi^2 \geq \chi^2_{\min}$, the rule is regarded as the important rule and then stored into the ruleset.

Each individual is evaluated by the fitness function defined by

$$\text{Fitness} = \sum_{r \in R} \left\{ \chi^2(r) + 10(n_{\text{ante}}(r) - 1) + \alpha_{\text{new}}(r) \right\}, \qquad (8)$$

where R in $r \in R$ is the set of suffixes of the extracted important rules from the individuals, $n_{\text{ante}}(r)$ is the number of attributes in the antecedent part of rule r, and $\alpha_{\text{new}}(r)$ is the additional constant shown as (9)

$$\alpha_{\text{new}}(r) = \begin{cases} \alpha_{\text{new}}, & \text{when rule } r \text{ is newly extracted,} \\ 0, & \text{otherwise.} \end{cases}$$

$$(9)$$

Therefore, the fitness function of GNP is concerned with importance, complexity, and novelty of rule r.

The pseudocode of evolving class association rules by GNP is summarized in Algorithm 1.

4. Simulations

4.1. Data Set. Owing to the lack of realistic data sets of smart human care services, the benchmark data set NSL-KDD is used to verify the validity of the proposed method

(a) DR and ACC

(b) NFR and PFR

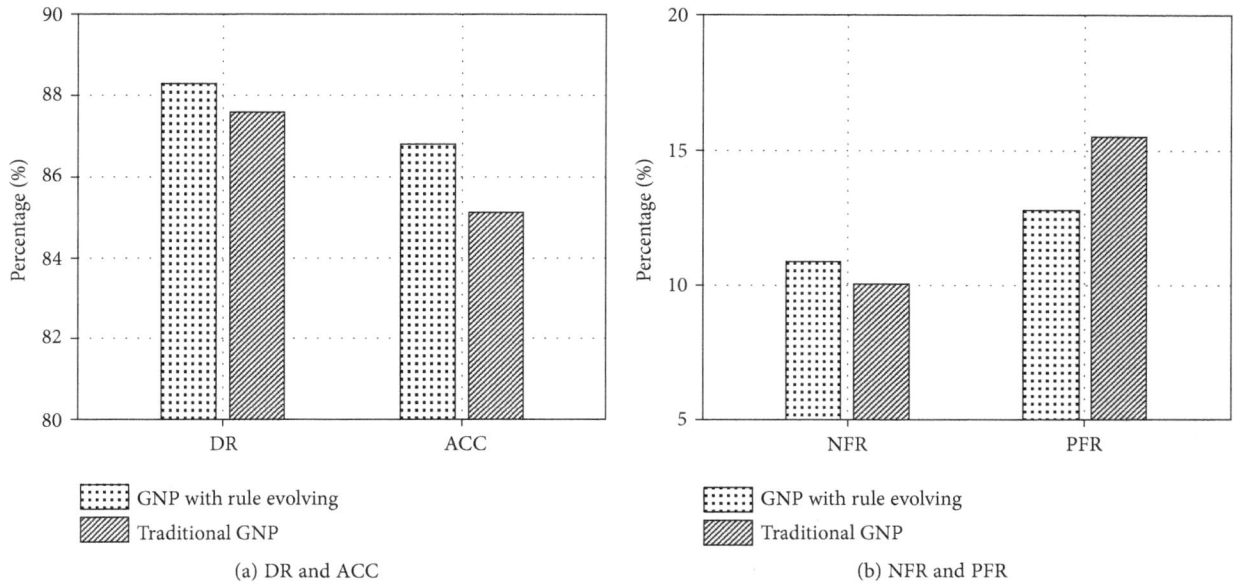

FIGURE 5: Detection performance comparisons of the traditional GNP and GNP with rule evolving.

TABLE 3: The accuracy comparisons of the traditional GNP and GNP with rule evolving (%).

Classifier	Traditional GNP (%)	GNP with rule evolving (%)
SVM	72.89	75.98
BP	71.87	74.82
MLP	73.53	74.91
KNN	75.69	77.30
Logic regression	73.56	74.83
Decision tree	79.63	80.22
AdaBoost	75.23	73.50
Naive Bayes	76.31	74.72
Cluster Gaussian	85.07	86.78
Average	**75.97**	**77.00**

[23, 24]. NSL-KDD is a new version of KDD CUP 1999 data set [25]. Both NSL-KDD and KDD CUP 1999 include a wide variety of intrusions simulated in a military network environment, which is difficult for a self-build simulation environment to acquire such diversified categories of intrusions. However, the NSL-KDD data set is different from KDD CUP 1999, which is composed of the most difficult detected data evaluated by the classical classification methods. Thus, NSL-KDD is a challenge data set for evaluating the intrusion detection methods. Moreover, compared to the KDD CUP 1999 data set, the intrusion detection performance on the NSL-KDD data set will not be biased towards the intrusions easily detected.

Each audit data in NSL-KDD consists of 41 attributes including continuous and discrete ones and one class label. Except for normal audit data, there are four types of attacks in this dataset, which are denial of service (DOS), probe, user to root (U2R), and root to local (R2L).

4.2. Parameter Settings. We use $\text{support}_{\min}(N) = 0.1$, $\text{support}_{\min}(I) = 0.075$, $\text{confidence}_{\min} = 0.8$, and $\chi^2_{\min} = 6.64$, where N and I indicate normal and intrusion, respectively. Class association rules are extracted for each class using GNP. The population size of GNP is 120. The number of processing nodes and judgment nodes are 10 and 100, respectively. In addition, the crossover rate is 1/5. The mutation rate for P_{m1} is 1/3 and for P_{m2} is 1/3, in which P_{m1} and P_{m2} mutate the connections of the branches and the contents of the nodes, respectively. The condition of termination is 1000 generations.

4.3. Result Analysis. First, we randomly select 2000 normal data and 2000 intrusion data as the training set. The testing set consists of 9711 normal data and 12,833 intrusion data. Both the training set and the testing set are from NSL-KDD, which avoids redundant records and improves the difficulty level of KDD Cup 1999. GNP-based class association rule mining is conducted on the prepared training set. The proposed method is compared with the traditional GNP-based class association rule mining shown in Figure 4. Different from GNP with rule evolving, the traditional GNP has no action of automatic selection of useful rules, which always extracts a great number of class association rules. After 1000 generations, the traditional GNP increasingly generates rules, while the proposed method has been converged already. We can conclude that GNP with rule evolving has the strong ability to reduce the rule quantity in rule mining. It can be also regarded as an online rule pruning scheme.

Furthermore, the detection performance of the proposed method on the NSL-KDD data set is investigated. Here, we use the classifier of cluster Gaussian referred to the literature [26]. The cluster Gaussian classifier utilizes

TABLE 4: The accuracy comparisons of CBA, CMAR, and GNP with rule evolving (%).

Classifier	Decision tree	SVM	KNN	AdaBoost	Cluster Gaussian
CBA	70.64	74.47	71.75	72.83	86.09
CMAR	72.39	72.28	72.99	72.19	85.48
GNP with rule evolving	80.22	75.98	77.30	73.50	86.78

the information of known normal and intrusion data to find the extract boundary of normal and known intrusions. In terms of the data distribution, it clusters the similar data which are supposed to have the similar network behaviors. And the classifier further uses Gaussian functions to find the cluster boundaries and data distribution to determine the cluster number. Table 2 shows the confusion matrix of the detection results. "A" is the actual class of the data and "C" is labeled by the classifier. From the confusion matrix, detection rate (DR), accuracy, positive false rate (PFR), and negative false rate (NFR) are calculated. DR indicates the rate of the data that are correctly classified into normal or intrusion. ACC (accuracy) is the rate of the data that are accurately classified as normal, misuse intrusion, or anomaly intrusion. PFR represents that the classifier identifies the normal data as misuse intrusion or anomaly intrusion. NFR represents that misuse and anomaly intrusions are identified as normal.

Therefore, according to Table 2, DR, accuracy, PFR, and NFR are calculated as follows:

$$DR = \frac{8202 + 4776 + 6202 + 17 + 546}{22544} = 87.58\%,$$

$$Accuracy = \frac{8202 + 4776 + 6202}{22544} = 85.08\%,$$

$$PFR = \frac{31 + 1478}{9711} = 15.54\%,$$

$$NFR = \frac{0 + 1292}{5322 + 7511} = 10.07\%.$$

(10)

From the results, misuse intrusion and normal are easy to distinguish by the evolved rules. Though most of anomaly intrusions have been identified, a lot of anomaly intrusions are still difficult to detect. Furthermore, we compare the traditional GNP with the proposed method on DR, accuracy, NFR, and PFR. As shown in Figure 5, GNP with rule evolving obtains higher DR and ACC and lower PFR, but NFR is a little bit higher in GNP with rule evolving than in the traditional GNP. Therefore, anomaly intrusions are still difficult to distinguish. The similarities between anomaly intrusions and normal patterns account for this phenomenon.

In order to further demonstrate the proposed method, classical machine learning algorithms are taken as comparative classifiers, including support vector machine (SVM), back propagation (BP) neural network, multilayer perception (MLP), k-nearest neighbor (KNN), logic regression, decision tree, AdaBoost, naive Bayes, and cluster Gaussian.

Table 3 shows the detection accuracy of the traditional GNP and GNP with rule evolving based on different

classifiers. By evolving 1000 generations, GNP extracts 33,723 rules and the proposed GNP extracts 436 rules. Then, 9 classifiers are used to evaluate the intrusion detection performance based on the two GNP. Among them, 7 classifiers on GNP with rule evolving acquire higher accuracy than those on the traditional GNP. In addition, the average accuracy of GNP with rule evolving is also better than that of the traditional GNP. The results demonstrate that the proposed method can evolve better rules for intrusion detection. Besides, we evaluate the proposed method by comparing it with the classical classification rule mining methods such as classification based on associations (CBA) [27] and classification based on multiple association rules (CMAR) [28]. Both CBA and CMAR contain the rule pruning procedure. With the default classifiers, CBA and CMAR obtain accuracies of 74.63% and 72.17%, respectively, which are lower than the average accuracy of 77% obtained using GNP with a rule selection mechanism. In addition, we select some of the classical classification methods to evaluate the effectiveness of the proposed method, which are the decision tree, SVM, KNN, AdaBoost, and cluster Gaussian. Table 4 illustrates the accuracy comparisons of CBA, CMAR, and GNP with rule evolving. As shown in Table 4, GNP with rule evolving has higher classification accuracies than the other rule-based methods. Thus, it is necessary to consider the rule evolving technique in the rule mining. And the rule evolving is capable of selecting useful rules and reducing the redundant and irrelevant rules.

5. Conclusions

In this paper, an intrusion detection system based on evolving class association rules is proposed as a security solution for smart human care services. In general, it utilizes a class association rule evolving strategy to construct the intrusion detection system. As a data mining solution, GNP with rule evolving can generate diversified class association rules and control the quantity of the rules simultaneously. For intrusion detection, the significance test is performed to ensure the importance of generated rules. In order to generate the more discriminative class association rules, the Jaccard distance is modified to measure the similarity between rules and different rule sets. In this way, the distance of the rules in the rule set with the same class is minimized and the distance between rules in the rule sets with different classes is maximized. The simulations conducted on the NSL-KDD dataset theoretically verify that GNP with rule evolving efficiently controls the quantity of generated rules and improves the detection performance by reducing the redundancy of the rules. In the future, we plan to verify the effectiveness of the intrusion detection system on the self-build simulation environment.

Conflicts of Interest

The authors declare that they have no conflicts of interest.

Acknowledgments

This work was supported in part by Basic Scientific Research Fund for Central Universities under Grant no. 2015QNB20, in part by the Natural Science Foundation of Jiangsu Province under Grant no. BK20150204, in part by China Postdoctoral Science Foundation under Grant no. 2015M581884, and in part by National Natural Science Foundation of China under Grant no. 51734009 and no. 51504255.

References

[1] I. Butun, S. D. Morgera, and R. Sankar, "A survey of intrusion detection systems in wireless sensor networks," *IEEE Communications Surveys & Tutorials*, vol. 16, no. 1, pp. 266–282, 2014.

[2] M. A. Faisal, Z. Aung, J. R. Williams, and A. Sanchez, "Data-stream-based intrusion detection system for advanced metering infrastructure in smart grid: a feasibility study," *IEEE Systems Journal*, vol. 9, no. 1, pp. 31–44, 2015.

[3] S. S. Wang, K. Q. Yan, S. C. Wang, and C. W. Liu, "An integrated intrusion detection system for cluster-based wireless sensor networks," *Expert Systems with Applications*, vol. 38, no. 12, pp. 15234–15243, 2011.

[4] S. Pan, T. Morris, and U. Adhikari, "Developing a hybrid intrusion detection system using data mining for power systems," *IEEE Transactions on Smart Grid*, vol. 6, no. 6, pp. 3104–3113, 2015.

[5] D. E. Denning, "An intrusion-detection model," *IEEE Transactions on Software Engineering*, vol. SE-13, no. 2, pp. 222–232, 1987.

[6] J. P. Anderson, "Computer security threat monitoring and surveillance," Tech. Rep., James P. Anderson Co., Fort, Washington, PA, USA, 1980.

[7] S. X. Wu and W. Banzhaf, "The use of computational intelligence in intrusion detection systems: a review," *Applied Soft Computing*, vol. 10, no. 1, pp. 1–35, 2010.

[8] A. A. Aburomman and M. B. I. Reaz, "A novel weighted support vector machines multiclass classifier based on differential evolution for intrusion detection systems," *Information Sciences*, vol. 414, pp. 225–246, 2017.

[9] E. Kabir, J. Hu, H. Wang, and G. Zhuo, "A novel statistical technique for intrusion detection systems," *Future Generation Computer Systems*, vol. 79, pp. 303–318, 2018.

[10] C. Guo, Y. Ping, N. Liu, and S. S. Luo, "A two-level hybrid approach for intrusion detection," *Neurocomputing*, vol. 214, pp. 391–400, 2016.

[11] J. Cannady, "Artificial neural networks for misuse detection," in *Proceedings of the 1998 National Information Systems Security Conference*, pp. 443–456, Arlington, VA, USA, 1998.

[12] D. Parikh and T. Chen, "Data fusion and cost minimization for intrusion detection," *IEEE Transactions on Information Forensics and Security*, vol. 3, no. 3, pp. 381–389, 2008.

[13] W. Hu, W. Hu, and S. Maybank, "Adaboost-based algorithm for network intrusion detection," *IEEE Transactions on*

Systems, Man, and Cybernetics, Part B (Cybernetics), vol. 38, no. 2, pp. 577–583, 2008.

[14] W. W. Cohen, "Fast effective rule induction," in *Proceedings of the Twelfth International Conference on Machine learning*, pp. 115–123, Tahoe City, California, 1995.

[15] A. Tajbakhsh, M. Rahmati, and A. Mirzaei, "Intrusion detection using fuzzy association rules," *Applied Soft Computing*, vol. 9, no. 2, pp. 462–469, 2009.

[16] T. Ozyer, R. Alhajj, and K. Barker, "Intrusion detection by integrating boosting genetic fuzzy classifier and data mining criteria for rule pre-screening," *Journal of Network and Computer Applications*, vol. 30, no. 1, pp. 99–113, 2007.

[17] C. H. Tsang, S. Kwong, and H. Wang, "Genetic-fuzzy rule mining approach and evaluation of feature selection techniques for anomaly intrusion detection," *Pattern Recognition*, vol. 40, no. 9, pp. 2373–2391, 2007.

[18] J. V. Hansen, P. B. Lowry, R. D. Meservy, and D. M. Mcdonald, "Genetic programming for prevention of cyberterrorism through dynamic and evolving intrusion detection," *Decision Support Systems*, vol. 43, no. 4, pp. 1362–1374, 2007.

[19] D. Song, M. I. Heywood, and A. N. Zincir-Heywood, "Training genetic programming on half a million patterns: an example from anomaly detection," *IEEE Transactions on Evolutionary Computation*, vol. 9, no. 3, pp. 225–239, 2005.

[20] E. Lazcorreta, F. Botella, and A. Fernández-Caballero, "Towards personalized recommendation by two-step modified apriori data mining algorithm," *Expert Systems with Applications*, vol. 35, no. 3, pp. 1422–1429, 2008.

[21] S. Mabu, K. Hirasawa, and J. Hu, "A graph-based evolutionary algorithm: genetic network programming (GNP) and its extension using reinforcement learning," *Evolutionary Computation*, vol. 15, no. 3, pp. 369–398, 2007.

[22] M. Levandowsky and D. Winter, "Distance between sets," *Nature*, vol. 234, no. 5323, pp. 34-35, 1971.

[23] A. Tavallaee, E. Bagheri, W. Lu, and A. A. Ghorbani, "A detailed analysis of the kdd cup 99 data set," in *2009 IEEE Symposium on Computational Intelligence for Security and Defense Applications*, pp. 53–58, Ottawa, ON, Canada, 2009.

[24] http://nsl.cs.unb.ca/NSL-KDD/.

[25] http://kdd.ics.uci.edu/databases/kddcup99/.

[26] N. Lu, S. Mabu, Y. Li, and K. Hirasawa, "Classification with clustering and Gaussian functions in intrusion detection system," *IEEJ Transactions on Electronics Information and Systems*, vol. 134, no. 12, pp. 1908–1915, 2014.

[27] B. Liu, W. Hsu, and Y. Ma, "Mining association rules with multiple minimum supports," in *KDD '99 Proceedings of the Fifth ACM SIGKDD International Conference on Knowledge Discovery and Data Mining*, pp. 337–341, San Diego, California, USA, 1999.

[28] W. Li, J. Han, and J. Pei, "Cmar: accurate and efficient classification based on multiple class-association rules," in *Proceedings 2001 IEEE International Conference on Data Mining*, pp. 369–376, San Jose, CA, USA, 2001.

18

Implementation of a Low-Cost Energy and Environment Monitoring System based on a Hybrid Wireless Sensor Network

Dong Sik Kim,[1] Beom Jin Chung,[2] and Sung-Yong Son[3]

[1]*Department of Electronics Engineering, Hankuk University of Foreign Studies, Yongin-si, Gyeonggi-do 17035, Republic of Korea*
[2]*Smart Green Home Research Center, Gachon University, Gyeonggi-do, Republic of Korea*
[3]*Department of Electrical Engineering, Gachon University, Gyeonggi-do, Republic of Korea*

Correspondence should be addressed to Sung-Yong Son; xtra@gachon.ac.kr

Academic Editor: Stefania Campopiano

A low-cost hybrid wireless sensor network (WSN) that utilizes the 917 MHz band Wireless Smart Utility Network (Wi-SUN) and a 447 MHz band narrow bandwidth communication network is implemented for electric metering and room temperature, humidity, and CO_2 gas measurements. A mesh network connection that is commonly utilized for the Internet of Things (IoT) is used for the Wi-SUN under the Contiki OS, and a star connection is used for the narrow bandwidth network. Both a duty-cycling receiver algorithm and a digitally controlled temperature-compensated crystal oscillator algorithm for frequency reference are implemented at the physical layer of the receiver to accomplish low-power and low-cost wireless sensor node design. A two-level temperature-compensation approach, in which first a fixed third-order curve and then a sample-based first-order curve are applied, is proposed using a conventional AT-cut quartz crystal resonator. The developed WSN is installed in a home and provides reliable data collection with low construction complexity and power consumption.

1. Introduction

Indoor temperatures are traditionally controlled using thermostats and their corresponding actuators or valves, as shown in Figure 1(a). In order to sense the room temperatures of everyday living spaces efficiently, wireless sensors utilizing the industrial, scientific, and medical (ISM) 2.4 GHz radio band with wireless link distances that are usually restricted within a single room, as shown in Figure 1(b), have recently been developed.

It is essential to monitor the indoor air quality of residential and commercial apartment complex areas efficiently based on measured room temperatures, humidity levels, CO_2 gas amounts, and other factors to control the indoor environmental conditions properly. A particular monitoring system usually depends on the existing building automation system (BAS), as shown in Figure 2, as well as on the data collected by the numerous sensors in the building and stored in a BAS server. The exploitation of wireless sensor network (WSN) technology can provide further efficiency

and flexibility in data collection approaches [1, 2]. Energy usage data, such as those collected from electric metering in the advanced metering infrastructure (AMI), can also be collected efficiently through collaboration with environment data collection. Furthermore, a WSN can deliver energy information to the in-home display (IHD) system efficiently [3]. The sensor node, which is considered in this paper, can deliver electrical energy usage information and measure temperature and humidity values. The proposed WSN can transfer and collect both energy and environment data.

For mass deployment in residential spaces or office rooms, both the construction cost and the complexity of a WSN should be minimized while still maintaining reliable wireless link margins [4, 5]. Based on an Internet of Things (IoT) technique using IPv6, we can construct a WSN. However, instead of implementing the IoT WSN for all sensors, including the end nodes, which could result in a high cost, partially adopting the IoT WSN can reduce the implementation cost and decrease the power consumption. In order to build a low-cost IoT-based WSN, a hybrid

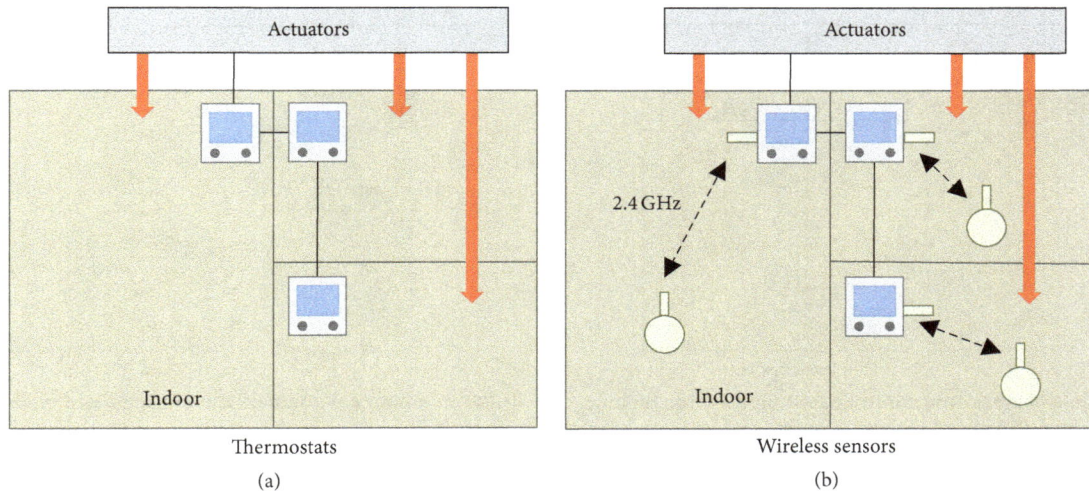

FIGURE 1: Indoor temperature control based on temperature sensors. (a) Traditional thermostats and actuators. (b) Wireless sensors for accurate temperature sensing.

WSN that exploits both high and low data-rate bands is developed [6]. In the hybrid WSN, the high-rate wireless link controls a mesh network in order to collect and transmit data to a server based on the IoT, and the low-rate wireless link collects data directly through a star-connected narrow bandwidth (NB) communication channel at low data rates in order to guarantee long-lasting wireless links [7]. The Korean regulations for NB radio stipulate an 8.5 kHz bandwidth with 12.5 kHz channel spacing, while the high-rate wireless link can have a much broader bandwidth.

In this paper, a low-cost hybrid WSN system utilizing the 917 MHz and 447 MHz radio bands is developed and implemented for energy and environmental monitoring [8]. The RFID/USN, or Wi-SUN (IEEE802.15.4 g), 917 MHz radio band with 200 kHz channel spacing is used to achieve a more stable RF link margin than the one that the 2.4 GHz-ISM band can achieve. In apartment and home environments, transmitted RF signals can reach other rooms due to diffraction of the signals. Hence, a lower frequency of 917 MHz is advantageous to a 2.4 GHz frequency for a high data-rate network [7, 9]. In a low data-rate network, the use of NB radios utilizing the 447 MHz band can significantly reduce the number of gateways. In order to achieve both low-power and low-cost goals, a duty-cycling algorithm [10] is developed at the physical layer of the receiver. Also, a digitally controlled temperature-compensated crystal oscillator (DCTCXO) [11] in which a low-cost microcontroller unit (MCU) controls the oscillation frequency by utilizing several functions of the RF integrated circuit (RFIC) is developed.

This paper is organized as follows. In Section 2, the architecture of the proposed low-cost hybrid WSN is described. In Section 3, an implementation of the network in residential indoor environment monitoring is introduced. In Section 4, a low-cost implementation of an RF radio component containing a DCTCXO is discussed. The paper is concluded in the final section.

2. The Low-Cost Hybrid Wireless Sensor Network

In this section, a low-cost implementation of the hybrid WSN for indoor environment monitoring is introduced.

The IoT technique, which utilizes Wi-Fi or Wi-SUN, can be employed to gather the indoor temperature data for temperature control. The distance ranges for most commonly used Wi-Fi (IEEE802.11) are 20–70 meters and 100–250 meters for indoor and outdoor environments, respectively. Hence, each house should have an individual IoT gateway, as shown in Figure 3. The sub-1 GHz Wi-SUN band can also implement IoT-based sensors, and each gateway can collect more data from several houses than it can in the Wi-Fi case because of the narrower receiver bandwidth and better diffraction properties. The wider collection capability of Wi-SUN can reduce the implementation cost when compared to that of the Wi-Fi case. However, a much lower implementation cost is required for practical residential environment monitoring.

As illustrated in Figure 2, a 447 MHz band NB radio can be used in a customized network to gather sensed data. In this case, a simple gateway can collect the data from numerous sensors in several houses similarly to the way it can in a Wi-SUN implementation, as shown in Figure 4. However, because the customized network is simpler than the network used in the Wi-SUN case, the installation cost of the WSN, which is based on the use of NB radios in a customized network, is significantly lower than the costs of both the Wi-Fi and Wi-SUN installations. The simple gateway in Figure 4 can be connected to Ethernet through a Wi-Fi interface in order to transmit the collected data to a data server.

If a small number of sensors are scattered over a wide area (e.g., over a whole building), then the LoRa modulation technology is an appropriate candidate for NB radio because of its long-range capability and its low data rate of 18 bps

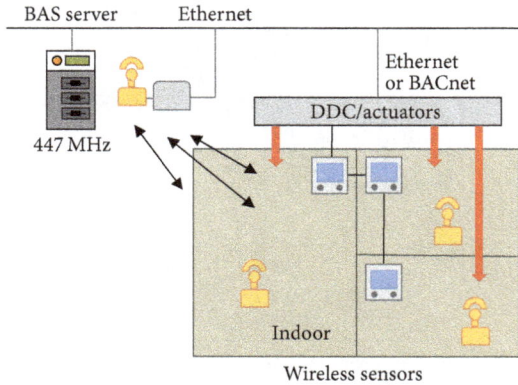

FIGURE 2: Indoor temperature control based on both the building automation system (BAS) and wireless sensors.

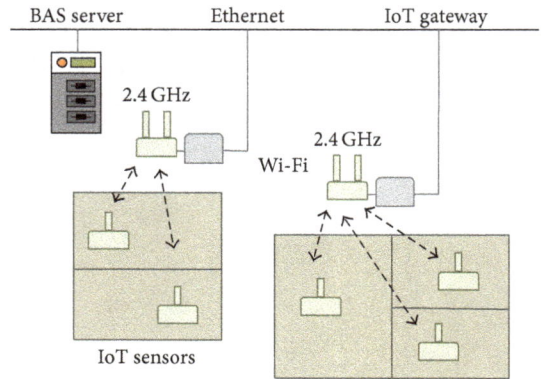

FIGURE 3: Indoor environment monitoring based on the Internet of Things (IoT).

operating at a 7.8 kHz receiver bandwidth, as shown in Figure 4. However, assuming that the number of sensors is large, collecting all of the data from a single gateway under the very low data rate of LoRa technology requires a considerably long time. In order to alleviate the low data-rate problem occurring in NB radios while achieving a long wireless distance, a hybrid WSN system [6], as illustrated in Figure 5, is employed in this study. To collect the sensed data efficiently, the hybrid WSN includes both a low data-rate network based on a customized star connection and a high data-rate network based on a standard mesh connection. The low data-rate network is constructed using NB radios to ensure that the RF link has a long range because it is generally believed to minimize transmission power and protocol while achieving a very low implementation cost. On the other hand, the high data-rate network collects sensed data from the sensors corresponding to the low data-rate star connection, gathering them through the high data-rate mesh connection, to send them to a server. The power-consumption constraint is not as significant for the sensor nodes in the high data-rate network as it is for those in a low data-rate one.

3. Implementation of the Hybrid Wireless Sensor Network

In this section, an implementation of the hybrid WSN in indoor environment monitoring is introduced. In the implemented network, the low data-rate network transmits sensed data periodically through the NB radios under a nonslotted ALOHA protocol. On the other hand, the high data-rate network operates through radios with relatively wider bandwidths under a constrained application protocol (CoAP) using the Contiki OS.

Figure 5 shows an overview of the implemented hybrid WSN system in which three types of components are used: simple wireless sensors, CoAP wireless sensors, and a CoAP gateway. Simple wireless sensors can measure room temperature and humidity without consuming much power and can transmit the sensed data to the corresponding collector, the CoAP wireless sensor, through an NB air link utilizing the 447 MHz band. This process corresponds to that of the

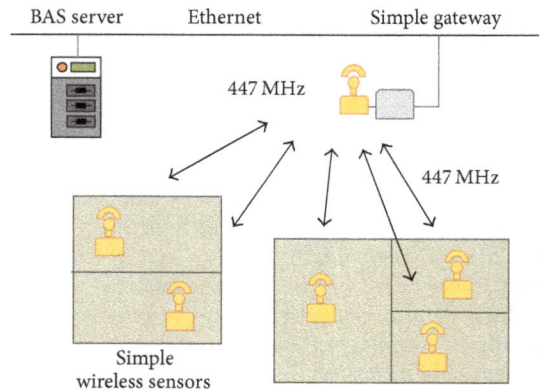

FIGURE 4: Indoor environment monitoring based on the use of NB radio modules under the 447 MHz band.

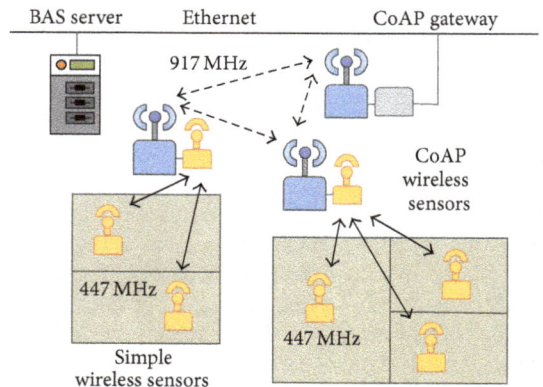

FIGURE 5: Implemented indoor environment monitoring system based on the use of the hybrid WSN under the 917 MHz and 447 MHz bands.

low data-rate star connection in the hybrid WSN. A mesh network utilizing a 917 MHz band RF link under the routing protocol for low-power and lossy networks (RPL) as well as the low-power IPv6 network (6LowPAN) protocol using the Contiki OS is constructed among the CoAP wireless sensors. The CoAP wireless sensors (CoAP server) can measure the

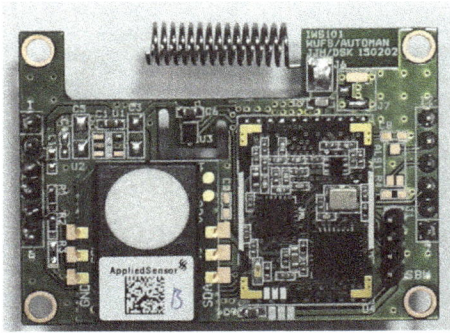

FIGURE 6: Integrated wireless sensor module (IWSM) containing a 447 MHz band RF module with a helical antenna, temperature and humidity sensor, and optional CO_2 sensor.

amount of present CO_2 gas as well as the local temperature and humidity. The CoAP gateway (CoAP client), which is an Ethernet-based mesh network gateway, transmits collected data to the BAS server through Ethernet under the CoAP RESTful application-layer protocol.

In order to manufacture the hybrid WSN system efficiently, an integrated wireless sensor module (IWSM) is developed, as shown in Figure 6. The IWSM contains an RF radio module, which is composed of a low-cost MCU (TI, MSP430G2553), an RFIC (Silicon Labs, Si4463) with a frequency reference, and a temperature and humidity sensor. The option of adding a CO_2 gas sensor is available. The CO2 sensor, which is based on the metal-oxide-semiconductor (MOS) technology (IAQ-CORE), is employed for low-power consumption with a small size instead of the conventional nondispersive infrared (NDIR) type sensors. The NDIR-based CO_2 sensor can usually provide accurate measurements of the CO_2 gas amount. However, the sensor size is relatively big and requires high powers with a long standby time. On the other hand, the MOS-based CO_2 sensor can efficiently measure the CO_2 gas amount in an indirect method with low-power consumption. The low-cost MCU has a flash memory of 16 kB and a RAM of 512 B but does not have any hardware multiplier. In the standby and active modes, the MCU consumes currents of $0.5\,\mu A$ and $330\,\mu A$, respectively, at an operating voltage of 3.0 V. An NB radio for use in an industrial temperature range from $-40°C$ to $85°C$ requires a temperature-compensated crystal oscillator (TCXO) component with a frequency error of ± 2 ppm to ensure communication. Even though the present study only requires sensing indoor temperatures in commercial temperature range of $0°C–70°C$, the industrial specification is applied to the NB radio design. Operated at a data rate of 2.4 kbps with an 8.5 kHz receiver bandwidth and a modulation index of 0.5 based on GFSK modulation, the receiver sensitivity of the proposed IWSM design is -118 dBm with a bit error rate less than 0.1% and a transmission power of 10 dBm. The typical accuracies of temperature and humidity sensors are $\pm 0.3°C$ and $\pm 2\%$ RH, respectively.

As shown in Figure 7(a), the simple wireless sensor is composed of an IWSM without a CO_2 sensor and is

powered by two "AA" batteries for ease of installation and maintenance in residential areas. Hence, both its power consumption and implementation cost should be very low as mentioned in the properties of the low-cost hybrid WSN. A lithium iron disulfide (Li-FeS$_2$) battery can be used to sense outdoor temperatures as low as $-40°C$. The sensors in each room of several houses can transmit measured data to the corresponding CoAP wireless sensors via star connections at the same low data rates as those of 447 MHz band NB communications.

The CoAP wireless sensor, which is based on a Cortex-M3 processor, collects measured data from simple wireless sensors through the 447 MHz band RF module of the IWSM. Then, it shares the data with other CoAP wireless sensors and with the CoAP gateway via a mesh wireless network utilizing the IPv6 network protocol under both the Contiki OS [12] and CoAP through the 917 MHz band RF module, as shown in Figure 7(b). The CoAP sensor uses AC power and can measure the amount of CO_2 gas present. The CoAP gateway contains a Cortex-A8 platform (Beaglebone Black) to communicate through Ethernet with the BAS server under the Linux OS, as shown in Figure 8. The gateway also forms a mesh network with the CoAP wireless sensors under the 917 MHz band. For a network with ten CoAP wireless sensors and a CoAP gateway, the average transmission delay time was 13.8 ms at a transmission rate of 99.9%.

4. Low-Power and Low-Cost Wireless Sensors

For a low-cost implementation of the simple wireless sensor with low-power consumption, the RF module should consume little power in both receiving and transmitting modes. A superregenerative receiver can significantly reduce both the power consumption and implementation cost under a low operating voltage [13]. In this paper, the superheterodyne receiver of a commercially available RFIC is used to achieve a longer range than that of a superregenerative receiver. In order to reduce the power consumption of the RFIC, a duty-cycling algorithm is employed based on a fast direct preamble detection algorithm. The DCTCXO algorithm is then implemented.

4.1. Duty-Cycling Receiver Algorithm. In order to reduce the average power consumption of the RF module receiver, an efficient duty-cycling algorithm is employed in which the receiver is turned on and off periodically to detect the transmitted preamble signal. Here, the duty cycle is proportional to the average operating current of the receiver. The receiver checks the existence of the preamble signal over a short period of activity. If the preamble signal is detected, then the receiver remains on until the reception of the entire data packet is complete. If there is no preamble signal, the receiver is turned off immediately. In order to reduce the average operating current, a fast detection scheme is developed to decrease the preamble detection time. In the scheme, the MCU of the RF module observes each directly received signal to verify whether the signal is correct or not while decoding the packet. Without a duty-cycling algorithm, the average current is 16.7 mA, which can be reduced to

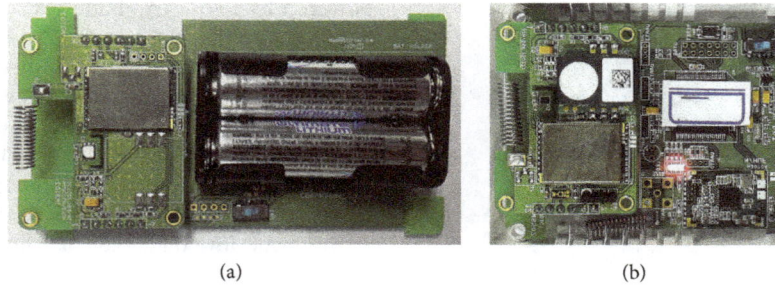

FIGURE 7: Simple and CoAP wireless sensors. (a) Simple wireless sensor circuit board. (b) CoAP wireless sensor circuit board.

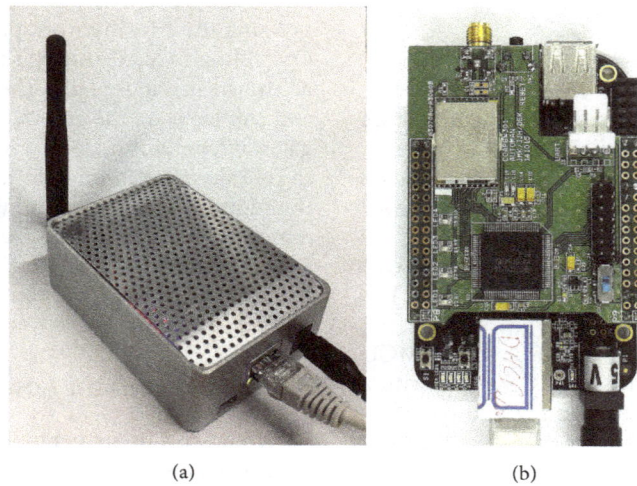

FIGURE 8: CoAP gateway. (a) Exterior of CoAP gateway. (b) CoAP gateway circuit board with a 917 MHz band RF module and a Cortex-A8 processor.

10.4 mA using a conventional duty-cycling algorithm with synchronized detection. However, the average current can be further reduced to 5.6 mA by applying the proposed duty-cycling algorithm based on fast preamble detection.

4.2. Digitally Controlled Temperature-Compensated Crystal Oscillator. In order to obtain a sufficient RF link margin, an NB RF module with a TCXO component can be employed, as shown in Figure 9(a). The operating current of the TCXO component is 1.8 mA. Therefore, using a conventional crystal resonator instead of the TCXO component can reduce the average operating currents in both the receiver and transmitter modes. Furthermore, replacing the expensive TCXO component with a conventional crystal resonator can significantly reduce the implementation cost. However, the frequency error, which depends on variations in temperature, increases the receiver bandwidth, and thus both the receiver sensitivity and selectivity are degraded, even though the automatic frequency control (AFC) function is active. For a crystal resonator with a frequency error of ±20 ppm, the receiver bandwidth is 36.6 kHz. Hence, the receiver is highly susceptible to adjacent channel transmissions under 12.5 kHz channel spacing. On the other hand, the frequency error of the TCXO shown in Figure 9(a) is ±2.0 ppm with a receiver bandwidth of 6.5 kHz. Narrowing the bandwidth increases the receiver sensitivity, and thus the transmission power can be reduced while still achieving a similar RF link margin. Consequently, the power consumption can be reduced [12, 13].

In this study, in order to achieve a similar receiver sensitivity performance to that of the TCXO with a frequency error of ±2.0 ppm, a DCTCXO algorithm is implemented based on a digital technique using the inherent functions of both the MCU and RFIC of the RF module for the AT-cut quartz crystal resonator, as shown in Figure 9(b). Digital techniques have been employed to design higher-precision TCXO than that which analog techniques are able to achieve. Buroker and Frerking [11] and Mroch and Hykes [14] used the available memory, the digital-to-analog converters (DACs), and the varactor diodes to compensate the crystal frequencies with both coarse analog and fine digital techniques with respect to temperature. Azcondo et al. [15] proposed microcomputer temperature compensation of a crystal oscillator by utilizing both the dual mode operation and the varactor diodes. The

(a)   (b)

FIGURE 9: Digitally controlled temperature-compensated crystal oscillator (DCTCXO). (a) NB RF module with a TCXO component (±2.0 ppm). (b) NB RF module with both an AT-cut quartz crystal resonator (±20 ppm) and a DCTCXO algorithm implementation for a ±2.0 ppm frequency error.

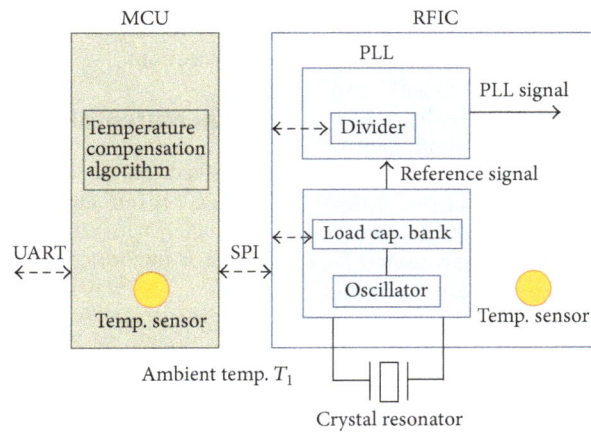

FIGURE 10: Frequency control portion of the RF module with the AT-cut quartz crystal resonator in the proposed DCTCXO.

capacitor banks were also used to trim the crystal frequency [16, 17]. The purpose of employing the digital technique, which is introduced to the proposed DCTCXO, is to achieve the same performance as that of the analog TCXO while minimizing the production procedure by not adding any components.

A block diagram of the frequency control portion of the designed DCTCXO is illustrated in Figure 10. Instead of using the DACs and the varactor diodes, the crystal oscillation frequency can be easily controlled by the load capacitor bank, which can be selected by the MCU. From an externally connected frequency counter, the channel frequency f_0 is tuned to a temperature of 25°C by adjusting the RFIC load capacitors to provide a coarse resolution of 400 Hz/div (<1 ppm trimmable). The phase locked loop (PLL) divider register values are then calculated from the measured temperatures with a fine resolution of 15 Hz/div to compensate the crystal frequency for temperature. Here, the crystal's temperature can be measured by the temperature sensor of either the MCU or the RFIC. The accuracy of the temperature sensors operating within the temperature range

from −40°C to 85°C is usually as low as ±2°C. Furthermore, because of the heat that emanates from the RFIC when it operates in either receiver or transmitter mode, measurement of the crystal's temperature results in transient responses with biases [17]. Hence, the measured temperatures should be compensated for according to the mode in which the RF is operating.

The proposed DCTCXO algorithm is now introduced. For the channel frequency f_0, the relative frequency change of the AT-cut quartz crystal resonator can be represented by a third-order curve, where the frequency deviation at the inflection temperature T_0 is zero, as follows:

$$\frac{\Delta f}{f_0} = A_1 (T - T_0) + A_2 (T - T_0)^2 + A_3 (T - T_0)^3, \quad (1)$$

where the numerator $\Delta f := f - f_0$ is the frequency change. In (1), the parameters A_1, A_2, and A_3 are constants that depend on the physical properties of the crystal unit, including the angle of its cut, ratios of its dimensions, and order of its overtone. The inflection temperature of the AT-cut crystal is approximately equal to 25°C. If we can obtain the parameters for a set of crystal resonators, then we can compensate the crystal frequency for temperature by using the temperature curve in (1). However, the parameters of the crystal resonators can vary during manufacturing due to nonuniform cut angles, and these parameters are determined by taking samples. Hence, the accuracy of temperature compensation performed by using a single compensation curve with fixed parameters can exhibit errors that are quite large. Therefore, a simple technique for measuring the nonuniformity of a crystal resonator sample is necessary to conduct temperature compensation more accurately. Let $\Delta\theta := \theta - \theta_0$ denote the cut-angle difference between the intended angle θ and the zero-temperature coefficient-reference angle θ_0, measured in degrees. For the AT-cut crystal resonator, Bechmann [18] found a change rate of $-0.08583 \times 10^{-6} \Delta\theta$ for the first parameter A_1 with respect to the cut-angle change $\Delta\theta$. In contrast, A_2 and A_3 had relatively lower change rates of $-0.07833 \times 10^{-9} \Delta\theta$ and $-0.033 \times 10^{-12} \Delta\theta$, respectively. Hence, the cut-angle change $\Delta\theta$ dominates the value of A_1 but not the values of A_2 and A_3. In other words, various crystal resonator samples with different values of $\Delta\theta$ can have approximate third-order curves of (1) in which the values of A_1 vary, while A_2 and A_3 do not change significantly. Therefore, the difference between the two third-order curves of (1) can be approximated by a first-order curve. After applying temperature compensation using a fixed third-order curve of (1), the relative change of the resultant frequency f' can be given as a first-order curve that is equal to zero at the inflection temperature 25°C, as

$$\frac{\Delta f'}{f_0} \approx a_1 \left(T - T_0 \right), \tag{2}$$

where the constant a_1 is defined as $a_1 := A_1 - A_1'$ and A_1' is the first parameter of the crystal resonator sample to be temperature compensated. In (2), the numerator $\Delta f' := f' - f_0$ is the resultant frequency change. By measuring the parameter a_1 in (2), the third-order compensated frequency curve can be compensated once again to achieve an even more accurate frequency reference. In the proposed DCTCXO, the calculation of a_1, which depends on the crystal resonator sample, is automatically performed using the frequency-offset measurement ability of the RFIC's AFC block. As shown in Figure 11, a reference-frequency RF signal originating from a high-precision RF signal generator enters the RF module, and $\Delta f'$, the frequency offset of the crystal resonator sample at an ambient temperature of T_1, which is different for 25°C, is calculated by the AFC. The parameter a_1 can then be obtained from the frequency-offset register of the AFC. The proposed DCTCXO algorithm can be summarized as follows.

4.2.1. DCTCXO Algorithm. First, conduct third-order temperature compensation using a fixed third-order curve of (1). Second, conduct first-order temperature compensation using the first-order curve of (2), which is obtained from the crystal resonator sample to be temperature compensated.

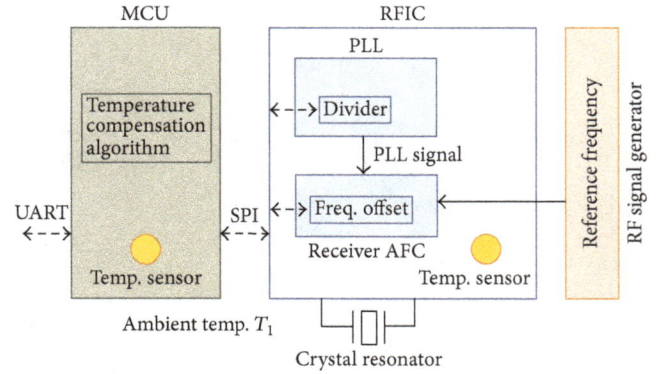

FIGURE 11: Frequency-offset measurement portion of the RF module for first-order compensation in the proposed DCTCXO.

A more detailed description of the algorithm and practical experimental observations will be introduced in the results section.

5. Experimental Results

In this section, an experiment performed on the proposed DCTCXO is introduced first. Then, a low-cost hybrid WSN, constructed with a low-cost RF module and implemented in a home, is introduced.

5.1. Proposed DCTCXO for the Low-Cost RF Module. The relative frequency changes versus the environment temperatures of five RF module samples were obtained experimentally, as shown in Figure 9(b). AT-cut crystal resonators were used to generate the RF reference frequency of 32.0000 MHz. In Figure 12(a), the frequency errors of the five RF module samples, given in parts per million (ppm), are shown and fitted to third-order curves. Using the fitted third-order curves, the results of temperature compensation performed on the five samples are shown in Figure 12(b). The compensation reduced the maximum frequency error from ±9.3 ppm to ±2.8 ppm. As addressed in (2), the compensated curves are first-order curves whose shapes depended on the crystal resonator samples. As shown in Figure 13(a), first-order fitting was conducted on each compensated result appearing in Figure 12(b). The second iteration of temperature compensation was then performed using the first-order curve corresponding to each crystal resonator sample. The final compensated curves are shown in Figure 13(b), revealing that the maximum frequency error was reduced significantly to ±0.68 ppm.

Unfortunately, first-order fitting using all the crystal sample data, shown in Figure 13, is not practical. Hence, a simple, practical technique, in which a frequency-offset measuring technique was employed by the RFIC's AFC block, was used to find the first-order curve of the crystal sample, as shown in Figure 11. In Figure 14, the frequencies measured using the frequency-offset register of the AFC are depicted versus the true frequency offsets. Biases existed depending on the input frequency offsets, and a first-order curve η was

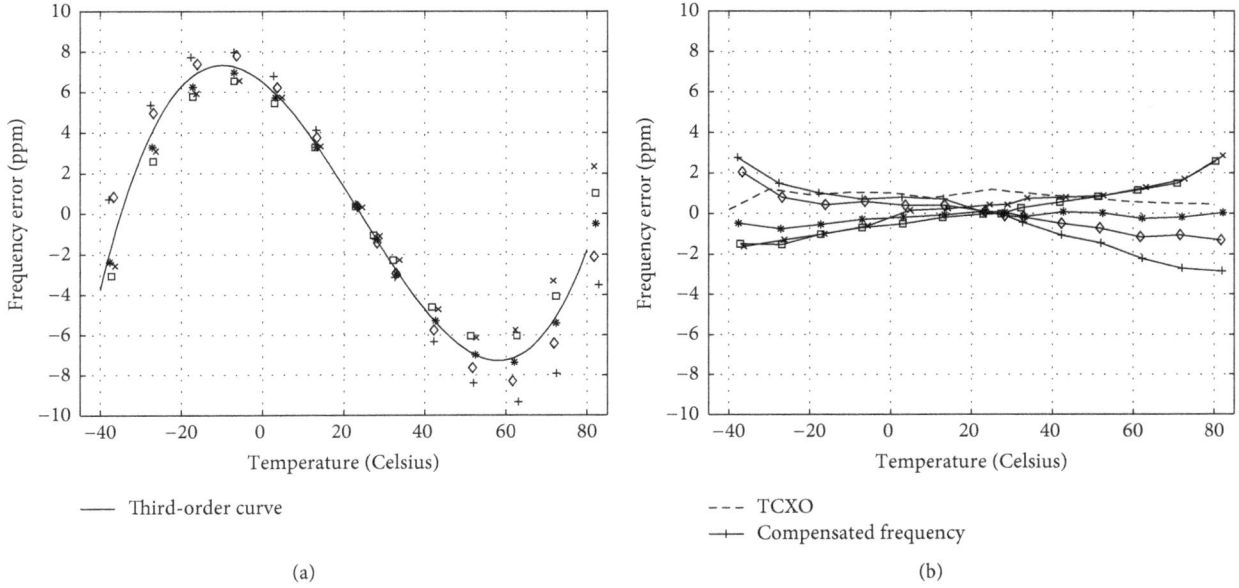

— Third-order curve

(a)

--- TCXO

—+— Compensated frequency

(b)

FIGURE 12: Temperature compensation using a fixed third-order curve. (a) Frequency errors (in ppm) versus ambient temperatures with third-order curves fitted to the five RF model samples (±9.3 ppm). (b) Temperature-compensated results using the fitted third-order curve (±2.8 ppm).

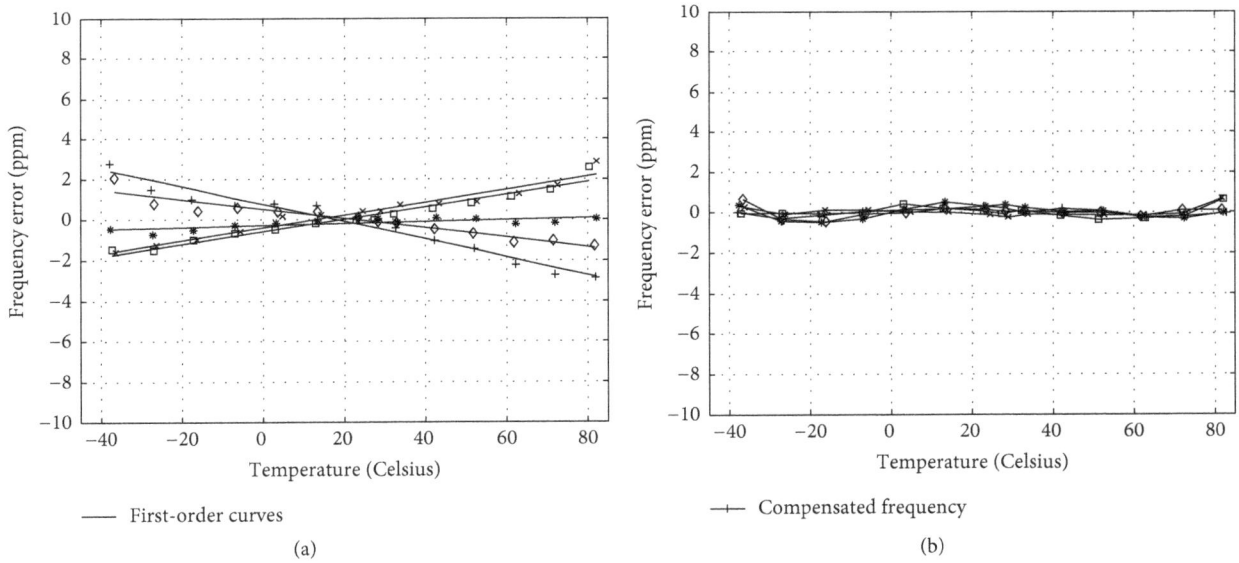

— First-order curves

(a)

—+— Compensated frequency

(b)

FIGURE 13: Temperature compensation based first on the fixed third-order curves and then on the individual first-order curves. (a) First-order curve fitting for each RF module sample using all of the third-order compensated data of Figure 12(b). (b) Temperature-compensated results using first the fixed third-order curves and then the individual first-order curves (±0.68 ppm).

used to describe the bias curve. Hence, in order to use the frequency f_{offset} measured from the frequency-offset register of the AFC at a specific ambient temperature T_1 that is higher than the inflection temperature T_0, compensation using the inverse function of the bias curve η^{-1} should be applied. A curve of (2) obtained this way is then given by the parameter a_1 as

$$a_1 = \frac{\eta^{-1}\left(f_{\text{offset}}\right)}{T_1 - T_0}. \tag{3}$$

The first-order curves obtained from both the frequency-offset registers and (3) are depicted in Figure 15(a), and

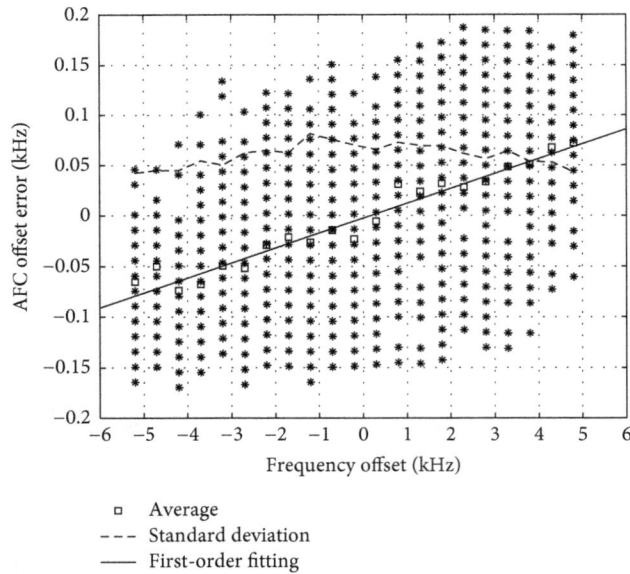

FIGURE 14: Frequency-offset measurement from the AFC frequency-offset register of RFIC. The compensation curve is obtained by fitting a first-order curve to the measured frequency offsets from the frequency-offset register of AFC.

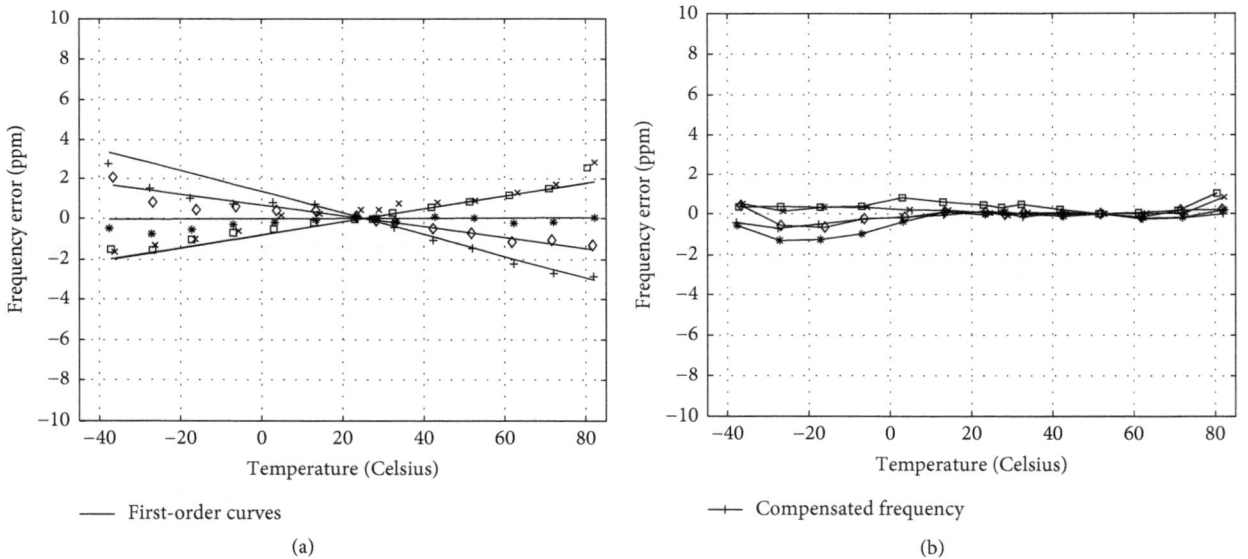

FIGURE 15: Temperature compensation in the proposed DCTCXO. (a) First-order curve fitting for each RF model sample using (3) at 50°C. (b) Temperature-compensation results using the fixed third-order curve and then the individual first-order curve (±1.3 ppm).

the compensated results are shown in Figure 15(b). The fitting increased the maximum frequency error slightly to ±1.3 ppm from that of Figure 13(a). In the experiment shown in Figure 15, the ambient temperature was $T_1 \approx 50°C$. For a practical RF module design, it is important to measure one frequency error at a specific temperature and to conduct automatic compensation, even though the temperature-compensation performance is degraded slightly, especially at a low temperature range. The memory required to implement

the proposed DCTCXO was 520 B, and the execution time of the 1 MHz-clock MCU was 3.5 ms.

The NB RF module of the 477 MHz band was tested at a bit rate of 2.4 kbps in a home environment and compared with a wider bandwidth receiver operating at the same rate. The experimental results are illustrated in Figure 16. The transmitter was located on the eighth floor, and the receiver was moved around the space to measure both the RSSI values and the packet error rates (PERs). The narrower bandwidth

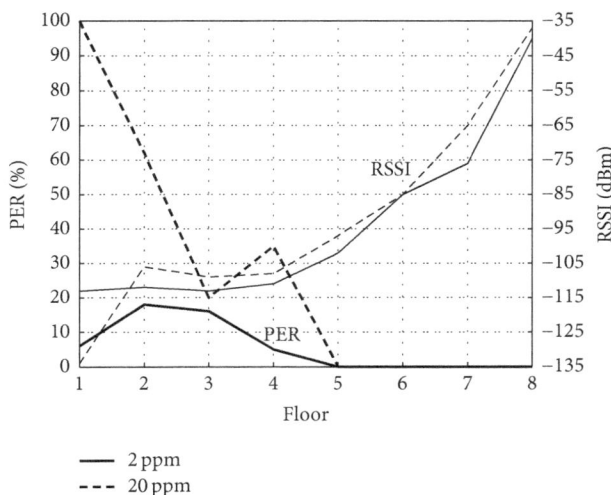

FIGURE 16: Comparison of receiver bandwidths operated under the same bit rate of 2.4 kbps for which the transmitter was located on the eighth floor of the building. The frequency errors of ±2.0 ppm and ±20 ppm had receiver bandwidths of 6.5 kHz and 36.6 kHz, respectively.

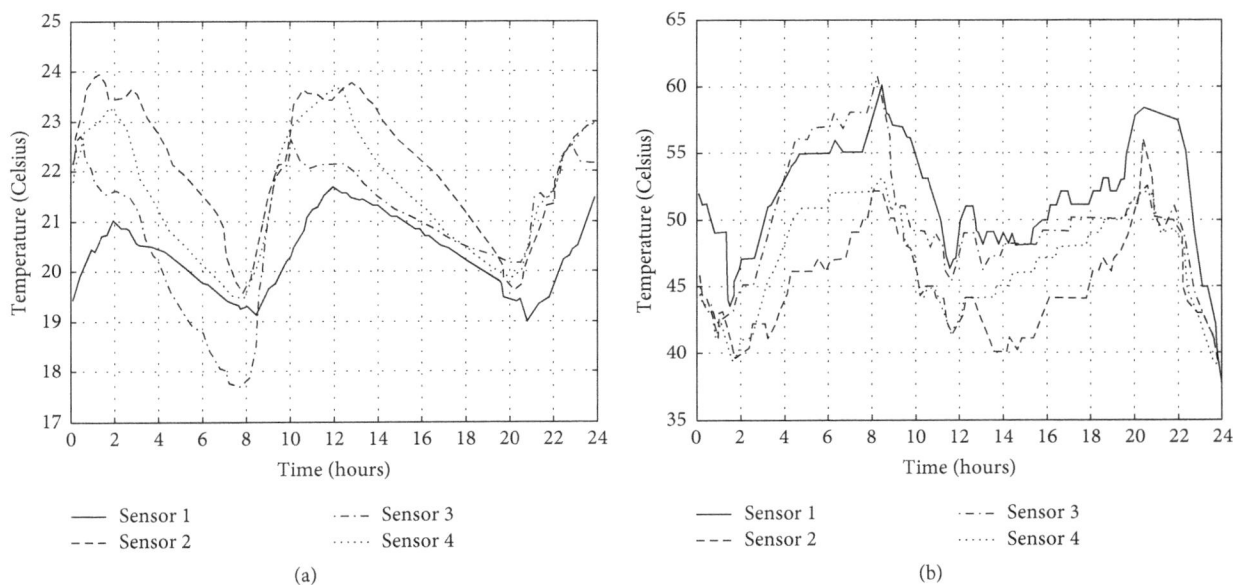

FIGURE 17: Comparison of receiver bandwidths operated under the same bit rate of 2.4 kbps for which the transmitter was located on the eighth floor of the building. The frequency errors of ±2.0 ppm and ±20 ppm had receiver bandwidths of 6.5 kHz and 36.6 kHz, respectively.

receiver demonstrated better packet error rates and could provide longer transmission ranges than the wider bandwidth receiver could. From the results shown in Figure 16, it is clear that the CoAP wireless sensor could cover more than five of the building's stories.

5.2. Experiment Performed on the Hybrid WSN. The proposed hybrid WSN system was constructed in a five-story library building. Each story had one CoAP gateway, two CoAP wireless sensors, and eight simple wireless sensors to collect temperature and humidity data. The data measured in one day, depicted in Figure 17, provided meaningful

information with respect to the rooms' locations. In Figure 18, the MOS-based CO_2 sensor of IWSM was compared with the conventional NDIR-based CO_2 sensors in an office environment with moving people. We can notice that the MOS-based CO_2 sensor can efficiently measure the CO_2 gas amount change below 2000 ppm similarly to the NDIR-based sensor cases.

6. Conclusion

A low-cost hybrid wireless sensor network utilizing the 917 MHz and 447 MHz RF bands is implemented for indoor

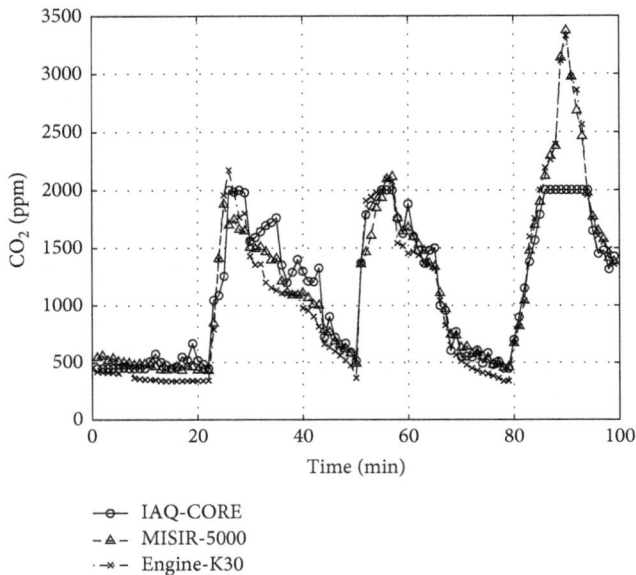

FIGURE 18: Comparison of the MOS-based CO_2 sensor of IWSM (IAQ-CORE) with the conventional CO_2 sensors, which are based on NDIR (MISIR-5000 and Engine-K30).

environment monitoring. A duty-cycling receiver algorithm and a digitally controlled temperature-compensation crystal oscillator (DCTCXO) are implemented to achieve low-power and low-cost wireless sensor nodes. In the proposed DCTCXO, a two-step compensation is performed, in which a fixed third-order curve is applied first and a first-order curve is applied after that, to achieve a high-precision frequency reference. The first-order compensation curve for a given crystal resonator sample is automatically obtained from the frequency-offset measurement conducted in the AFC block. By implementing both the duty-cycling algorithm and the proposed DCTCXO, the average operating current of the RF module is reduced from 16.7 mA to 4.0 mA.

Conflicts of Interest

The authors declare that there are no conflicts of interest regarding the publication of this paper.

Acknowledgments

This work was supported by Korea Institute of Energy Technology Evaluation and Planning (KETEP) and by the Ministry of Trade, Industry & Energy (MOTIE) of the Republic of Korea (20151210200240 and 20141010501840).

References

[1] J. Lozano, J. I. Suarez, P. Arroyo, J. M. Ordiales, and F. Alvarez, "Wireless sensor network for in door air quality monitoring," *Chemical Engineering Transactions*, vol. 30, pp. 319–329, 2012.

[2] S. Bhattacharya, S. Sridevi, and R. Pitchiah, "Indoor air quality monitoring using wireless sensor network," in *Proceedings of the 6th International Conference on Sensing Technology (ICST '12)*, pp. 422–427, IEEE, Kolkata, India, December 2012.

[3] D. S. Kim, S.-Y. Son, and J. Lee, "Developments of the in-home display systems for residential energy monitoring," *IEEE Transactions on Consumer Electronics*, vol. 59, no. 3, pp. 492–498, 2013.

[4] S. Abraham and X. Li, "A cost-effective wireless sensor network system for indoor air quality monitoring applications," *Procedia Computer Science*, vol. 34, pp. 165–171, 2014.

[5] T. Torfs, T. Sterken, S. Brebels et al., "Low power wireless sensor network for building monitoring," *IEEE Sensors Journal*, vol. 13, no. 3, pp. 909–915, 2013.

[6] A. Kim, J. Han, T. Yu, and D. S. Kim, "Hybrid wireless sensor network for building energy management systems based on the 2.4 GHz and 400 MHz bands," *Information Systems*, vol. 48, pp. 320–326, 2015.

[7] F. Liang and B. Liu, "Research of sub-GHz wireless sensor network and its application in grain monitoring system," *Journal of Computational Information Systems*, vol. 7, no. 10, pp. 3491–3498, 2011.

[8] J. Han, J. Lee, E. Lee, and D. S. Kim, "Development of a low-cost indoor environment monitoring system based on a hybrid wireless sensor network," in *Proceedings of the IEEE International Conference on Consumer Electronics (ICCE '16)*, pp. 461-462, January 2016.

[9] M. N. Halgamuge, T.-K. Chan, and P. Mendis, "Experiences of deploying an indoor building sensor network," in *Proceedings of the 3rd International Conference on Sensor Technologies and Applications (SENSORCOMM '09)*, pp. 378–381, June 2009.

[10] G. Ghidini and S. K. Das, "An energy-efficient markov chain-based randomized duty cycling scheme for wireless sensor networks," in *Proceedings of the 31st International Conference on Distributed Computing Systems (ICDCS '11)*, pp. 67–76, July 2011.

[11] G. Buroker and M. Frerking, "A Digitally Compensated TCXO," in *Proceedings of the 27th Annual Symposium on Frequency Control*, pp. 191–198, Cherry Hill , NJ, USA, 1973.

[12] W. Dargie and C. Poellabauer, *Fundamentals of Wireless Sensors Networks: Theory and Practice*, John Wiley Sons, 2010.

[13] B. Otis and J. Rabaey, "Ultra-low power wireless technologies for sensor networks," *Ultra-Low Power Wireless Technologies for Sensor Networks*, pp. 1–184, 2007.

[14] A. Mroch and G. Hykes, "A miniature high stability TCXO using digital compensation," in *Proceedings of the 30th Annual Symposium on Frequency Control*, pp. 292–300, 1976.

[15] J. C. Blanco, F. J. Azcondo, and J. Peire, "New Digital Compensation Technique for the Design of a Microcomputer Compensated Crystal Oscillator," *IEEE Transactions on Industrial Electronics*, vol. 42, no. 3, pp. 307–315, 1995.

[16] Q. Huang and P. Basedau, "Design considerations for high-frequency crystal oscillators digitally trimmable to sub-ppm accuracy," *IEEE Transactions on Very Large Scale Integration (VLSI) Systems*, vol. 5, no. 4, pp. 408–416, 1997.

[17] D. S. Kim, E. Lee, J. Han, and J. Lee, "Digitally controlled temperature-compensated crystal oscillator in developing low-cost radio frequency modules," *Information*, vol. 18, no. 7, pp. 3157–3166, 2016.

[18] R. M. Cerda, *Understanding Quartz Crystals and Oscillators*, Artech House Microwave Library, London, UK, 2014.

Permissions

All chapters in this book were first published in JS, by Hindawi Publishing Corporation; hereby published with permission under the Creative Commons Attribution License or equivalent. Every chapter published in this book has been scrutinized by our experts. Their significance has been extensively debated. The topics covered herein carry significant findings which will fuel the growth of the discipline. They may even be implemented as practical applications or may be referred to as a beginning point for another development.

The contributors of this book come from diverse backgrounds, making this book a truly international effort. This book will bring forth new frontiers with its revolutionizing research information and detailed analysis of the nascent developments around the world.

We would like to thank all the contributing authors for lending their expertise to make the book truly unique. They have played a crucial role in the development of this book. Without their invaluable contributions this book wouldn't have been possible. They have made vital efforts to compile up to date information on the varied aspects of this subject to make this book a valuable addition to the collection of many professionals and students.

This book was conceptualized with the vision of imparting up-to-date information and advanced data in this field. To ensure the same, a matchless editorial board was set up. Every individual on the board went through rigorous rounds of assessment to prove their worth. After which they invested a large part of their time researching and compiling the most relevant data for our readers.

The editorial board has been involved in producing this book since its inception. They have spent rigorous hours researching and exploring the diverse topics which have resulted in the successful publishing of this book. They have passed on their knowledge of decades through this book. To expedite this challenging task, the publisher supported the team at every step. A small team of assistant editors was also appointed to further simplify the editing procedure and attain best results for the readers.

Apart from the editorial board, the designing team has also invested a significant amount of their time in understanding the subject and creating the most relevant covers. They scrutinized every image to scout for the most suitable representation of the subject and create an appropriate cover for the book.

The publishing team has been an ardent support to the editorial, designing and production team. Their endless efforts to recruit the best for this project, has resulted in the accomplishment of this book. They are a veteran in the field of academics and their pool of knowledge is as vast as their experience in printing. Their expertise and guidance has proved useful at every step. Their uncompromising quality standards have made this book an exceptional effort. Their encouragement from time to time has been an inspiration for everyone.

The publisher and the editorial board hope that this book will prove to be a valuable piece of knowledge for researchers, students, practitioners and scholars across the globe.

List of Contributors

Xiujuan Du and Chunyan Peng
Computer Department, Qinghai Normal University, Xining 810008, China
Key Laboratory of the Internet of Things of Qinghai Province, Xining 810008, China

Meiju Li
Computer Department, Qinghai Normal University, Xining 810008, China
Key Laboratory of the Internet of Things of Qinghai Province, Xining 810008, China
College of Physics and Electronic Information Engineering, Qinghai Nationalities University, Xining 810007, China

Yujia Sun, Xiaoming Wang, Jiyan Yu and Yu Wang
Ministerial Key Laboratory of ZNDY, Nanjing University of Science and Technology, Nanjing 210094, China

Wenzhao Feng, Junguo Zhang, Chunhe Hu, Yuan Wang and Hao Yan
School of Technology, Beijing Forestry University, Beijing 100083, China

Qiumin Xiang
Chongqing Mobile Communications Limited Company, Chongqing 404100, China

Mengxing Huang, Yong Bai, Zhuhua Hu and Yanfang Deng
State Key Laboratory of Marine Resource Utilization in South China Sea, College of Information Science & Technology, Hainan University, Haikou 570228, China

Mingshan Xie
State Key Laboratory of Marine Resource Utilization in South China Sea, College of Information Science & Technology, Hainan University, Haikou 570228, China
College of Network, Haikou University of Economics, Haikou 571127, China

Enjie Ding
IoT Perception Mine Research Center, China University of Mining and Technology, Xuzhou 221008, China

Yanjun Hu
School of Information and Control Engineering, China University of Mining and Technology, Xuzhou 221008, China

Xiansheng Li, Tong Zhao and Lei Zhang
IoT Perception Mine Research Center, China University of Mining and Technology, Xuzhou 221008, China
School of Information and Control Engineering, China University of Mining and Technology, Xuzhou 221008, China

Abebe Belay Adege, Getaneh Berie Tarekegn and Yirga Yayeh Munaye
Department of Electrical Engineering and Computer Science, National Taipei University of Technology, Taipei, Taiwan

Hsin-Piao Lin and Lei Yen
Department of Electronic Engineering, National Taipei University of Technology, Taipei, Taiwan

Shuang Jia and Lin Ma
School of Electronics and Information Engineering, Harbin Institute of Technology, Harbin 150080, China

Danyang Qin
Department of Communication Engineering, Heilongjiang University, Harbin 150080, China

Stelios A. Mitilineos, Stelios M. Potirakis, Nicolas-Alexander Tatlas and Maria Rangoussi
Department of Electrical and Electronics Engineering, University of West Attica, Campus 2, 250 Thivon and P. Ralli, Aigaleo, 122 44 Athens, Greece

Anand Paul and Hameed Pinjari
School of Computer Science and Engineering, Kyungpook National University, Daegu, Republic of Korea

Won-Hwa Hong and Hyun Cheol Seo
School of Architectural, Civil, Environmental and Energy Engineering, Kyungpook National University, Daegu, Republic of Korea

Seungmin Rho
Department of Media Software, Sungkyul University, Anyang, Republic of Korea

Hongyu Chen and Xin Xu Fangling Pu
School of Electronic Information, Wuhan University, Wuhan, Hubei 430072, China Collaborative Innovation Center of Geospatial Technology, Wuhan University, Wuhan, Hubei 430079, China

Zhaozhuo Xu
Electrical Engineering Department, Stanford University, Palo Alto, CA 94305, USA

Nengcheng Chen
State Key Laboratory for Information Engineering in Surveying, Mapping and Remote Sensing (LIESMARS), Wuhan University, Wuhan, Hubei 430079, China

Sheeraz Ahmed
Iqra National University, Peshawar, Pakistan Career Dynamics Research Centre, Peshawar, Pakistan

Mujeeb Ur Rehman
Career Dynamics Research Centre, Peshawar, Pakistan

Atif Ishtiaq
Iqra National University, Peshawar, Pakistan

Sarmadullah Khan
School of Computer Science and Informatics, De Montfort University, Leicester LE1 9BH, UK

Armughan Ali
COMSATS Institute of Information Technology, Attock, Pakistan

Shabana Begum
Islamia College University, Peshawar, Pakistan

Maher Ibrahim Sameen, Biswajeet Pradhan and Omar Saud Aziz
School of Systems, Management and Leadership, Faculty of Engineering and Information Technology, University of Technology Sydney, Building 11, Level 06, 81 Broadway, Ultimo, NSW 2007, Australia

Yan Wang, Xuehan Wu and Long Cheng
Department of Computer and Communication Engineering, Northeastern University, Qinhuangdao 066004, China

Fang Zhu
School of computer and communication engineering, Northeastern University at Qinhuangdao, Northeastern University, Qinhuangdao 066004, China

Junfang Wei
School of Resources and Materials, Northeastern University at Qinhuangdao, Northeastern University, Qinhuangdao 066004, China
Key Laboratory of Dielectric and Electrolyte Functional Material Hebei Province, School of Resource and Materials, Northeastern University at Qinhuangdao, Qinhuangdao 066004, China

Jian Chen, Yu Tan and Yongjun Zheng
College of Engineering, China Agricultural University, 17 Tsinghua East Rd, Beijing 100083, China

Peng Li and Gangbing Song
Department of Mechanical Engineering, University of Houston, 4800 Calhoun Rd, Houston, TX 77204, USA

Yu Han
College of Water Resources & Civil Engineering, China Agricultural University, 17 Tsinghua East Rd, Beijing 100083, China

Nannan Lu, Yanjing Sun and Song Li
School of Information and Electrical Engineering, China University of Mining and Technology, Xuzhou, Jiangsu, China

Hui Liu
Institute of Information Photonics and Optical Communications, Beijing University of Posts and Telecommunications, Beijing, China

Dong Sik Kim
Department of Electronics Engineering, Hankuk University of Foreign Studies, Yongin-si, Gyeonggi-do 17035, Republic of Korea

Beom Jin Chung
Smart Green Home Research Center, Gachon University, Gyeonggi-do, Republic of Korea

Sung-Yong Son
Department of Electrical Engineering, Gachon University, Gyeonggi-do, Republic of Korea

Index

www.ingramcontent.com/pod-product-compliance
Lightning Source LLC
Chambersburg PA
CBHW082041190326
41458CB00010B/3431